W0227964

S. Imura · N. Akamatsu · H. Azuma
K. Sawai · S. Tanaka (Eds.)

Hip Biomechanics

With 276 Figures

Springer-Verlag Berlin Heidelberg GmbH

Shinichi Imura, M.D.
Fukui Medical School, 23 Shimoaizuki, Matsuoka-cho, Yoshida-gun, Fukui, 910-11
Japan

Noriya Akamatsu, M.D.
Yamanashi Medical College, 1110 Shimokato, Tamaho-cho, Nakakoma-gun,
Yamanashi, 409-38 Japan

Hirohiko Azuma, M.D.
Saitama Medical School, 38 Morohongo, Moroyama-machi, Iruma-gun, Saitama, 350-04
Japan

Kazuhiko Sawai, M.D.
Aichi Medical University, 21 Iwasaku, Nagakute-cho, Aichi-gun, Aichi, 480-11 Japan

Seisuke Tanaka, M.D.
Kinki University School of Medicine, 377-2 Ohno-Higashi, Sayama, Osaka, 589 Japan

ISBN 978-4-431-68239-4   ISBN 978-4-431-68237-0 (eBook)
DOI 10.1007/978-4-431-68237-0

Printed on acid-free paper

Library of Congress Cataloging-in-Publication Data
Hip biomechanics / S. Imura ... [et al.] (eds.). p. cm. "Hip Biomechanics
Symposium, held on November 19, 1992, in Fukui city, Japan"—Pref. Includes
bibliographical references and index. ISBN 978-4-431-68239-4
                1. Hip joint—Mechanical properties—Congresses. 2. Hip
joint—Surgery—Simulation methods—Congresses. 3. Artificial hip joints—Congresses. I. Imura, Shinichi. II. Hip Biomechanics Symposium, (1992 : Fukui-shi,
Japan), [DNLM: 1. Hip Prosthesis—congresses. 2. Hip—physiology—congresses.
3. Hip—physiopathology—congresses. 4. Hip—surgery—congresses. 5. Hip Joint
—surgery—congresses. 6. Biomechanics—congresses. WE 855 H6673 1993],
RD549.H52 1993, 617.5'81—dc20, DNLM/DLC, for Library of Congress, 93-5020

© Springer-Verlag Berlin Heidelberg 1993
Originally published by Springer-Verlag Berlin Heidelberg New York in 1993
Softcover reprint of the hardcover 1st edition 1993

This work is subject to copyright. All rights are reserved, whether the whole or part of the material
is concerned, specifically the rights of translation, reprinting, reuse of illustrations, recitation,
broadcasting, reproduction on microfilms or in other ways, and storage in data banks.
The use of registered names, trademarks, etc. in this publication does not imply, even in the
absence of a specific statement, that such names are exempt from the relevant protective laws and
regulations and therefore free for general use.
Product liability: The publisher can give no guarantee for information about drug dosage and
application thereof contained in this book. In every individual case the respective user must check
its accuracy by consulting other pharmaceutical literature.

Table of Contents

III Hip Joint Surgery and Simulation

A   Total Hip Arthroplasty

B   Osteotomy

Preface

It was a great honor for us to preside over the Hip Biomechanics Symposium, held on November 19, 1992, in Fukui City, Japan. We believe it was the first interdisciplinary conference on biomechanical approaches to hip disorders; this volume contains the presentations made at the conference.

The information found here will be of value not only for orthopedic surgeons but also for researchers. The advent of biomechanical approaches to hip disease has made possible the understanding of fundamental hip pathologies, an understanding which enables us to institute the appropriate non-surgical or surgical treatment. Indeed, efficient surgical planning is not possible without taking these approaches.

This comprehensive volume consists of six major sections: loading, gait analyses, THA, osteotomies, motion analyses, and stem designs for stability. Each of these sections brings together many of the leading researchers in this field.

We hope this volume will be a spur to future understanding of and on fruitful research on hip disorders.

The Editorial Committee wish to express their deep gratitude to the contributors and the publisher.

Shinichi Imura
Noriya Akamatsu
Hirohiko Azuma
Kazuhiko Sawai
Seisuke Tanaka

I

Hip Joint and Loading

1

Two-Dimensional Finite Element Analysis for Stress Distribution in Normal and Dysplastic Hips

Nobuo Konishi[1]

Summary. Two-dimensional finite element analysis was achieved for stress distribution in the normal hip model, with a center-edge (CE) angle (Wiberg) [1] of 30°, and in the dysplastic hip models with CE angles of 15° and 0°. Under loads of 5° or 15° slanted resultant force, stress concentration was seen in dysplastic hip model at the lateral portion of the subchondral bone of femoral head and at the joint cartilage just beneath the lateral edge of acetabulum. If the resultant force has nearly vertical direction, the acetabular coverage of 30° CE angle may not be sufficient to avoid stress concentration at the area of the femoral head that faces the lateral edge of the acetabulum.

Key words: Hip dysplasia—Finite element analysis—Stress in joint cartilage

Introduction

Osteoarthritic change in the dysplastic hip is considered to be developed by increased mechanical stress on a reduced area for load transmission [2]. The area necessary to maintain the stress on the hip within the physiological limit has not yet been determined. Although many factors, such as patient age, body weight, activity, and the physical or physiological condition of bone and cartilage, are involved in the development of secondary osteoarthritis, these, as yet undetermined, etiologic factors are not considered here. Our purpose is only to determine the level of acetabular coverage sufficient to prevent secondary osteoarthritis. For this purpose, stress analysis by the two-dimensional finite element method was achieved.

---

[1] Department of Orthopaedics, Aichiken-Saiseikai Hospital, 1-17 Kamisaradouri, Nishi-ku, Nagoya, 451 Japan

3

Materials and Methods

Three types of hip joint model were prepared for the finite element analysis. The first was a model of the normal hip joint with a CE angle (Wiberg) [1] of 30°. The second was a model of mild dysplastic hip with a CE angel of 15°, and the third was a model of severe dysplastic hip, with a CE angle of 0°. The normal model was made by tracing the outline of a hip joint on a radiograph of normal female hips (Fig. 1). The traced outline was modified so that its CE angle was just 30° and its femoral head had a regular circular shape. The traced outline was then reconstructed to the finite element model, which had 216 nodes and 211 elements in the case of the normal hip (Fig. 2 left). The models of mild and severe dysplastic hip joints (Fig. 2 right) were made by removing some elements from the normal hip joint model so that their CE angles were just 15° or just 0°. In any type of hip joint model, there are three parts; the acetabular, femoral head, and joint cartilage components. The acetabular and femoral head components were adopted to have a uniform quality of cortical bone with an elastic module of $120\,000\,\text{kg/cm}^2$ and a thickness of 20 mm. The joint cartilage component was constructed from two types of truss element;

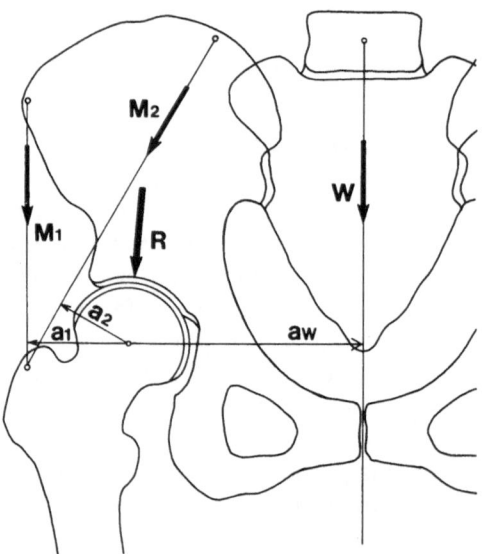

Fig. 1. Prototype of the analytic model used for the finite element method. The outlines of the pelvis and femur were traced from a radiograph of a normal female subject. The body weight vector (W) was loaded at the median line of the pelvis. With the body weight vector, the force of the gluteus medius muscle (M_1) and the force of the gluteus maximus (M_2) were arranged to yield the resultant force (R) which was slanted medially 5° (load case 1) or 15° (load case 2) from the vertical line. The distances from the center of the femoral head to the body weight vector, to the gluteus medius, and to the gluteus maximus are expressed as a_w, a_1, and a_2, respectively

CE Angle= 30°   CE Angle= 0°

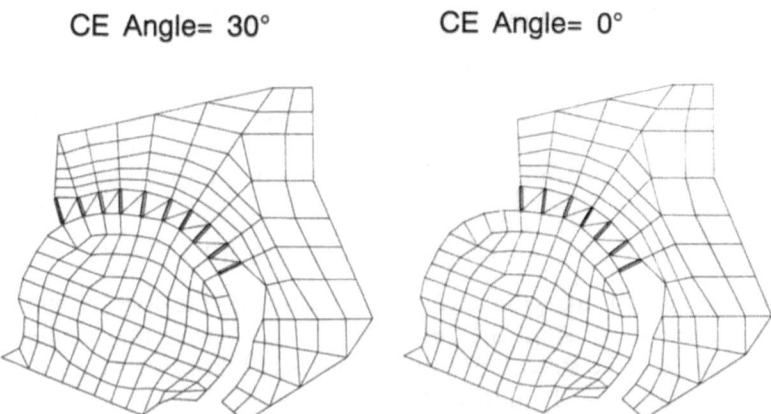

Fig. 2. Normal hip joint model with center-edge (*CE*) angle of 30° (*left*) and severe dysplastic hip model with CE angle of 0° (*right*) used for the finite element analysis. The acetabulum and femoral head components were constructed from planar elements. The joint cartilage component was constructed from two types of truss elements; vertical and bias truss

vertical and bias trusses. The vertical truss, which is arranged vertically to the joint surface, has an elastic module of cartilage (120 kg/cm^2) and cut surface area so that it corresponds to the surface area of the joint cartilage. The bias truss, which is a bias linkage between a point of the acetabular joint surface and a point of the femoral head, has a very small elastic module that is only 1/100 of the elastic module of the vertical truss. These special arrangement allow us to transmit only a load vertical to the joint surface. The distal end of the femoral head is fixed.

The forces loaded on the model are a set of body weight and muscle contractile forces. Their direction and quantity were determined via the following considerations. According to Blount [3] and Pauwels [4], the resultant force of the hip joint is medially slanted 15° from the vertical direction. Recent investigations based on instrumented hip endoprosthesis have reported a more vertical orientation of the resultant force [5,6]. Our investigation, based on a mathematical model of the hip joint, gave a three-dimensional direction of the resultant force that had a medial slant of 4° and a forward slant of 7° [7]. From these considerations, we adopted two cases of load conditions; one (load case 1) was arranged to have a medially slanted resultant force of 5° and the other (load case 2) a resultant force of 15°. In each of these load cases, the body weight load, which was 5/6 of the total body weight, or 50 kg, was loaded at the median line of the pelvis. The two abductors; the gluteus medius and gluteus maximus muscles, were arranged to equalize with the body weight and to have a resultant force with a medially-slanted orientation of 15° or 5°. Their force was calculated by the following equations (for the symbols used in the equations see Fig. 1).

From the equalization of moments around the hip joint,

$$M_1a_1 + M_2a_2 = Wa_W$$

where M_1 and M_2 are the contractile forces of the gluteus medius and gluteus maximus, W is the body weight load, and a_1, a_2, and a_w are the lever-arm lengths of, or distances from the center of the femoral head to M_1, M_2, and W.

From the equation of the vertical component of forces,

$$M_1 + M_2 \sin 60° + W = R \sin 85°$$

Where the axis of the gluteus medius is vertical, the axis of the gluteus maximus is slanted 30° from the vertical line. The axis of the resultant force (R) was assumed to be slanted 5° from the vertical line in load case 1.

From the equation of the horizontal component of forces,

$$M_2 \cos 60° = R \cos 85°$$

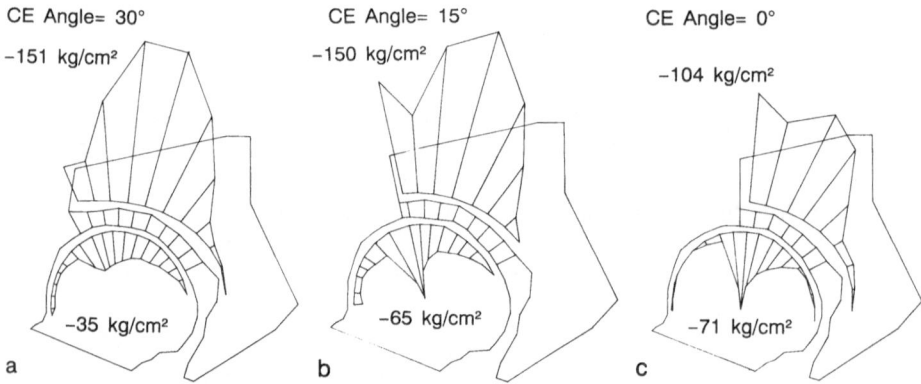

Fig. 3. a In the normal hip joint model for load case 1, the center portion of the subchondral region of the acetabulum has greater stress (151 kg/cm²) than the peripheral portion of the same region. The subchondral region of the femoral head has a more uniform tendency of stress distribution (35 kg/cm² at the maximum) than the subchondral region of the acetabulum. **b** In the mild hip dysplasia model for load case 1, the center and the lateral peripheral portion of the subchondral region of the acetabulum have greater stress than the medial portion of the same region (150 kg/cm² at the central portion). The lateral portion of the subchondral region of the femoral head that confronts the lateral edge of the acetabulum has a stress concentration of 65 kg/cm². **c** In the severe hip dysplasia model for load case 1, the lateral portion of the subchondral region of the acetabulum has greater stress (104 kg/cm²) than the medial portion, although the maximum stress in this region is less than the maximum stress in the same region of the normal hip model. A more pronounced stress concentration (71 kg/cm²) occurs at the lateral portion of the subchondral region of the acetabulum than in the mild hip dysplasia model

Solving these equations yields the following: a resultant force of 169 kg; gluteus medius force, 92.8 kg; and the gluteus maximus force, 29.5 kg in load case 1; and values of 181, 43.8, and 93.8 kg, respectively, in load case 2.

The three types of hip joint model with two types of load case in each model were analyzed by a finite element analysis program (FEM3-program; Fujitsu, Tokyo, Japan), using a Nagoya university computer (FACOM-M200; Fujitsu, Tokyo, Japan).

Results

Load Case 1 (Resultant Force of 5° Medial Slant)

In the normal hip joint model (Fig. 3a), the center portion of the subchondral region of the acetabulum has greater stress than the peripheral portion of the same region. The subchondral region of the femoral head has a more uniform tendency of stress distribution than the subchondral region of the acetabulum. The joint cartilage has a uniform tendency of stress distribution, although the center has greater stress than the peripheral portion (Fig. 4a).

In the mild hip dysplasia model (Fig. 3b), the center and the lateral peripheral portion of the subchondral region of the acetabulum have greater stress than the medial portion of the same region. The lateral portion of the subchondral region of the femoral head that confronts the lateral edge of the acetabulum has a stress concentration. The center and the lateral portions of the joint cartilage have greater stress than the medial portion (Fig. 4b).

In the severe hip dysplasia model (Fig. 3c), the lateral portion of the subchondral region of the acetabulum has greater stress than the medial portion, although the maximum stress in this region is less than the maximum stress

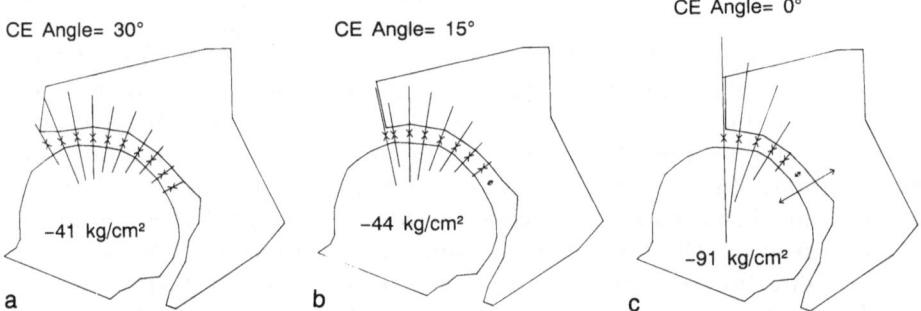

Fig. 4. a In normal hip joint model for load case 1, the joint cartilage has a uniform tendency of stress distribution, although the center has greater stress (41 kg/cm²) than the peripheral portion. **b** In the mild hip dysplasia model for load case 1, the center and the lateral portions of the joint cartilage have greater stress (44 kg/cm²) than the medial portion. **c** In the severe hip dysplasia model for load case 1, the stress concentration (91 kg/cm²) is seen in the lateral edge of the joint cartilage

in the same region of the normal hip model. There is more pronounced stress concentration at the lateral portion of the subchondral region of the acetabulum than in the mild hip dysplasia model. The stress concentration is also found in the lateral edge of the joint cartilage (Fig. 4c).

Load Case 2 (Resultant Force of 15° Medial Slant)

The stress distribution findings in load case 2 are similar to the findings in load case 1. The subchondral region of the acetabulum has greater stress at the center portion in the normal hip model and has greater stress at the lateral portion in the hip dysplasia models. The stress concentration at the lateral portion that confronts the lateral edge of the acetabulum is found in the subchondral region of the hip dysplasia models. The joint cartilage has a stress concentration only in the severe hip dysplasia model.

Discussion

There is some difficulty in performing finite element analysis of a joint surface which transmits only a vertical compressive load and allows free rotational movement around the joint. We therefore adopted the arrangement of two types of truss element, vertical and bias truss, with the elastic module of the bias truss being much smaller than that of the vertical truss. Selective trans-mitting capacity for the compressive load only and not for the tensile load is usually achieved by a recalculating technique: if a model has the calculated result of tensile stress in the joint cartilage element, the modified model, in which the elements with tensile stress are eliminated, is recalculated through finite element analysis. This recalculation is repeated until no element of the joint cartilage has tensile stress. However, this recalculating technique was not adopted in our study, because it requires much cost and time.

It was a feasible result that the stress concentration was found at the lateral edge of the acetabulum or at its opponent area in the femoral head and joint cartilage in the hip dysplasia model. In load case 1, with resultant force of 5° medial slant, the mild hip dysplasia model showed a stress concentration at the lateral portions of the femoral head and joint cartilage, whereas in load case 2, the same model showed no stress concentration at the lateral edge portion of the joint cartilage. These observations indicate that acetabular coverage over 35° to the direction of resultant force is sufficient to prevent a stress concentration at the lateral edge portion of the acetabulum, femoral head, and joint cartilage. An acetabular coverage over 20° around the direction of resultant force was accompanied by a stress concentration at the lateral edge portion of all hip joint components. Acetabular coverage of over 30° around the direction of the resultant force had an intermediate result, between those of acetabular coverage of over 35° and those of 20°, in that the stress concentration was found at the lateral portion of the femoral head. The actual direction of the resultant force that acts in the hip of normal subjects is not known. If the resultant force has a nearly vertical direction, acetabular coverage of 30° CE angle may not be sufficient to avoid stress concentration at the area of femoral head that faces the lateral edge of the acetabulum.

Two peaks of large compressive stress were seen in the acetabulum, one being distributed at the central portion and the other at the lateral edge of the subchondral region in the acetabulum. This peculiar pattern of stress distribution in the acetabulum was difficult to interpret. It may be a characteristic pattern of load transmission between the two circular joint surfaces with the same center point. The interspace between the two circular surfaces decreases to an eccentric orientation under a compressive load and the interspace just below the compressive load becomes narrowest where the largest stress develops. Alternatively, this peculiar distribution may be due to insufficient simulation of the hip joint. In our finite element analysis, the acetabulum component was only part of the pelvis. Therefore, the two main downward loads, body weight and muscular force, were exerted at the most lateral and the most medial points of the acetabular component. These loads bend the acetabular component to yield compressive stress at the central portion of the component. For these reasons, the calculated stress value in the acetabular component may be overestimated and may differ from the actual stress in the acetabulum.

This study showed the maximum compressive stress in the femoral head in the severe dysplastic hip to be $71\,kg/cm^2$, and in the joint cartilage this was $91\,kg/cm^2$. The stress level that destroys cortical bone was reported to be $900\,kg/cm^2$ or greater [8,9], and this is more than ten times the estimated maximum stress values seen in the dysplastic hip. On the other hand, repeatedly loaded stress over a long period is considered to initiate bone destruction at a much lower level than that which destroys the bone in a single load [10]. If our estimated value for maximum stress in the severe dysplastic hip represents the actual stress value in the patient with hip dysplasia, the lower threshold of the stress that initiates degenerative change in the subchondral bone or joint cartilage is within the range of $60\text{–}90\,kg/mm^2$, since the maximum stress seen in the mild and severe dysplastic hip fall within this range. Again, since our model for finite element analysis is a two-dimensional configuration with a uniform thickness of $20\,mm$ in all components, and is constructed only from hard tissue components, ignoring soft tissue components such as the acetabular lip and the fibrous ligaments around the joint capsule, it is uncertain whether our estimated stress value represents the actual value in a living subject.

Acknowledgment. The author wishes to thank Dr. Takamasa Sakurai and Hidehiko Ninomiya, Department of Civil Engineering, Toyota Industrial College, for their work in finite element analysis.

References

1. Wiberg G (1939) Studies on dysplastic acetabula and congenital subluxation of the hip joint, with special reference to the complications of osteoarthritis. Acta Chir Scand [Suppl] 83:1–33
2. Maquet PGJ (1985) Biomechanics of the hip—as applied to osteoarthritis and related conditions. Springer, New York, pp 37–41
3. Blount WP (1956) Don't throw away the cane (presidential address, the American Academy of Orthopaedic Surgeons). J Bone Joint Surg [Am] 38-A:695–708
4. Pauwels F (1935) Schenklsbruch. Ein mechanisches Problem. Ferdinand Enke, Stuttgart (cited by Blount [3])
5. Davy DT, Kotzar GM, Brown RH, Heiple KG, Goldberg VM, Heiple KG Jr, Berilla J, Burstein AH (1988) Telemetric force measurements across the hip after total arthroplasty. J Bone Joint Surg [Am] 70-A:45–50
6. Hodge WA, Carlson KL, Fijan RS, Burgess RG, Riley PO, Harris WH, Mann RW (1989) Contact pressures from an instrumented hip endoprosthesis. J Bone Joint Surg [Am] 71-A:1378–1385
7. Konishi N (1988) Three-dimensional resultant force in the hip calculated on mathematical model by the non-linear programming method (in Japanese). Rinsho Seikei Geka (Clinical Orthopaedic Surgery) 23:1491–1497
8. Asenzi A, Bonucci E (1968) The compressive properties of single osteon. Anat Rec 161:377–392
9. Hert J, Kucera P, et al (1965) Comparison of the mechanical properties of both the primary and haversian bone tissue. Acta Anat 61:412–423
10. Seireg A, Kempke W (1969) Behavior of in vivo bone under cyclic loading. J Biomech 2:455–461

2

Stress Distribution at Hip Joint During Level Walking

Kazuhiko Sakamoto, Yoshinobu Hara, Akira Shimazu[1],
and Kenji Hirohashi[2]

Summary. In 1950, Bressler and Frankel [1] calculated dynamic hip joint force for the first time. Since then, researchers have reported the results obtained by various methods [2]. These methods are divided into two categories, based either on a mathematical model or on replacing the femoral head with a prosthesis. Of course the latter method is not universally acceptable. We adopted the model of Gilbert et al. [3], this being the simplest one, consisting of a system of two rigid segments and disregarding the change in center of pressure (COP) which moves from heel to toes. This model is acceptable under ideal special conditions in which only "bone to bone" force is calculated, that is, the pressure, based on the contraction of muscles which pass over joints, is disregarded. In other words, this model is suitable when there is ideally efficient walking with little contraction. Under this supposition, the force of the hip joint during level walking was calculated in the sagittal and frontal planes, using force platform, kinesiologic, and somatotype data for the stance phase. Two normals (a 29-year-old male; subject 1, and a 70-year-old female; 2) were examined in this study. Subject 1 walked rapidly and 2 walked slowly. We compared the components of the resultant force of the femoral head. In subject 1, the resultant force had a two-peak curve, the first peak reaching 2.7 times the body weight (bw); the second peak reached 2.2 times bw and the valley between the two peaks was 0.7 times bw. In subject 2, who had a slow gait, there was no finding of peaks as with subject 1, the shape being trapezoid; the maximal value was found in the latter half of the stance and was 0.88 times bw. In subject 1, the fore-aft components exceeded the vertical one, while in subject 2, the fore-aft component was larger than that of the ground reaction force, but did not exceed the vertical component at the femoral head. We depicted the stress distribution at the joint surface of the hip based

[1] Department of Orthopaedic Surgery, Osaka City Univerity, 1-5-7 Asahimachi, Abeno-ku, Osaka, 545 Japan
[2] Kanoya Gymnastic College, 1 Shiromizu-cho, Kanoya, Kagoshima, 891-23 Japan

11

on the calculated components, using a rigid body spring model (RBSM). Stress distribution was greatest posteriorly after heel contact; in the middle of the stance the stress to the vertical side was not so large. In subject 2, the stress was moderate in all phases of the stance. Since there is no way of verifying directly whether the three femoral components calculated above are acceptable, we attempted to verify these components indirectly, by comparing the vertical components calculated in the sagittal and frontal planes. In subject 1, the shape of the components was similar from heel contact to mid-stance, but suddenly after mid-stance the components in the sagittal plane decreased. In subject 2, the shape of the components was similar in all phases of the stance. Our findings could be useful for determining stress distribution at the intact hip joint surface of healthy subjects during level walking at a slow speed. However, in the latter half of the stance phase, with rapid walking, a more complete model, particularly in regard to the foot, would be preferable.

Key words: Gait—Hip joint—RBSM—Kinesiology—Biomechanics

Introduction

In the latter half of the nineteenth century, Braune and Fischer [4] had already tried to calculate the resultant force at the hip joint during level walking. However, their attempt was unsuccessful. In 1950, Bressler and Frankel [1] calculated dynamic hip joint force for the first time. In 1966, Rydell [5] succeeded in measuring dynamic hip joint pressure dircetly by replacing the femoral head with an endoprosthesis. This method is not acceptable from any point of view. The next year, in 1967, extended dynamic hip joint pressure was calculated by Paul [6], using techniques similar to those of Bressler and Frankel [1]. The results obtained by these latter methods have not been applied in a sustained fashion.

Studying pathologically deformed hip joints from the point of view of the resultant force and components is significant; Kawai's rigid body spring model (RBSM) [7] also provides a model by which we can simulate the condition of hip joint stress distribution more precisely. As a first step toward studying pathological hip joints in the near future, we applied the results obtained with a mathematical model in normal subjects to the RBSM.

Herein we describe the resultant force and components at the femoral head during level walking, calculated from force platform data, kinesiologic data for the lower extremities, and the mathematical model created by Gilbert et al. [3]. By putting the resultant forces into the RBSM, we followed the dynamic changes in stress distribution at the hip joint.

Materials and Methods

Our subjects were a 29-year-old man (subject 1) and a 70-year-old woman (subject 2). They were made to walk, as naturally as possible, on an 8-m path on which two platforms (Kistler Instruments, Hants, UK), of 2 m × 0.4 m, were mounted in parallel. Subject 1 walked at a cadence of 120, with a step length of 0.79 m and a speed of 94.8 m/min, this being a rapid gait. Subject 2 walked at a cadence of 96, with a step length of 0.39 m and a speed of 37.4 m/min, this being a slow gait.

The floor reaction force was measured three-dimensionally, and views of walking were recorded simultaneously from the frontal and lateral sides with a high-speed video system (NAC Instruments, Tokyo, Japan). White markers

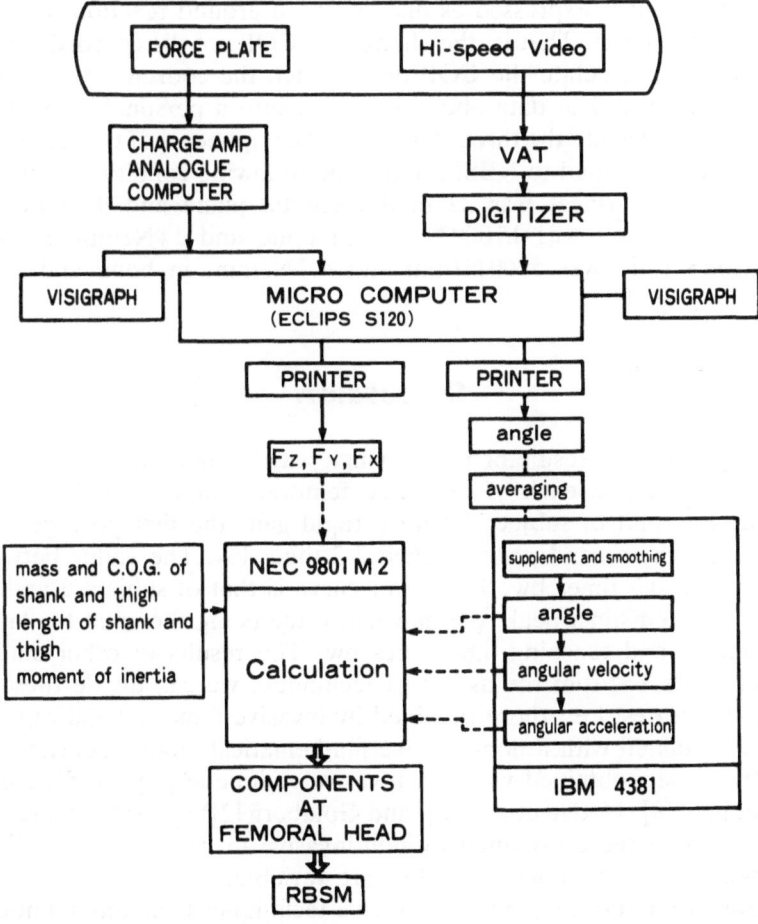

Fig. 1. Process for determining components at femoral head. *RBSM*, Rigid body spring model; *C.O.G.*, center of gravity; *VAT*, video action tracer

were patched at the relevant anatomical sites (i.e., major trochanter, lateral condyle and lateral malleolus in the sagittal plane; and the frontal skin of the hip joint, the patellar tendon, and the ankle joint space in the frontal plane) and they were digitized five times every 20 ms; the results were averaged mathematically to obtain angular changes. Filtering of raw data and supplementation of the data obtained each 40 ms, the angular velocity, and the acceleration were computed with spline function (with an IBM 4381).

Information regarding somatotype was obtained by X-ray of the lower extremities, determination of body weight, and by using Matsui's Table [8] of Japanese standard somatotypes. Based on this information, the moment of inertia, center of segmental gravity, and the weight of the segment were calculated.

We adapted the mathematical link model of Gilbert et al. [3]. This model may be the simplest one that consists of two segments. The foot is not regarded as it is in fact, but is expressed as one point, so ground reaction force always acts at the same point. That is, the changes of COP are disregarded. Thus, it is not necessary to calculate the COP or to match the coordinates of the space and force platform. The data above were put into a personal computer (PC-9801, NEC) to calculate the three-dimensional components of the femoral head. The results were applied to RBSM under the following conditions: the number of divided joint surfaces was 22 in the sagittal plane and 9 in the frontal plane; Young's ratio was 15 000 N/mm in bone and 14 N/mm in cartilage; and Poisson's ratio was 5000 N/mm and 4.7 N/mm, in bone and cartilage, respectively (Fig. 1).

Discussion

First, we studied the resultant force, which was obtained as the root of summation of the square each of the three femoral components. In Fig. 2, the upper curve is that of subject 1 with a rapid gait: the first peak reached 2.7 times bw and the second peak reached 2.2 times bw. The valley between the two peaks was 0.7 times bw. The lower curve is that of subject 2 with a slow gait: there is no distinct peak, the maximal value being shown at the latter half of the stance, and reaching 0.88 times bw. The results of other studies are shown in Table 1. From the aspect of technique, we can divide these results into two categories, i.e., those obtained by invasive femoral head replacement and those obtained with a non-invasive mathematical model. As shown in the Table, the results obtained with the invasive technique [Rydell 5], Kilvington and Goodman [9], Hodge et al. [10], and Goldberg [2]) appear to be reasonably consistent, while those obtained by non-invasive techniques (Paul [6], Seireg and Avikar [11], and Gilbert et al. [3]) are variable.

However we do not recommend invasive techniques from any point of view. Our result is smaller than most of these others, except for that of Gilbert et al. [3], whose model we adapted. This is why, as we have already mentioned

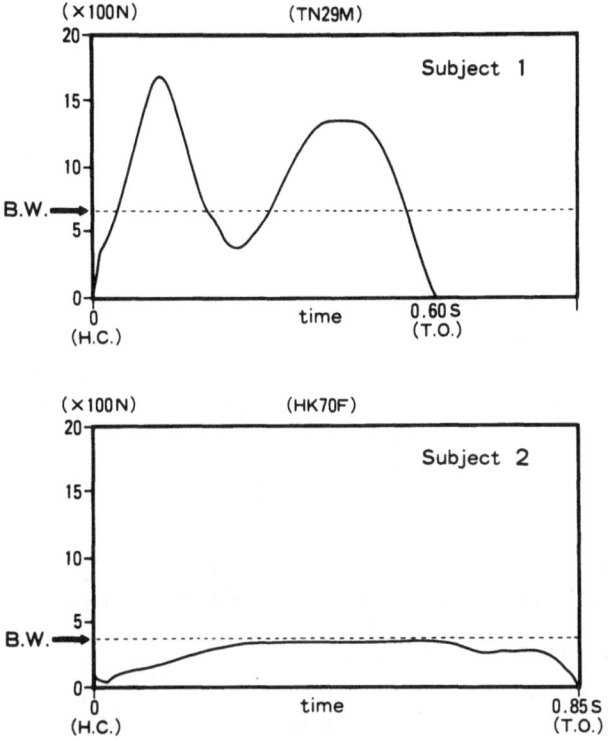

Fig. 2. Resultant force at femoral head. *BW*, Body weight; *T.O.*, toe off; *H.C.*, heel contact

Table 1. Maximal resultant force at hip joint in walking cycle.

Author	Times/bw	Method
Rydell (1965)	2.3, 2.8	Endoprosthesis
Paul (1967)	3.9	Mathematical model
Seireg and Arvikar (1975)	5.0	Mathematical model
Kilvington and Goodman (1981)	2.2	Endoprosthesis
Gilbert et al. (1984)	1.1	Mathematical model
Hodge et al. (1986)	2.6	Endoprosthesis
Goldberg (1988)	2.6, 2.8	Total hip replacement

above, we calculated the femoral components of an ideally efficient walker with little muscle contraction. Observing the motion of human walking, we assume that this action is completed by the actuators of the joint itself. Our result is calculated in the light of this supposition. Outwardly, we cannot see the muscle contraction that is the origin of these actuators. That is, the load produced by muscle contraction that passes over the joint is not included;

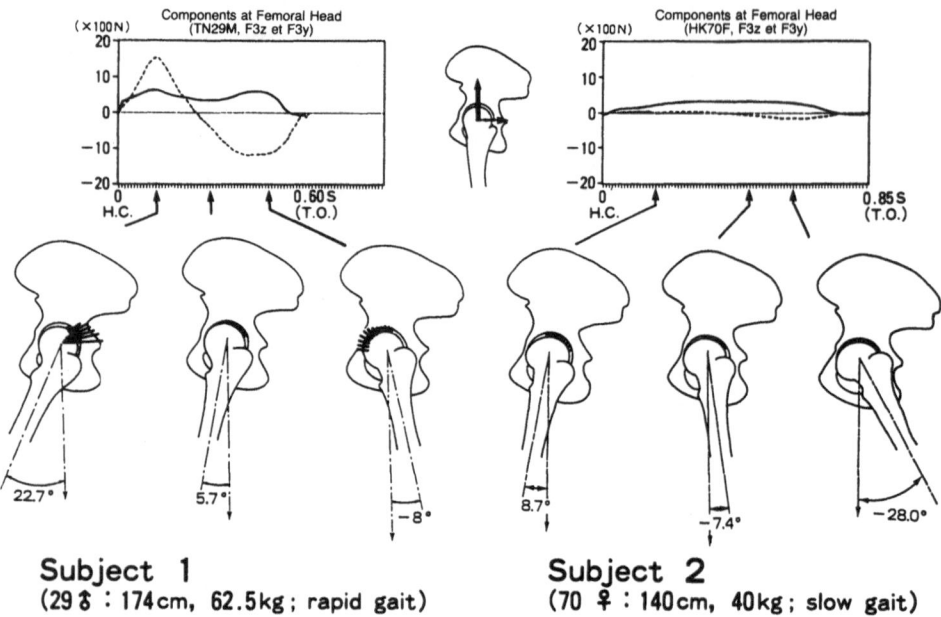

Subject 1
(29 ♂ : 174cm, 62.5kg ; rapid gait)

Subject 2
(70 ♀ : 140cm, 40kg ; slow gait)

Fig. 3. Stress distribution (sagittal plane)

the model calculates "bone-to-bone" force only. In addition, as vertical components, we used those calculated in the frontal plane.

Next, in terms of the supposition described, we simulated the stress distribution of the hip joint surface during level walking by means of the three components calculated at the femoral head in the sagittal and frontal planes.

In the sagittal plane (Fig. 3), the fore-aft and vertical components of both subjects are shown above, and the results simulated by RBSM are shown below.

In subject 1, a man with a rapid gait, the maximal value of the fore-aft components exceeds that of the vertical components by calculation, but since we have no means of verifying whether this result is acceptable, we apply the result, as it is, to RBSM. In the early phase of the stance, the stress distribution at the hip joint surface is significantly large at the posterior acetabular surface; in the midst of the stance, the stress toward the vertical direction is, unexpectedly, not so large; and at late stance, the stress distribution is large at the anterior acetabular surface, but is not as significantly large as in the early phase of the stance.

In subject 2, an elderly woman with a slow gait, the maximal value of the fore-aft components did not exceed that of the vertical components, but it exceeded that of the fore-aft component of the ground reaction force. In all phases of the stance, the stress was moderate.

In the frontal plane (Fig. 4), regarding subject 1, the upper curve shows that the lateral component is approximately biphasic, and the stress is large at

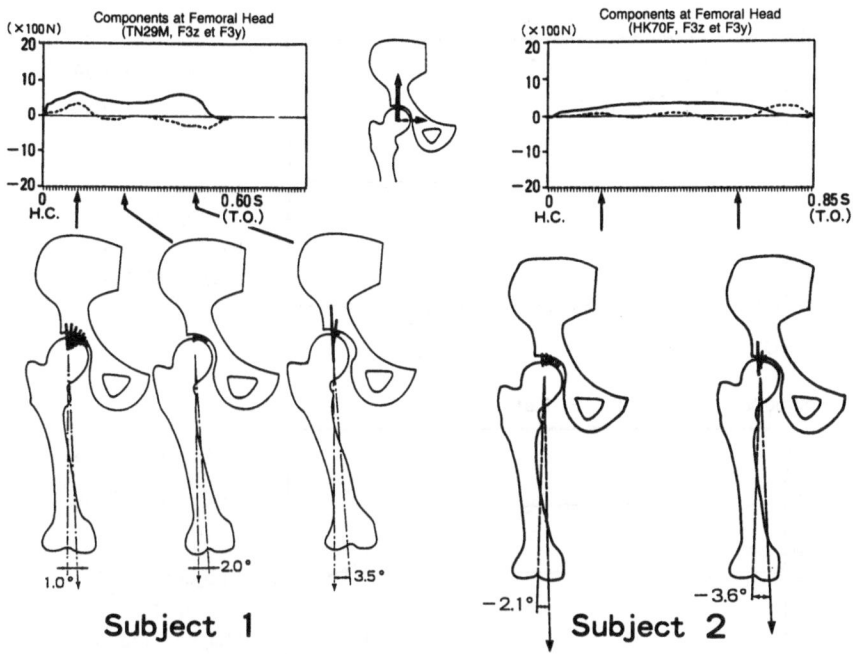

Fig. 4. Stress distribution (frontal plane)

the early phase of the stance and before the "toe-off", but in mid-stance the load and the stress is moderate. In subject 2, the lateral components have drifted polyphasically, but the resultant force is directed upward. We have no verification regarding whether this result is acceptable. Perhaps the only way to verify this is to measure the hip joint pressure by means of a femoral head endoprosthesis which is complete from any aspect, at the same time measuring the parameters, as described above, with a much more complete model that would overcome any problems, comparing the results obtained in two different ways.

Here we attempt to show whether the results we obtained are acceptable in the light of the supposition already outlined, not directly but indirectly. There is a contradiction inherent in the results which we obtained, since we calculated the components in terms of the supposition that the segments move only in one plane, i.e., in the sagittal or the frontal plane. That is, even is a healthy man or woman, the thigh flexes and extends, with a little accompanying abduction and adduction, and the shank flexes and extends, also with a little accompanying adduction and abduction, that is, the length of the segment is not constant but changes during walking. However, during the short period after heel contact, the knees are almost extended and the motion is moderate.

We compared the vertical components calculated in both the sagittal and frontal planes for the two subjects. That is, we examined this phenomenon,

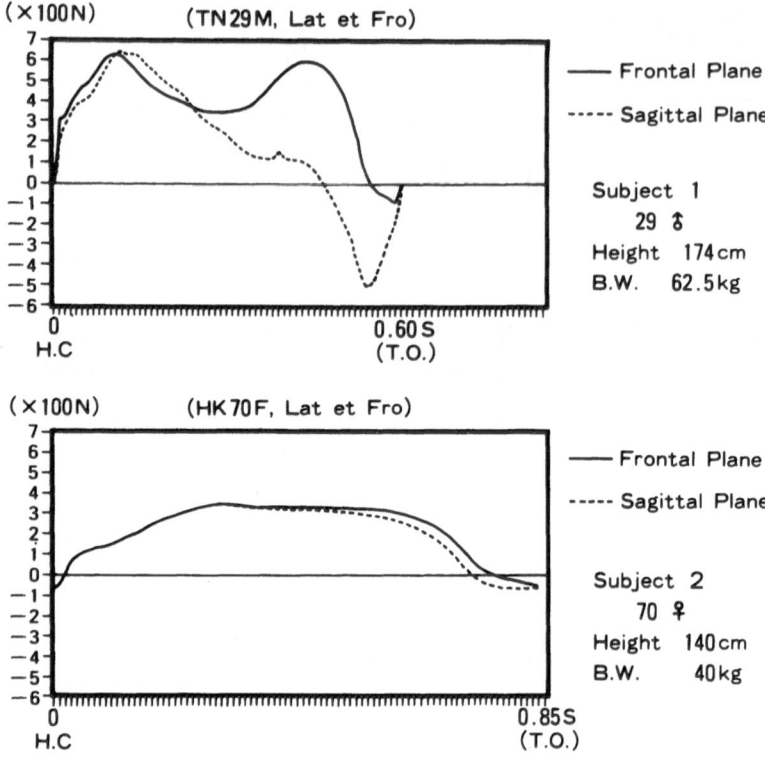

Fig. 5. Vertical component at femoral head. *Lat et fro*, lateral and frontal

walking, from two aspects, in the sagittal and frontal planes. The result would be acceptable if there were no distinct differences between the results for the two planes, even though they were calculated based on such different data for the motion of the thigh and shank, as well as the three components of the ground reaction force. The acceptability of the result would apply even if this result were to contradict the supposition mentioned above.

In subject 1 (Fig. 5, upper curves), in the first half of the stance phase the result may be acceptable, but in the latter half there is a distinct difference between the two results, the components in the sagittal plane decreasing suddenly at mid-stance and dropping below zero, and the result may not be acceptable.

In subject 2, the first half of the stance may be acceptable, and the latter half may also be acceptable, even though the result calculated in the sagittal plane is a little smaller than that calculated in the frontal plane.

We compared the findings for the lower extremities in both subjects and in both planes, and added the trajectories of the ground reaction force from the point of the markers and also from the point where the center of pressure was confirmed.

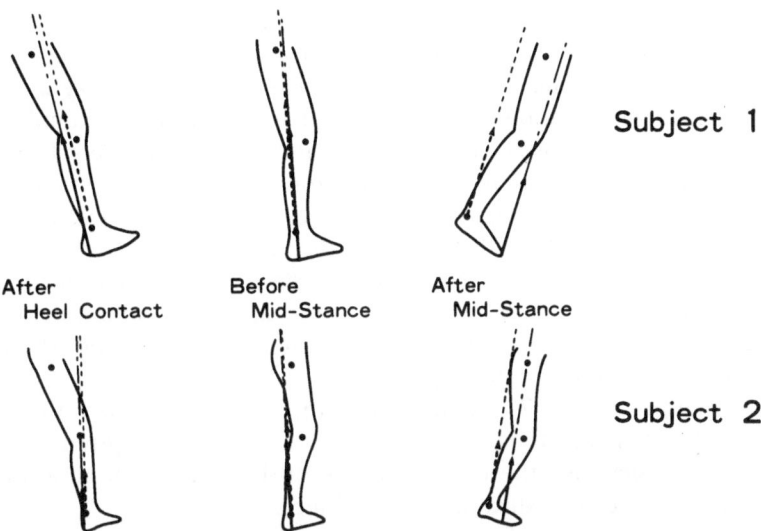

Fig. 6. Trajectories in sagittal plane. After mid-stance the distance between the two trajectories increased significantly in subject 1, who walked rapidly, using the tip of the hallux

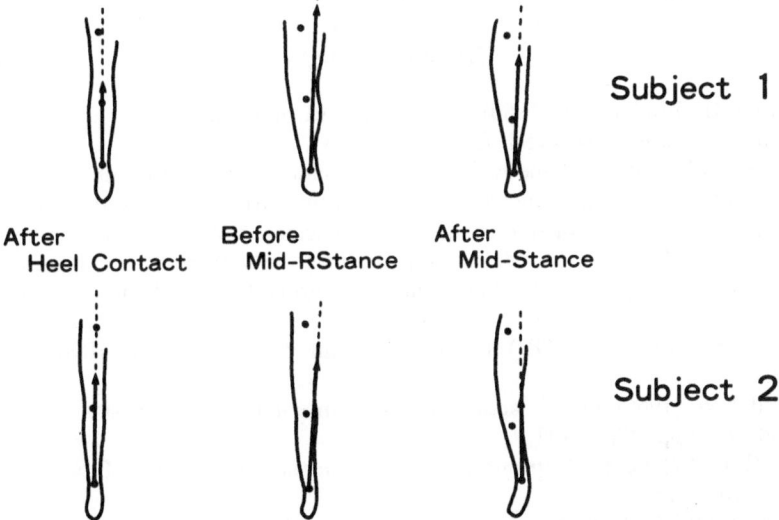

Fig. 7. Trajectories in frontal plane. The motion of the segments and the C.O.P. are small; there may be no significant difference between the trajectories in either subject

In the sagittal plane (Fig. 6), after heel contact until mid-stance there was no significant difference between the trajectories for the two subjects, but after mid-stance the distance between the two trajectories increased, more notably in subject 1, who walked rapidly, using just the tip of the hallux for body mass propulsion.

In fact, at the end stage of the stance, the ground reaction force acts at the tip of the hallux; however, in the two-segment model, this force acts near the site of the lateral malleolus. Thus, mathematically, this trajectory appears to create a large extension moment. The thigh is considered to be extended. That is to say, there is no muscle action to counteract gravity. This factor may explain why the result calculated in the sagittal plane seemed rather curious, the hip joint force being too small or negative, while the vertical component was noted at the latter half of the stance in this model.

In subject 2, even just before the "foot-off" stage, the ground reaction force has not yet reached and has not yet acted at the tip of the hallux; the distance between the two trajectories is thus not so great, and, in addition, the motion of thigh extension is not so rapid.

In the frontal plane (Fig. 7), the motion of the segments is small, the changes of COP also being small; thus there may be no significant difference between the trajectories which pass the frontal surface of the ankle joint and those that pass the true center of pressure in either subject.

We conclude that our findings could be applied to determine stress distribution at the intact hip joint surface in healthy subjects during level walking at slow speeds. In rapid walking, in the latter half of the stance, a more complete model, particularly in regard to the foot, would be preferable for determining stress distribution at the hip joint surface.

References

1. Bressler B, Frankel JP (1950) The force and moments in the leg during level walking. Trans Am Soc Mech Engrs 72:27–36
2. Goldberg M (1988) Telemetric force measurements across the hip after total arthroplasty (abstract of educational lecture) In: Proceedings of the 15th annual conference of the Orthopedic Biomechanics Society, Yokohama, September, 1988
3. Gilbert JA, Maxwell GM, McElhaney JH, et al. (1984) A system to measure the forces and moments at the knee and hip during level walking. J Orthop Res 2:281–288
4. Braune W, Fischer O (1987) The Human Gait. Springer, Berlin Heidelberg New York
5. Rydell NW (1966) Forces acting on the femoral head prosthesis. Acta Orthop Scand 37 [Suppl 88] 1–132
6. Paul JP (1967) Forces transmitted by joints in the human body. Proc Instn Mech Engrs 181:8–15
7. Kawai T (1977) A new element in discrete analysis of plane strain problem (in Japanese). Seisan Kenkyu 29:204–207
8. Matsui S (1967) Text for kinesiology (in Japanese). Kyorin-Shoin, Tokyo, p 112
9. Kilvington M, Goodman RMF (1981) In vivo hip joint forces recorded on a strain gauged "English" prosthesis using an implanted transmitter. Eng Med 10:175–188
10. Hodge WA, Fijan RS, Carlson KL, et al (1986) Contact pressures in the human hip joint measured in vivo. Proc Natl Acad Sci USA 83:2879–2883
11. Seireg A, Arvikar RJ (1975) The prediction of muscular load shearing and joint forces in the lower extremities during walking. J Biomech 8:89–102

3

Biomechanical Studies of the Hip Utilizing Computer Simulation

Takatoshi Ide, Yasuhiro Yamamoto, and Shigeru Tatsugi[1]

Summary. Estimation of the force distribution on the hip joint articulating surface was studied with stress analysis calculated by computer simulation based on Kawai's Rigid body spring model (RBSM) using plane radiographs. The result of the two-dimensional RBSM, modified to take into consideration the depth of the joint, showed that the maximum compressive stress on the contact surface was estimated at 0.031 body weight (BW)/mm in the normal hip. The maximum stress and the resultant force were decreased with an increase of the center-edge angle, but the change of the resultant force was not remarkable compared with the maximum stress. It was estimated that the maximum compressive stress was about five times with less than 0° of the center-edge angle compared with 27° in the normal hip. With the stress analysis using RBSM it was possible to estimate the biomechanical behavior of the hip precisely. This system is useful for clarifying osteoarthritis of the hip in biomechanical terms and for the prognosis of coxarthrosis in clinical cases.

Key words: Coxarthrosis—Rigid body spring model (RBSM)—Stress analysis—Center-edge angle

Introduction

The hip joint is an one of the most important weight-bearing joints in the human body. In order to investigate the origin and prognosis of coxarthrosis, it is essential to consider biomechanical factors in the anatomical features of the hip. The purpose of this study was to estimate force distribution on the hip joint articulating surface, using a computer simulation technique, and to determine the prognosis of coxarthrosis via biomechanical factors.

---

[1] Department of Orthopedic Surgery, Yamanashi Medical University, 1110 Shimokato, Tamaho-cho, Nakakoma-gun, Yamanashi, 409-38 Japan

Methods

Computer Simulation

The stress analysis was performed using Kawai's rigid body spring model (RBSM) [1]. A frontal section at the hip joint, consisting of the pelvis and femur segments, was used as the two-dimensional and two-element simulation model (Fig. 1). Virtual springs, which resist compressive and shear force, were positioned 1 mm apart along the joint between the acetabulum and the femoral head. Virtual muscle structure springs, which resist only tensile force, were positioned between the pelvis and the great trochanter of the femur as the abduction muscle. An external load of 60 kgf as the body weight was loaded at the center of the fifth lumbar spine and the femur element was fixed simultaneously. The center of gravity of the pelvis was positioned at the center of the fifth lumbar spine, and the femoral center of gravity was positioned at the mid point of the femoral shaft.

In the RBSM method, bone segments are generally assumed to be rigid, while articulating surfaces and musculo-ligamentous structures are simulated by springs. It is possible to classify the virtual spring characteristics into three types. The compressive spring resists compressive force, but it breaks when tensile force is loaded. The musculo-ligamentous spring has characteristics opposite to those of the compressive spring, and the shear spring is regarded as friction between adjacent bodies. The RBSM is formulated by considering the rigid body to be in equilibrium with external loading [2,3]. Reaction forces

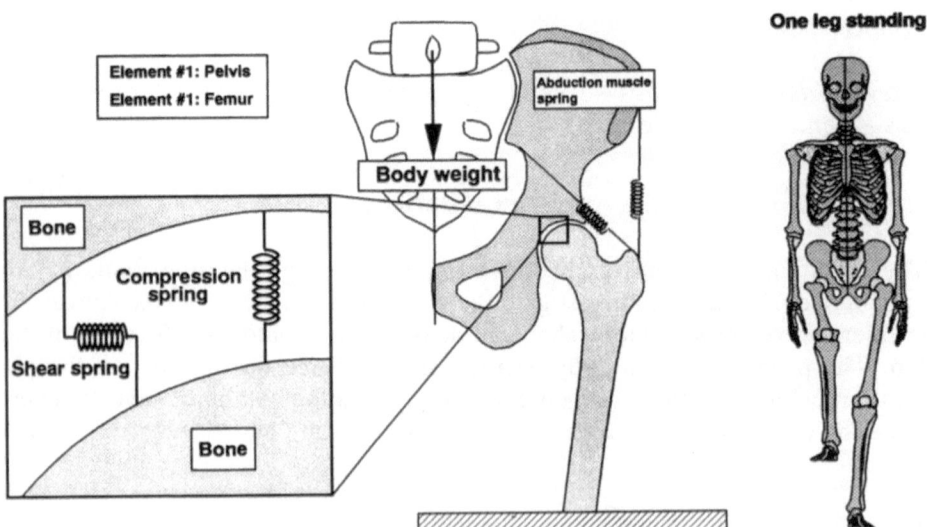

Fig. 1. Computer simulation model of the hip: Virtual springs were positioned along the joint, and muscle springs were established as the abduction muscle. The external load of 60 kgf as the body weight was applied through the 5th lumbar spine

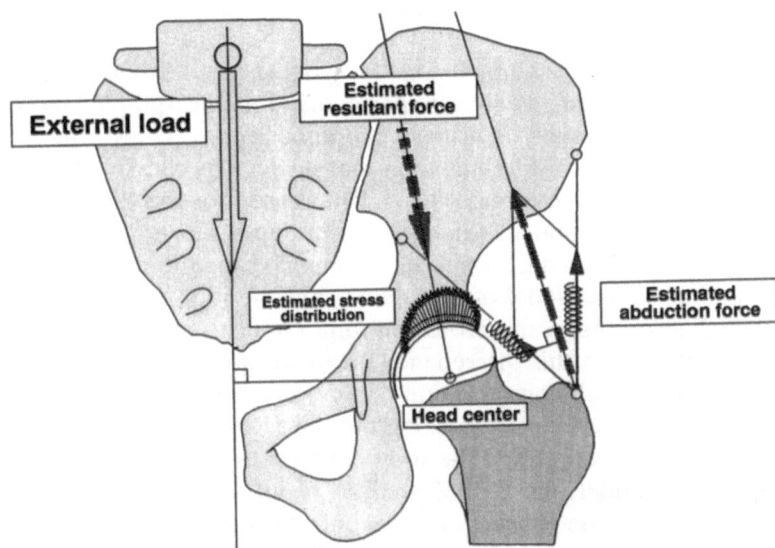

Fig. 2. Estimation of the resultant force: The resultant force acting on the femoral head center was estimated by equillibrium analysis with the external gravity load and the abduction force

between adjacent bodies are produced by the spring system distributed over the possible contact surface between the adjacent bodies. The weight which is added to these rigid bodies is transmitted through these springs. The calculation using RBSM must rely upon an iterative process in order to achieve the equilibrium condition, in contrast to the difficulties encountered with contact stress problems or with the use of the finite element method (FEM) technique.

From the analysis using RBSM, the stress distribution on the hip joint surface and the abduction force were estimated in the one leg standing situation. The abduction force was determined by taking the moment of the abductor muscle force and the body gravitational force with respect to the hip joint. The resultant force acting though the center of the femoral head toward the acetabulum can be calculated using the equilibrium equations with the lever arm theory (Fig. 2).

Material Constant and Modified Two-Dimensional RBSM Model

In the RBSM method, the material constant of the cartilage, ligament, and muscle should be included in the analysis as the spring constant. In the past, many authors have reported marked differences in elastic modulus values from 10 to 50 MPa as estimated by different experimental studies [4,5]. However, these material constants were not derived for the purpose of formulating computational mechanics problems. Hence, it is not always necessary to describe their precise biomechanical behavior in vivo precisely for the purpose of conducting FEM analysis.

According to a previous comparison study between an experimental loading test with frest frozen human cadaver specimens of metacarpo-phalangeal joints (MP joints) and stress analysis using a two-dimensional RBSM, the estimated displacement was 6.7 times greater than that of the experiment while the peak pressure was only 0.6 times greater [6,7]. The reason for this discrepancy could be explained by a plain model where the joint depth was not considered. This can be remedied by reducing a three-dimensional shaped joint and projecting it into a two-dimensional model of unit depth, but the joint cartilage springs will be reassigned according to the projected joint contact surface depth.

To find a solution to this discrepancy problem, a modified two-dimensional (2D) RBSM model with consideration of the joint depth was created. If many of the virtual springs are lined up along the depth of the joint, the spring constant can be adjusted according the joint contact surface depth (Fig. 3). Using the adjusted 2 D model, the RBSM analysis result was found to approach the experimental data. The elastic modulus of the cartilage was estimated as 25.8 MPa and the spring constant of the cartilage was 22.6 N/mm per unit area after displacement matching by the modified 2 D RBSM [8]. These material constants were used in the present study.

Anatomical Joint Area and Weight Bearing Area

It is obvious that the radiographically-defined joint contact area will not be responsible for the load transmission mechanism under weight bearing. In contrast, certain areas will carry a high pressure while other parts may become separated and thus carry tensile forces. Therefore, the entire joint area was defined as the anatomical area of joint contact, and the area related to the load transmission was defined as the actual weight-bearing area (Fig. 4). The actual

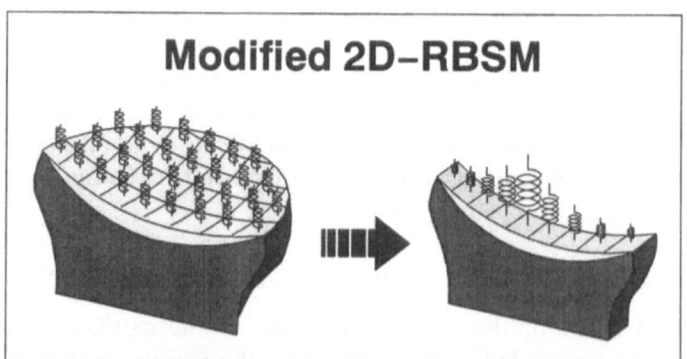

Fig. 3. Concept of the modified 2D-RBSM: The joint contact area was divided into small unit subsquares and multiplied in the transverse direction. The number of units of the spring created non-uniform two dimensional springs across the length of the contact line. *2D-RBSM*, two-dimensional rigid body spring model

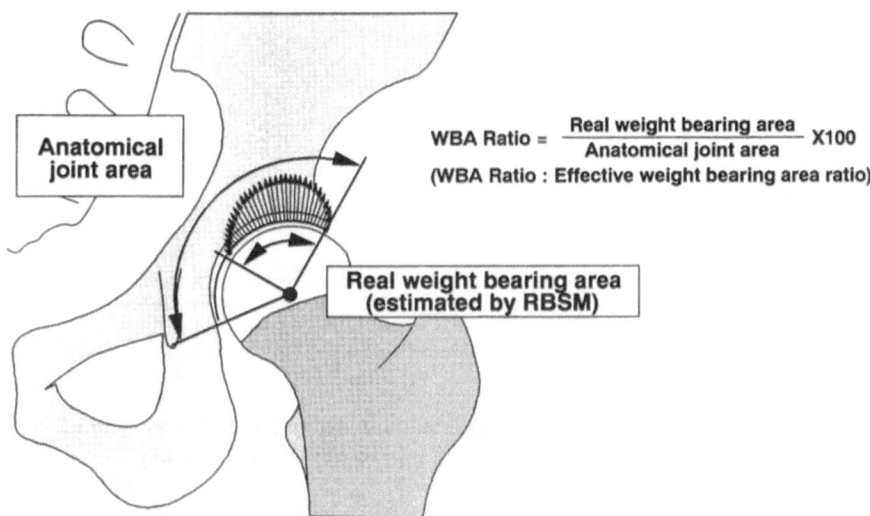

Fig. 4. Definition of effective weight-bearing ratio (*WBA ratio*). The real weight-bearing area and WBA ratio was estimated with RBSM. The center-edge angle was defined as 27° and the anatomical joint area had 140° of the open angle in the standard hip mode

weight-bearing area expressed as a percentage of the anatomical area of joint contact was defined as the effective weight-bearing area ratio (WBA ratio).

Standard Model of the Hip

A standard model of the hip for the stress analysis using RBSM was made with a Japanese standard physical constitution (Fig. 5). The standard normal model was defined as that having a radius of the femoral head of 46 mm, excluding the cartilage thickness, and an acetabulum radius of 50 mm. Therefore the radius of the joint contact surface was 48 mm, the cartilage thickness was 4 mm, the center-edge angle (CE angle) was 27°, and the Sharp angle was 41°. The anatomical joint area of contact in the standard normal model has a 140° open angle which is bounded by a line from the femoral head center to the tear drop and a line through the acetabulum edge from the head center. The stress analysis using RBSM was performed with this standard model, and the force and pressure distribution were determined analytically subject to variations of these anatomical parameters.

Analysis of Clinical Cases

A total of 89 clinical cases of secondary osteoarthritis with acetabular hypoplasia plus 16 normal hips were analyzed using the computer simulation technique with RBSM. In order to identify the key landmarks for the origin and insertion point of the abduction muscle on the great trochanter, the inner and lateral

T. Ide et al.

1	Radius of femoral head (without cartilage layer)	46mm
2	Radius of acetabulum	50mm
3	Radius of the joint surface (included acetabulum and femoral head side)	48mm
4	Thickness of cartilage layer	4mm
5	Center–edge angle(CE angle)	27 deg.
6	Sharp angle	41 deg.
7	Acetabular–head index:AHI(a/b%)	84.0
8	Open angle of the anatomical joint area	140 deg.

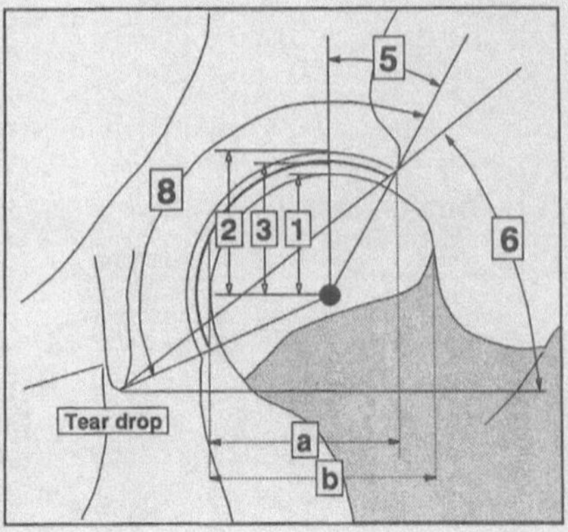

Fig. 5. Standard model of the hip: The standard normal model was defined for the stress analysis using RBSM with Japanese standard physical constitution

edge of the pelvis and the center of the fifth lumbar spine were selected. To determine the joint contact line, a computer graphics X-ray image analyzing system was used. The radiographic image of the hip was scanned and entered into a personal computer through a video scanner (Epson: GT-6000), and key landmarks were digitized on the cathode ray tube (CRT) screen. A joint contact line was composed with several segments of arcs which could be made from clear landmarks on the joint articulating surface, and the center of the femoral head was approximated using the mean value of these arc centers.

Results

Standard Model of the Hip

It was estimated that the resultant force for the femoral head was 2.74 times the body weight and the abduction force was 1.81 times that in the standard normal hip model under one-legged standing conditions. The maximum compressive stress on the contact surface was estimated at 0.031 body weight (BW)/mm ($BW = body\ weight$), the mean value was 0.02 BW/mm, and the WBA ratio was estimated at 65.65% of the available area of contact. The stress distribution of the standard hip model is shown in Fig. 6.

Standard Model and CE Angle

Estimations of the stress distribution with various values of the CE angle were performed with RBSM. The analysis showed stress concentration on the joint contact surface (Fig. 7), the maximum compressive stress being increased remarkably with decreasing CE angle (Fig. 8). The maximum compressive stress with a CE angle of $-23°$ was 5.87 times that with the normal hip model. Changes of the stress distribution and the maximum compressive stress were

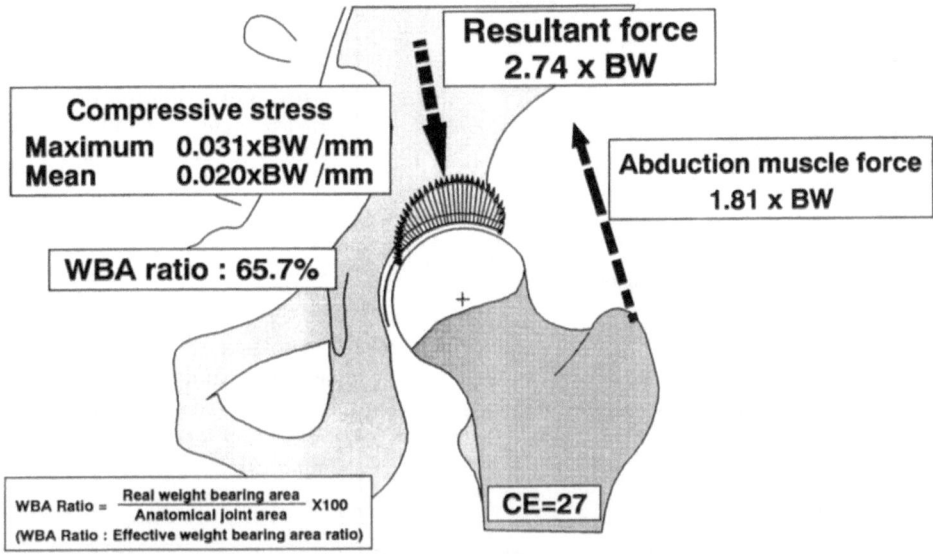

Fig. 6. Results for the standard normal hip model. *BW*, body weight

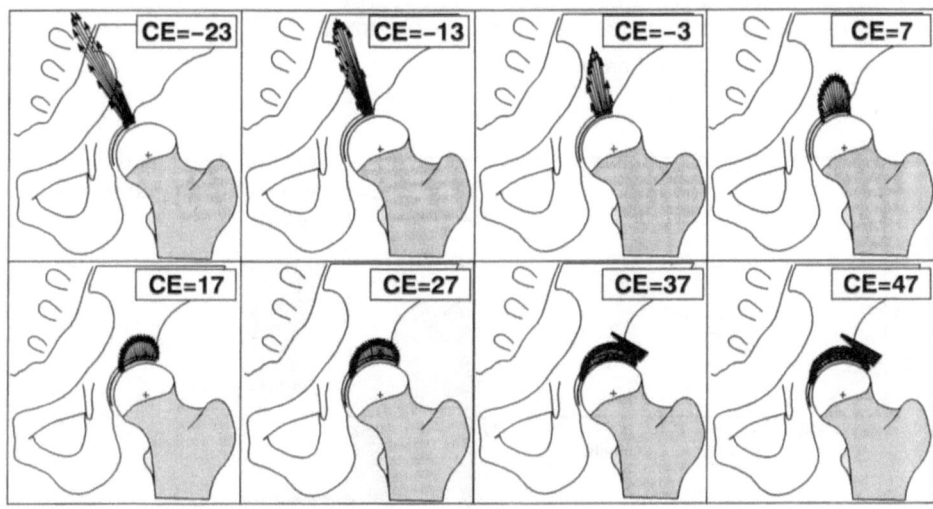

Fig. 7. Estimated stress distribution for each CE angle

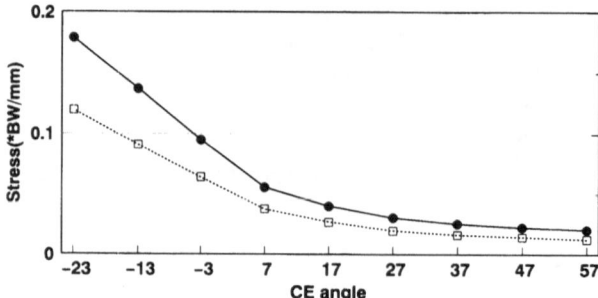

Fig. 8. Correlation between center-edge angle and estimated maximum compressive stress. *Open squares*, mean; *closed circles*, maximum

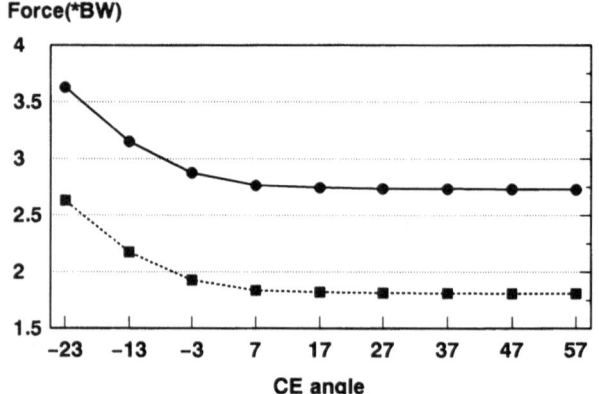

Fig. 9. Estimated resultant force (*closed circles*) and abduction force (*closed squares*)

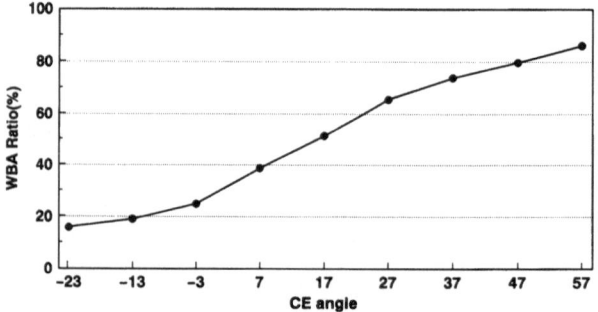

Fig. 10. Estimated effective weight bearing ratio (*WBA ratio*)

not significant when the CE angle was larger than 27°. The resultant force was increased with changes of the CE angle (Fig. 9); however, the change of resultant force according to the CE angle was not sensitive compared with the maximum stress. The effective weight-bearing area ratio was diminished linearly with decreasing CE angle (Fig. 10), with only 15.92% being estimated as the WBA ratio when the CE angle was −20°.

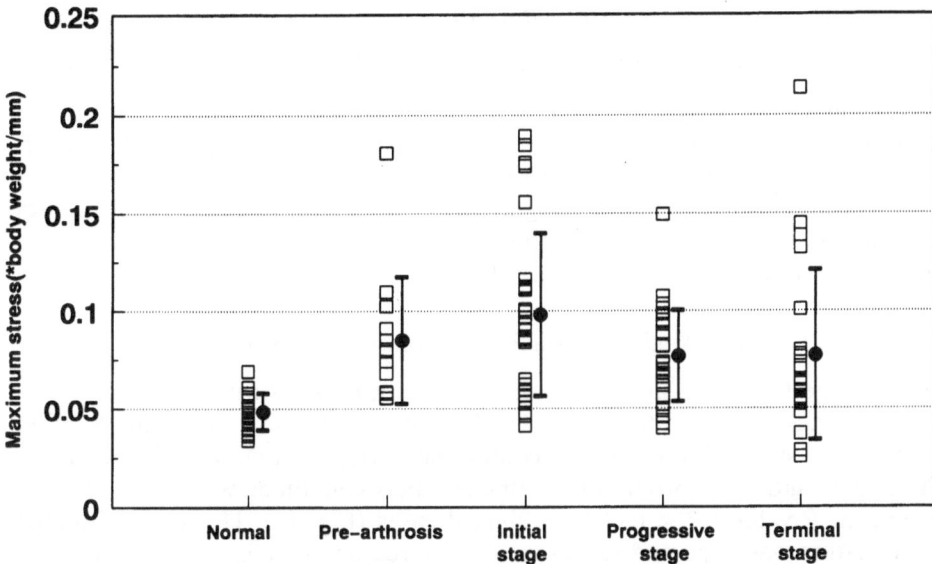

Fig. 11. Maximum compressive stress according to Japanese coxarthrosis criteria

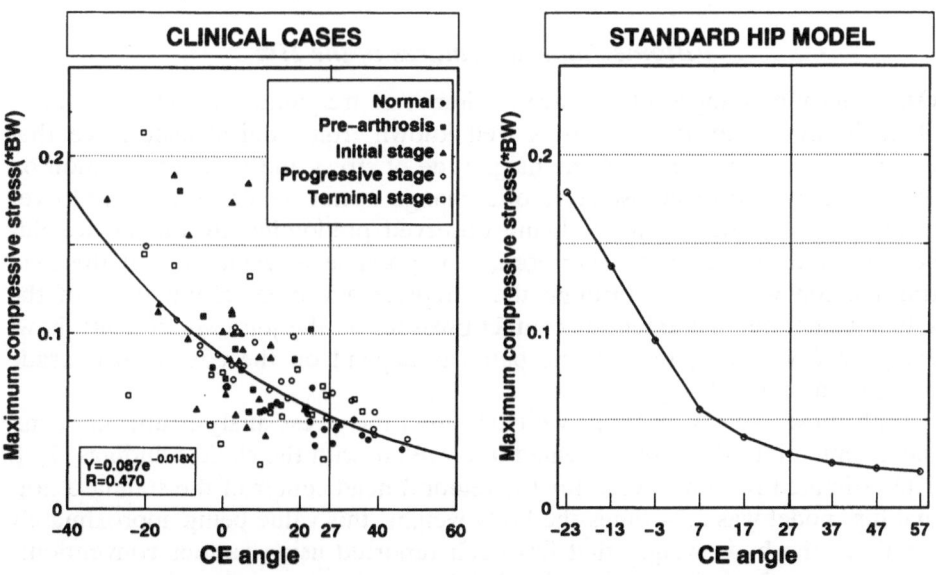

Fig. 12. Correlation between CE angle and maximum compressive stress in clinical cases: The trend of maximum compressive stress associated with the CE angle was similar when compared to the result in the standard normal hip model. *Closed circles*, normal; *closed squares*, pre-arthrosis; *closed triangles*, initial stage; *open circles*, progressive stage; *open squares*, terminal stage. $Y = 0.087\,e^{-0.018X}$; $r = 0.470$

Rsults in Clinical Cases

Stage of Coxarthrosis and Maximum Compressive Stress

In clinical cases, the mean value of the estimated maximum compressive stress was 0.048 BW/mm in the 16 normal hips studied. The estimated maximum compressive stress was 1.75 times the normal value in pre-arthrosis of coxarthrosis according to the staging criteria of the Japanese Orthopedics Association, being 2.03 times in the initial stage, and about 1.6 times in the progressive and terminal stage (Fig. 11).

CE Angle and Maximum Compressive Stress

The average CE angle in the 16 normal cases was 26.78°; it was 6.85° in pre-arthrosis, 2.20° in the initial stage, 11.97° in the progressive stage, and 8.90° in the terminal stage. The maximum compressive stress increased with decreasing CE angle, and the correlation coefficient between them was 0.470, with a significant P value ($P < 0.005$) in clinical cases (Fig. 12). However, although the variation was large and values were scattered among the cases, the trend of maximum compressive stress associated with the CE angle was similar when compared to the result in the standard normal hip model.

Discussion

Biomechanical Features of the Hip

Osteosclerotic changes of the acetabulum can frequently be observed on a plane X-ray in coxarthrosis. It is well known from clinical experience that osteosclerosis corresponds to the magnitude of stress and to its distribution on the hip joint. However osteosclerotic change is not generally observed over whole areas of the hip joint, being observed predominantly on the weight-bearing area of the joint. Therefore, it is possible to estimate that the load transmission would pass through the osteosclerotic area. Estimations of the actual weight-bearing extent or contact pressure on the joint are difficult, however, since various osteosclerotic patterns depend on the coxarthrosis grade recognized on the X-ray.

With these aspects in mind, we performed the stress analysis utilizing computer simulation; the results seemed to correlate with the clinical findings [7,9]. The estimated resultant force for the femoral head center of the standard normal hip model was 2.74 times the body weight, this value being approximately 2.8 times the body weight that has been reported using various conventional analytical methods in the past [4,5]. The maximum compressive stress was estimated as 0.031 BW/mm, and it was equivalent to 1.86 kgf/mm^2 of the contact pressure loaded through the joint contact area with a body weight of 60 kg. The actual weight-bearing area was estimated to take approximately 60% of the anatomical joint contact area in the standard normal hip model.

Coxarthrosis Prognosis

The estimated maximum compressive stress in clinical cases seemed to be correlated with decreasing CE angle ($r = 0.47$), although the variation between cases was large. It was considered that estimation of coxarthrosis progression using these factors could be determined with only 47% probability if the CE angle were used as a prognosis indicator. Both the maximum compressive stress and the resultant force for the femoral head increased with decreasing CE angle in the standard normal hip model, but changes in the resultant force were not as remarkable as those in the maximum compressive stress. For example, the maximum compressive stress in pre-arthrosis with 6.85° CE angle on average was 1.75 times that in the normal, with 26.78° CE angle in clinical cases, while the resultant force was 1.27 times normal. In the standard hip model, this trend was more remarkable. The maximum stress with 7° CE angle was 1.83 times the normal with 27°, while the resultant force was only 1.01 times normal. Determination of the femoral head center for the estimation of the resultant force, using equilibrium analysis or CE angle is measurement, is considered to be difficult in coxarthrosis with a deformed femoral head, using a conventional technique. On the other hand, determination of the femoral head center is not necessary for stress analysis when using the RBSM method. Therefore, it was concluded that, with stress analysis utilizing RBSM, it was possible to correctly estimate the biomechanical behavior of the hip joint. This method is thus useful for determining the prognosis of osteoarthritis of the hip.

Hypoplasia of the Acetabulum and Compressive Stress

The maximum compressive stress was increased steeply with reducing CE angle (under 27°) in the analysis of the standard normal hip model. However, changes in the maximum stress were not significant with a CE angle over 27°. If the stress value was assumed to be 1.00 with a CE angle of 27°, it was roughly estimated that the value was nearly 0.75 times this with a CE angle of more than 27°, two times between 0° and 27°, and five times at less than 0°. The

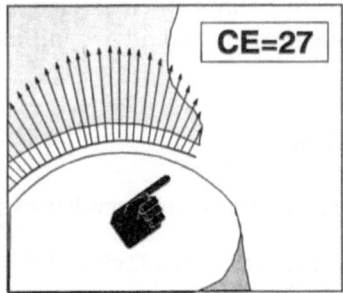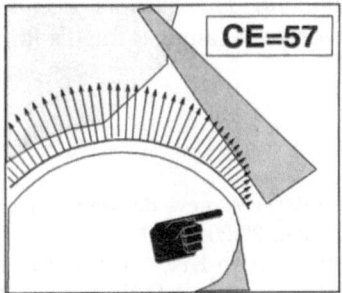

Fig. 13. Stress distribution on the bone segment of the shelf operation: The compressive stress on the edge of the bone block was small compared with the normal

compressive stress on the acetabular edge with a CE angle of more than 27°
was less than the stess in the standard normal hip model (Fig. 13). It was
estimated that the compressive stress on the lateral edge of the bone segment
would be less than that in the normal hip when a shelf operation was performed.
Therefore, bone absorption after the operation could be estimated with this
lesser compressive stress on the lateral edge of the bone block. It was considered that these results could provide some possible solutions to the problem
of how to decide the size of a bone fragment when acetabulum plasty is
performed using a shelf operation for hypoplasia of the hip. The size of the
bone transplant should be selected so that the CE angle will be around 27°. It is
not necessary to perform a bone block transplantation when the CE angle is
more than 27° in patients with a normally shaped femoral head, because the
range of hip motion may be limited and there is a possibility of bone absorption
after the operation. However, if the selection of bone segment size could be
determined by the CE angle, it would be more beneficial, in cases which
involve deformities of the femoral head, to select bone size using computer
simulation based on RBSM.

Conclusions

1. The maximum compressive stress on the contact surface was estimated at
   0.031 BW/mm, and the effective weight-bearing area ratio was estimated at
   65.6% in the computer simulation of the standard normal hip model.
2. The maximum compressive stress and the resultant force decreased with
   increasing CE angle; however, the change in resultant force was not as
   remarkable as the change in maximum stress.
3. Hip joint contact stress analysis using RBSM allowed the estimation of the
   biomechanical behavior of this joint. It is possible to use these results in
   determining the prognosis of coxarthrosis in clinical cases, without the need
   to determine the femoral head center.

Acknowledgment. The authors would like to express their deep gratitude to
Professor Noriya Akamatsu for his kind suggestions and advice in this study.

References

1. Kawai T (1977) A new discrete model for analysis of solid mechanics problems.
   Seisan Kenkyu 29:208–210
2. Garcia-Elias M, An KN, Cooney WP, Linscheid RL, Chao EYS (1989) Transverse
   stability of the carpus. J Orthop Res 7:738–743
3. Horii E (1990) Effect on force transmission across the carpus in procedures used to
   treat Kienbock's disease. J Hand Surg 15:393–400

4. Askew MJ, Mow VC (1978) The biomechanical function of the cartilage fibril ultrastructure of articular cartilage. J Biomech 100:105–115
5. Kempson GE (1980) The mechanical properties of articular cartilage. The joints and synovial fluid. Academic, New York, pp 177–238
6. Ide T, Hara T, An KN, Chao EYS (1990) Stability and pressure distribution of joint articulating surfaces. In: Furg YC (ed) Abstracts of the 1st world congress of biomechanics, vol 11. UCSD Price Center, San Diego, p 57
7. Ide T, Yamamoto Y, Tatsugi S, Akamatsu N, Chao EYS (1990) Comparison studies between computer simulation and experimental measurements using the metacarpophalangeal joint (in Japanese). Proceedings of the 1990 annual meeting of the Japanese Society for Orthopaedic, Biomechanics JSOB 12:119–124
8. Ide T, Yamamoto Y, Tatsugi S, Akamatsu N, Chao EYS (1990) Determination of spring constants for the rigid body spring model (in Japanese). Proceedings of the 1990 annual meeting of the Japanese Society for Orthopaedic Biomechanics JSOB 12:125–131
9. Ide T, Chao EYS (1990) Computer-aided implant selection and placement analysis. In: Furg YC (ed) Abstracts of the 1st world congress of biomechanics, vol 11. UCSD Price Center, San Diego, p 192

4

—Overview—
Estimation of Femoral Head Resultant Force with Three-Dimensional Rigid Body Spring Model

Shinkichi Himeno[1]

Summary. Since 1981, when this author introduced the rigid body spring model (RBSM), it has been used for contact pressure analyses of the joint. This method has now been expanded three-dimensionally, and in this study it was applied for the estimation of the contraction forces of the muscles around the hip joint, and the resultant force vector of the femoral head. Every muscle is modeled as a spring whose stiffness is proportional to its cross-sectional area. The reaction forces for each spring, under various hip positions and loading conditions, are calculated for related muscle forces. The resultant force vector of the femoral head in the one-leg standing position is calculated as 2.93 times body weight in magnitude, and as 14.5° medially and 10.4° anteriorly in direction. The proximal part of the gluteus maximus muscle, the gluteus minimus muscle, and the gluteus medius muscle, play an important role in hip abduction. Varus osteotomy of 30° at the femoral neck decreases the magnitude of the resultant force vector to 0.8 times body weight, and increases medial inclination to 5° in direction. Co-contact of the hip muscles for shock absorption increases the magnitude of the resultant force, without directional change. In the subluxated hip, the magnitude of the resultant force is increased; however, the medial inclination of the force is also increased, and this reduces the hip instability. This method of resultant force estimation is expected to be useful in the biomechanical evaluation of the pre- and postoperative hip joint, and in facilitating improvements in joint prostheses.

Key words: Hip joint—Resultant force—Estimation—RBSM—Computer Simulation—FEM—Three-dimensional

[1] Division of Orthopedics, Himeno Hospital, Niishiro 2316, Hirokawa-machi, Yame-gun, Fukuoka, 834-01 Japan

Introduction

Estimation of the resultant force vector of the femoral head provides essential information for the pre- and postoperative biomechanical evaluation of the hip joint, and for improving prosthetic designs in arthroplasty. It is also valuable for rehabilitational planning to know how muscles work in each motion phase. For this reason, the resultant force vector of the femoral head has been estimated by a number of authors. Pauwels [1] performed two-dimensional balance analysis of the gluteus medius muscle and body weight, and estimated the resultant force as three times body weight in magnitude, and 16° medially in direction. Williams [2] created a pseudo three-dimensional model of the muscles, and Merchant [3] obtained a value of five times body weight by making direct measurements of chains that were analogous to muscles. Rydell [4] embedded strain gauges in hip prostheses to make direct measurements of the resultant force. Recent advances in microelectronics have now made it possible to perform in vivo measurements, even over periods of several years, as Davy et al. [5] have done.

Mathematical hip models have also been made to allow for the calculatation of muscle forces with a computer. Seireg and Arvikar [6] adopted linear programming, while Crowninshield and Brand [7] and Konishi [8] used non-linear programming, to estimate the resultant force. Such mathematical approaches are expected to be of great value because of their non-invasiveness and flexibility.

Kawai and Toi [9] developed a rigid body spring model (RBSM), a kind of finite element method, for application in civil engineering in areas such as sliding of land, mining problems, and so forth. In 1981, the author of this chapter introduced this technique to the field of biomechanics, principally to solve contact problems of the joint [10–16]. In this chapter, the application of this RBSM is expanded to allow for the estimation of the muscle forces of each muscle around the hip joint and their resultant force.

Materials and Methods

Non-Linear Programming

More than 30 muscle forces must be determined; however, only six equations are built, namely, translations and rotations of the x, y, and z axes. Thus, there are more unknown variables than equations, this situation being known as "statically indeterminate". In other words, there is an infinite number of solutions, i.e., patterns of muscle contractions, that satisfy the equations. Linear or non-linear programming is a mathematical method of selecting the best solution from among this infinite number of solutions, by evaluating each solution according to a criterion that is called the objective function. The validity of this method depends to a great extent, on the objective function.

Let F denote a muscle force, and S denote the cross-sectional area of a muscle. Seireg and Arvikar [6] adopted $\Sigma(F)$ for the objective function. Forces are then concentrated on quite a few muscles, those which have the longest lever arms to the related axes. This is obviously different from physiological observations. Crowninshield and Brand [7] adopted $\Sigma(F/S)^3$, and Konishi [8] adopted $\Sigma(F/S)^2$ as objective functions for each analysis. They did not clarify the physiological rationale of these criteria, and, worse, their methods had a serious disadvantage, in that the solution depended to a great extent on the method applied for muscle division. Suppose a muscle whose cross-sectional area is 3.0 generates a force of 3.0. When the muscle is virtually divided into a ratio of $1:2$, each muscle bundle should have forces of $1:2$. However, the criterion of Crowninshield and Brand [7] gives a value, of $0.8:2.2$, and that of Konishi [8] gives $0.6:2.4$. Herein, a more realistic objective function is given, in that it has a physiological basis.

Modelling of Muscle

The two following hypotheses are adopted in muscle modelling: (1) The muscle contraction force (F) is proportional to the cross-sectional area (S) of the muscle. This hypothesis is based on the physiological observation that muscle force is proportional to the number of actin/myosin filaments.

(2) The muscle contraction force is proportional to the elongated length produced ($\Delta\lambda$) when a load is applied. This force represents the function of the muscle spindle, a sensor of muscle length. The γ-fiber system gives the muscle spindle an objective length to be maintained. When the muscle is elongated by a load, the muscle spindle generates afferent pulses whose frequency is proportional to the difference between the objective length and the actual length. This signal is fed to α-neurons to produce contractions of the α muscle fiber.

The muscle contraction force (F) is considered to be proportional to the product of the cross-sectional area (S) of the muscle and its elongated length ($\Delta\lambda$) ... ($F = S * \Delta\lambda$). Compared with the balance equation of a spring ($F = k * x$), where K is the spring stiffness coefficient, and x is the elongation of the spring, the muscle is equivalent to a spring whose stiffness is proportional to its cross-sectional area (S). In the equilibrium state, the total energy stored in the spring takes the minimum value $\Sigma(k * x^2)$, namely $\Sigma(S * \Delta\lambda^2)$. After the substitution ($\Delta\lambda = F/S$) is made, the minimum energy stored in the spring system is $\Sigma(F^2/S)$. In conclusion, this muscle modelling is equivalent to taking $\Sigma(F^2/S)$ as the objective function in non-linear programming.

This objective function gives the correct solution ($1:2$) for the virtual muscle division problem noted above in the last paragraph of the section on non-linear programming.

Suppose two muscles work on the same joint axis. When a load is applied, the elongation of a muscle is proportional to its lever arm. Physiologically, muscle force is larger if a muscle has a larger cross-sectional area and larger

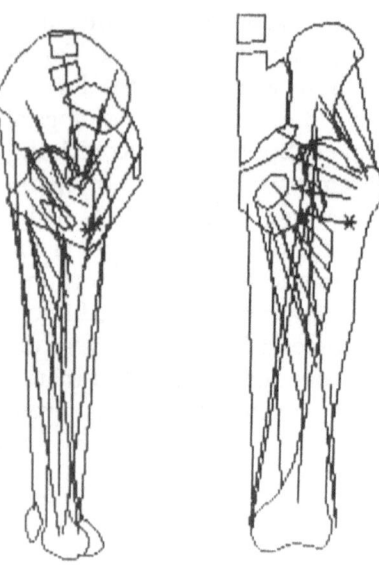

Fig. 1. Lateral view (*left*) and anteroposterior view (*right*) of the anatomical model employed in this analysis

lever arm. Although the model proposed here may look complicated, it is quite natural.

Cadaver Measurements

The three-dimensional coordinates of the origins and insertions of each muscle were measured as shown in Fig. 1. When the orgin or insertion of a muscle was spread widely, the muscle was divided into several bundles and measured.

In previous studies, the muscles were cut at the center of the muscle belly for direct measurement of their cross-sectional area [6–8]. This method of measurement is erroneous, and sometimes it is difficult to define the muscle belly center. To avoid this problem, a "cross-sectional index" (g/cm) was used; this was defined as muscle weight (g) divided by muscle fiber length (cm). The muscle weight related to muscle volume (cm^3) is divided by muscle length (cm). This index, accordingly, has the dimensions of area (cm^2).

Muscle Force Calculation with the RBSM

A muscle is modelled as an equivalent spring. Therefore, the optimum solution is obtained directly by finite element method-like calculations, without surveying the optimum solution in non-linear programming. As shown in the flowchart in Fig. 2, member stiffness matrices were calculated for each muscle spring, using the muscle origins, insertions, and cross-sectional indices. The femur is completely fixed, and only rotational degrees of freedom of the pelvic bone are allowed. The action line of the center of gravity is defined as the center of the symphisis pubis in the anteroposterior view and, the center of the

Fig. 2. Flowchart of the RBSM analysis

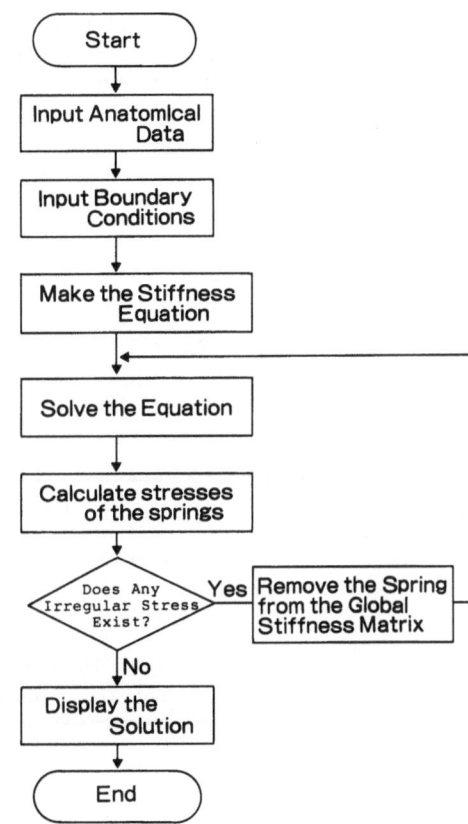

femoral head in the lateral view. Eighty-three percent of the body weight is loaded on the center of gravity.

After the initial solution is obtained, the stress of each spring is calculated. If compression is detected, the spring is removed from the system, because the muscle is slackened. Iterative calculations are continued until no compression is detected. When the calculation has converged, each muscle force, the resultant force of each muscle, and the body weight are calculated.

Results

Anterior Rotation of the Pelvis

The pelvis of the cadaver measured here was in a position of complete loss of anterior rotation. The first parametric study was made to calculate the resultant force vector of varying degrees of anterior rotation of the pelvis. Figure 3a shows that the resultant force had minimal magnitude, 2.93 times body weight, at 10° anterior rotation of the pelvis. In this position, the direction of the resultant force vector was 14.6° medially, and 10.4° anteriorly. This means that

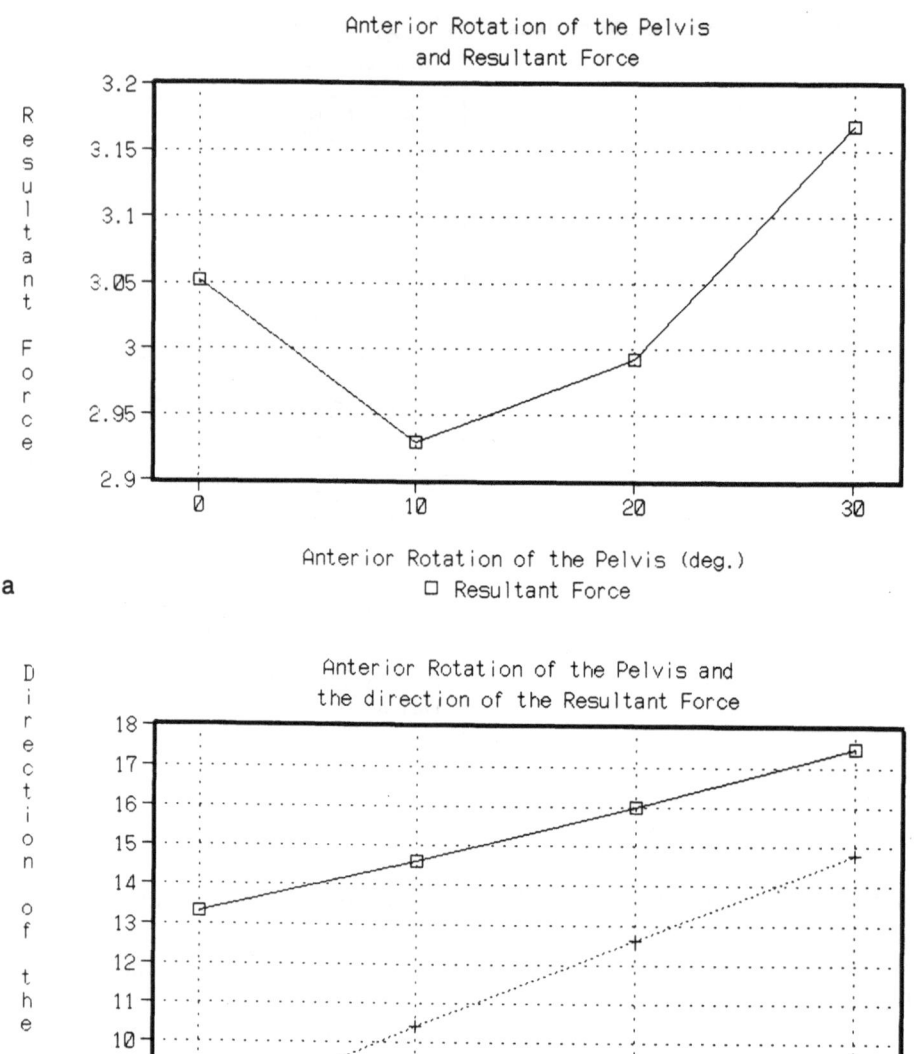

Fig. 3a,b. Anterior rotation of the pelvis and the resultant force vector; **a** magnitude (*open squares*, resultant force); **b** direction (*open squares*, medial angle; *crosses*, anterior angle)

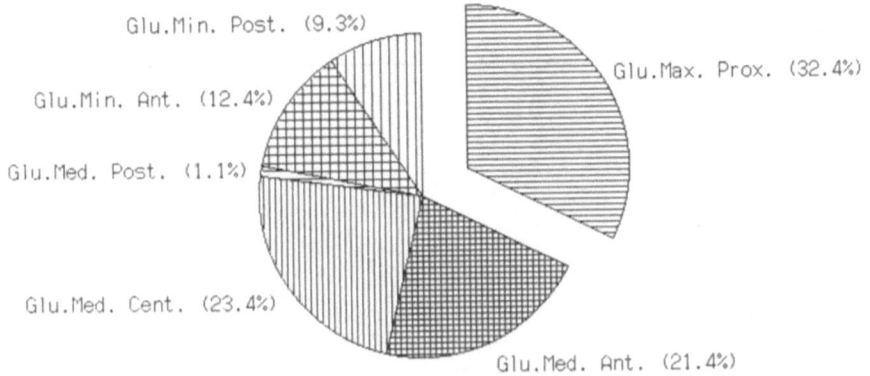

Glu.Min. Post. (9.3%)

Glu.Min. Ant. (12.4%)

Glu.Med. Post. (1.1%)

Glu.Med. Cent. (23.4%)

Glu.Max. Prox. (32.4%)

Glu.Med. Ant. (21.4%)   a

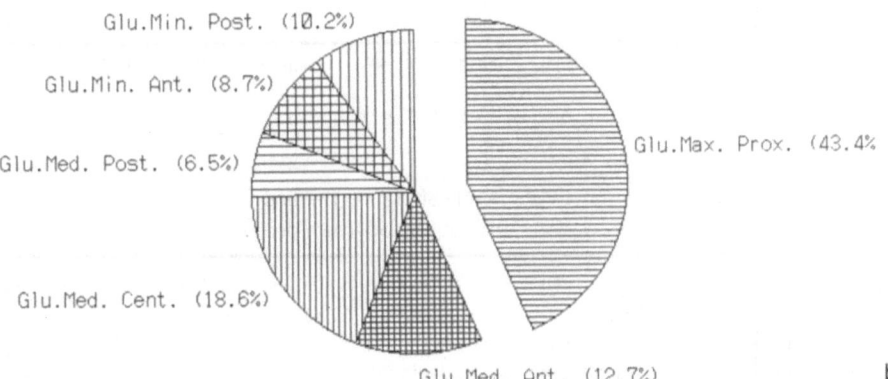

Glu.Min. Post. (10.2%)

Glu.Min. Ant. (8.7%)

Glu.Med. Post. (6.5%)

Glu.Med. Cent. (18.6%)

Glu.Max. Prox. (43.4%

Glu.Med. Ant. (12.7%)   b

Fig. 4a,b. Share of abduction moment; **a** 10° anteriorly rotated pelvis, **b** 20° anteriorly rotated pelvis. *Glu. min. post.*, gluteus minimus, posterior; *Glu. min. ant.*, gluteus minimus, anterior; *Glu. med. post.*, gluteus medius, posterior; *Glu. med. cent.*, gluteus mediums central; *Glu. max. prox.*, proximal part of the gluteus maximus; *Glu. med. ant.*, anterior part of the gluteus medius

the energy required for muscle contraction is minimum at the position of a 10° anteriorly-rotated pelvis from the lateral view shown in Fig. 1. In the following discussion, this is considered to be the standard position. When the anterior rotation of the pelvis increases, the medial and anterior angles of direction of the resultant force vector increase steadily.

The shares of the abduction moment to each muscle were analyzed, as shown in Fig. 4a. The proximal part of the gluteus maximus, the gluteus minimus, and the gluteus medius all play a role in hip abduction. When the pelvis is rotated anteriorly, the proximal part of the gluteus maximus plays a greater role in hip abduction, as shown in Fig. 4b. The gluteus maximus consists of the proximal part inserting the iliotibial tract, and the distal part

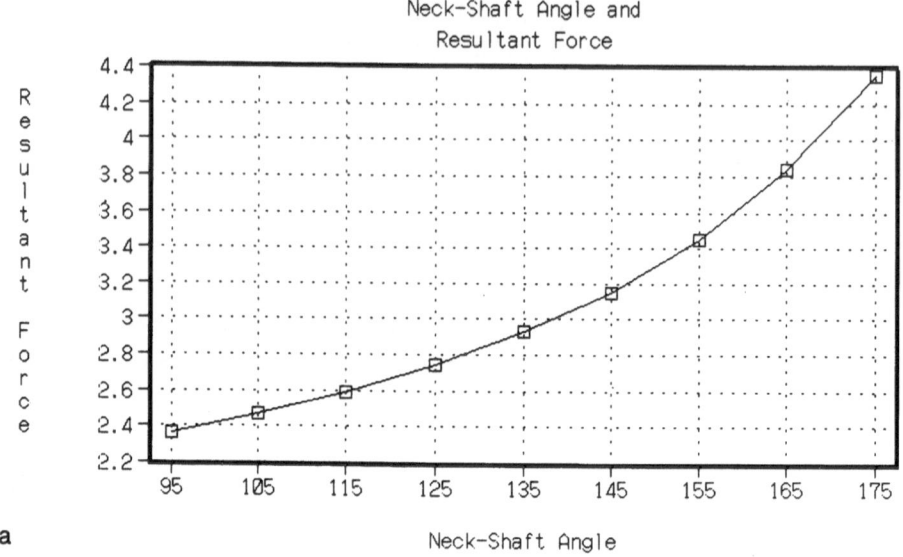

a

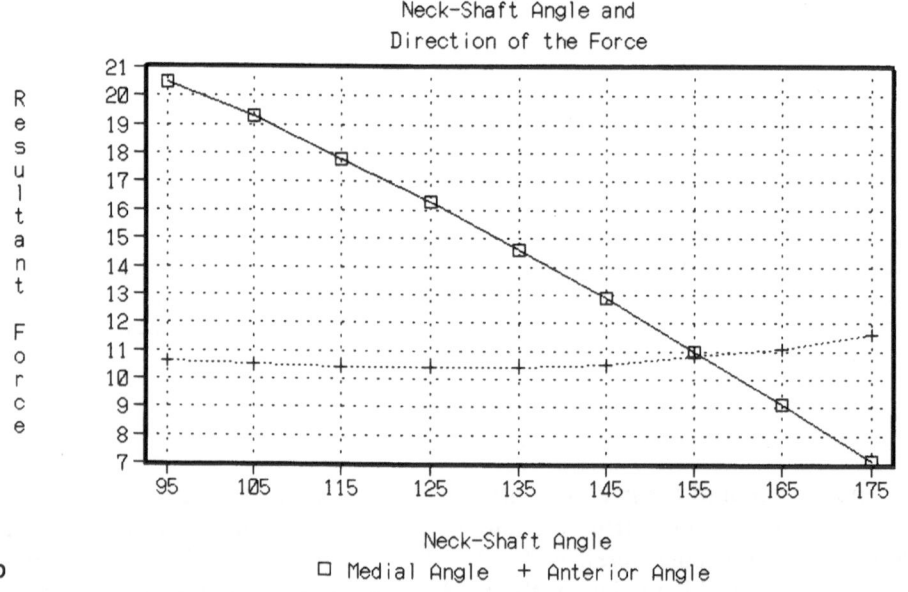

b

Fig. 5a,b. Neck-shaft angle and the resultant force vector; **a** magnitude *open squares*, resultant force **b**; direction (*open squares*, medial angle; crosses, anterior angle)

inserting the tuberositas glutea. The former is an abductor and the latter an extensor of the hip joint.

Neck-Shaft Angle of the Femur

The neck-shaft angle of this cadaver was 135°. This angle varies and the resultant vector is calculated as shown in Fig. 5a,b. When there is a varus deformity of the femoral neck, the magnitude of the vector decreases, the medial angle of the direction increases, and the anterior angle remains unchanged. With respect to varus osteotomy at the femoral neck, 30° varus osteotomy would be expected to reduce the magnitude of the resultant force vector to 0.8 times body weight, and to increase the medial angle of the direction to 5°.

Co-Contraction

In a previous study, carried out by Himeno (the author) and Tsumura [12], it was found that co-contraction of the flexor and the extensor in the finger dramatically reduced stress concentration at the margin of the finger joint in shock loading. The same mechanism is assumed to operate in the hip joint. When there is co-contraction of abductors and adductors of up to 50% of body weight, other than the body weight at the center of gravity, the magnitude of the resultant force vector is almost doubled (4.68 times body weight), while the direction of the vector remains unchanged (16.2° medially, and 11.3° anteriorly). When shock loading is assumed in the heel-strike phase of walking, preparatory co-contraction would be expected to reduce stress concentration at the margin of the acetabulum.

The Subluxated Hip Joint

The femoral head is subluxated to 50% of the head radius laterally and 25% proximally from the original location. The magnitude of the resultant force increases to 3.41 times body weight due to the elongation of the leverarm of the body weight. At the same time, however, the medial angle of the direction also increases to 19.1°, which factor compensates for the instability of the subluxated hip, while the anterior angle remains unchanged (11.7°).

Discussion

Validation by In Vivo Measurement

The reported direction of the resultant force vector, measured by strain gauges in prostheses, shows relatively little variation among authors (4° medially, and 10° anteriorly), while the magnitude of the vector, which varies according to the required extent of co-contraction, with minimal change occurring in vector direction, shows wide variations [4,5].

In earlier studies [4,5], the direction calculated on the basis of strain gauge measurement was assumed to be the direction of the resultant force vector itself. This is, however, erroneous; the direction calculated on this basis is the direction of the peak pressure, or, in other words, the direction of displacement. When an asymmetric load is applied, the displacement of the head deviates outward from the direction of the load itself [10,11,13,14]. If a load is applied 4° medially on the normal hip joint, the direction of displacement, or peak pressure, is located about 10° laterally. For the peak pressure to be located 4° medially, the resultant force should be inclined about 15° medially, which is comparable to the results presented here. In the sagittal plane, the acetabulum is deep enough to provide symmetrical load-bearing for the resultant force, and therefore, no dissociation is oberved between the direction of the load and the direction of displacement.

The resultant force vector presented here seems to contradict the measurements. However, previous studies did not address the dissociation of the loading and displacement directions in asymmetric load-bearing in a joint. Rather, previous findings can be explained by the resultant force vector estimated in this paper.

Subluxated Hip Joint

In the normal hip joint, the resultant force vector estimated here is comparable to that of Pauwels [1]. However, in the subluxated hip, the use of Pauwels' method leads to findings of increased magnitude and a less medially-inclined resultant force vector. Here, in contrast, the medial inclination of the resultant force increased quite substantially to stabilize the joint.

Pauwels' method is only two-dimensional, and neglects the gluteus minimus and the proximal part of the gluteus maximus muscle, which latter plays a more important role in the subluxated than in the normal hip. In fact, the contact pressure distribution of the subluxated hip, based on the resultant force estimated by Pauwels' method, is extremely eccentric in the marginal area of the acetabulum. If this eccentric distribution actually occurred, it would be very difficult for such patients to walk. However, most of them can walk without pain until they reach middle age. X-ray photographs of such patients show evenly distributed osteosclerosis along the acetabulum. Such findings can be more naturally explained in terms of the present results than in terms of previous findings.

Anterior Rotation of the Pelvis

When the pelvis is rotated anteriorly, the proximal part of the gluteus maximus muscle plays a more important role in hip abduction, and the resultant force vector inclines medially. A patient whose muscles are weakened tends to walk with an anteriorly-rotated pelvis, this involves the proximal part of the gluteus maximus muscle providing the increased abduction moment required for walking. Osteo-arthritic patients with acetabular dysplasia also tend to rotate the pelvis anteriorly. This is considered to compensate both for the instability of the hip, by providing a more medially-inclined joint force, and to compensate for anterior dysplasia of the acetabulum.

Indications and Limitations of Varus Osteotomy

When the femoral neck shows varus deformity, the magnitude of the resultant force decreases and the medial inclination of the force increases. For a stable ball and socket joint, it is necessary the joint force is located at least 30° medially from the joint edge [11,14]. As stated above, 30° varus osteotomy results in a 5° medial inclination of the resultant force. When the instability of the hip is relatively minor, and 5° medialization is sufficient to produce stability, this is a good indication for such osteotomy. If more than 5° medialization is required, however, joint stability cannot be recovered by this osteotomy alone.

Significance of Co-Contraction

The concept of co-contraction, i.e., the simultaneous contraction of agonist and antagonist muscles, has recently come to be regarded as important. In another study, Himeno (the author) and Tsumura [12] have analyzed this type of contraction. Co-contraction plays an essential role in improving anti-shock capability. Co-contraction of abductors and adductors, as shown here, increased the magnitude of the resultant force without directional change and thus was not at all hazardous to joint stability. Preparatory loading by such co-contraction relieves the stress concentration produced by a shock load.

Estimation of Muscle Forces and Contact Pressure

With the RBSM, the estimation of muscle force can be performed quickly and easily, as shown here. The simultaneous estimation of muscle force and contact pressure distribution was not discussed here; however, since the RBSM is one of the best tools available for the elucidation of contact problems, its use in such simultaneous estimations would be of value in elucidating contact problems in deformed and incongruent hip joints.

References

1. Pauwels F (1980) Biomechanics of the locomotor apparatus. Springer, New York, pp 76–105
2. Williams JF (1968) A force analysis of the hip joint. Biomed Eng 3:365–370
3. Merchant AC (1965) Hip abductor muscle force. An experimental study of the influence of hip position with particular reference to rotation. J Bone Joint Surg [Am] 47-A:472–476

4. Rydell NW (1966) Force acting on femoral head prosthesis. A study of strain gauge spplied prosthesis in living persons. Acta Orthop Scand [Suppl] 88:1–132
5. Davy DT, Katzar MS, Brown RH, et al (1988) Telemetric force measurements across the hip after total arthroplasty. J Bone Joint Surg [Am] 70-A:45–50
6. Seireg A, Arvikar RJ (1973) A mathematical model for evaluation of forces in lower extremities of musculo-skeletal system. J Biomech 6:313–326
7. Crowninshield RD, Brand RA (1981) A physiologically-based criterion of muscle force prediction in locomotion. J Biomech 14:793–801
8. Konishi N (1988) Estimation of three-dimensional resultant force of the hip by three-dimensional moment method and non-linear programming (in Japanese). Clin Orthop Surg (Rinsho-seikei-geka) 23:1491–1497
9. Kawai T, Toi Y (1977) A new Element in discrete analysis of plane strain problems. Seisan Kenkyu 29:204–207
10. An KN, Himeno S, Tsumura H, et al (1990) Pressure distribution on articular surfaces: Application to joint stability evaluation. J Biomech 23:1012–1020
11. Himeno S, Fujii T, Nishio A, et al (1981) Stability of the deformed hip (in Japanese). Clin Orthop Surg (Rinsho-seikei-geka) 16:746–753
12. Himeno S, Tsumura H (1983) The locomotive and control mechanism of the human finger and its application to Robotics. Proc international conference in advanced robotics 1:261–268
13. Himeno S (1983) Instability of the hip joint and its contact pressure. Biomechanics, 8-A:132–137
14. Himeno S, Tsumura H (1984) The role of the rotator cuff as the stabilizing mechanism of the shoulder. In: Bateman JE, Welsh RP, Decker BC (eds) Surgery of the shoulder. CV Mosby, St Louis, pp 17–20
15. Himeno S, Tsumura H (1985) Marginal stress distribution of the femoral component of THR and its mode of failure. In: Proceedings of the 31st annual meeting of the Orthopedic Research Society, Las Vegas, 21–24 Jan 1985
16. Tsumura H, Himeno S (1983) Load transmission and injury mechanism of the wrist joint. In: Proceedings of the ASME biomechanics symposium, July 8, 1983, Houston

II

Hip Joint and Gait Analysis

5

Quantitative Analysis of the Limp in Coxarthrosis

Yutaka Ohneda, Kenji Kawate, and Susumu Tamai[1]

Summary. To establish quantitative analysis of the limp in patients with coxarthrosis, the rotation of the pelvis during walking was measured in three dimensions. Fifty-two normal subjects and 46 patients with coxarthrosis were studied. Eight patients were measured again after total hip replacement and we compared the findings before and after surgery. The rotation of the pelvis during walking was measured in the frontal, sagittal, and horizontal planes during free walking, using three gas rate-sensors fixed to the pelvis. Then we drew waveforms, using a microcomputer with an analog-digital converter. In the normal subjects, the waveforms were regular, symmetrical, and highly reproducible on repeated gait cycles in the frontal plane. In contrast, compared with those of the normal subjects, the waveforms of the patients with coxarthrosis were irregular, asymmetrical, and poorly reproducible. These characteristics of the patients' waveforms were confirmed objectively when we calculated the symmetry, reproduction, and abnormality indices for the waveforms of the frontal plane, using the Fourier transform technique. In the sagittal plane, the pelvis was in anteversion at heel strike in each normal subject, while in most of the patients the pelvis inclined backward at heel strike of the affected side. In the horizontal plane, the waveforms were flat around the heel strike of the affected side in the majority of patients. The mean rotational angle of the patients was less than that of the normal subjects in the frontal plane, and larger in the sagittal plane. In all normal subjects, the rotational angle in the frontal plane was larger than that in the sagittal plane, while in 44 patients, the rotational angle in the frontal plane was less than that in the sagittal plane. We were therefore able to distinguish a patient from a normal subject by observation of the three indices of the frontal plane and by the ratio of the rotational angle in the sagittal plane to that in the frontal plane. The three indices, obtained by using the Fourier transform technique, and the ratio of the

[1] Department of Orthopaedic Surgery, Nara Medical University, 840 Shijo-cho, Kashihara, Nara, 634 Japan

rotational angle, were useful for quantitative analysis of the limp in patients
with coxarthrosis.

Key words: Pelvic rotation—Three-dimansional measurement—Human
walking—Limp—Quantitative assessment—Gas rate sensors—Fourier
transform

Introduction

Human walking, a peculiar activity of humans, is very difficult to understand.
This activity needs unity of the entire body and is affected by disability.

Gait analysis, which has been developed along with the development of photographic methods since the end of the nineteenth century, has been studied by
various techniques [1]. Levens et al. [2], using three cameras, took photographs of pins drilled into the cortices of bones. Inman et al. [3] measured
pelvic movements by photographing interrupted lights. Stokes et al. [4] used a
two-camera Selspot system interfaced with an HP1000 minicomputer. Thurston
et al. [5] observed the movements using a television-computer system, and
Van Leeuwen et al. [6] used precision potentiometers attached to a firmly
strapped external pelvis girdle. However, regarding the limp of the patient with
coxarthrosis, many mysteries remain to be solved.

The movement of the pelvis during walking shows a three-dimensional pattern which is altered by a variety of disease processes affecting the hip joint.
Therefore, we believe that we can gain some insight into the limp in patients
with coxarthrosis by measuring pelvic rotation. In this study, we measured
pelvic rotation during walking in the frontal, sagittal, and horizontal planes,
using three gas rate-sensors. The results for patients with coxarthrosis were
then compared with the results obtained from normal subjects.

Materials and Methods

We studied 52 normal subjects (10 males and 42 females) and 46 patients with
coxarthrosis (4 males and 42 females). The average age of the normal subjects
was 40 years (range, 20–82) and that of the patients was 47 years (range,
21–74). Eleven and 35 of the 46 patients showed bilateral and unilateral
involvement, respectively. The patients were further divided into two groups,
consisting of 17 early stage and 29 advanced and end-stage cases.

We used the gas rate-sensor system (rotational angle measurement system
G-2210; Anima, Tokyo, Japan). This is an angular accelerometer using hy-
drogen gas. We firmly attached three sensors to the pelvic girdle, one in each
dimension (frontal, sagittal, and horizontal; Fig. 1) [7]. The normal subjects
and patients then performed a total of six trials walking along an 8-m path into
which was incorporated a 250-cm-long force platform. The three-planar rotation
of the pelvis and the perpendicular component of the floor reaction force were

Fig. 1. A subject is fitted with a pelvic girdle on which three gas rate-sensors are installed (from [7] with permission)

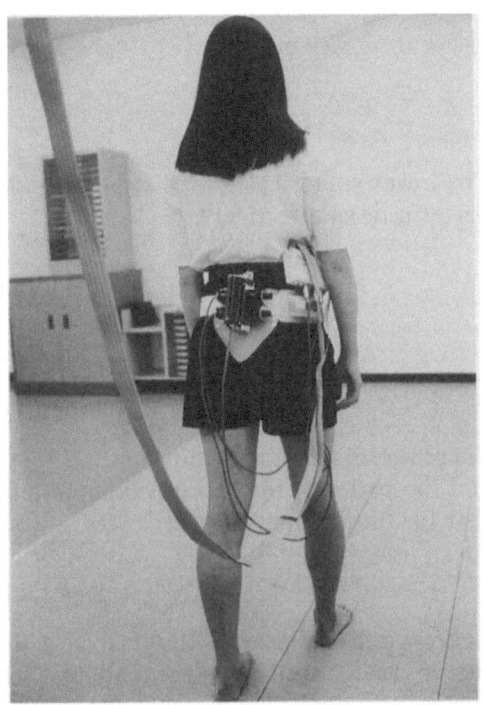

measured simultaneously. The analog data obtained were converted to digital form, and each individual waveform was drawn on the display, using a micro-computer. A measuring program was made, using C-language, with a sampling time of 3 ms.

The measured parameters were: rotational angle, waveform of the rotational angle, and frequency analysis of the waveform of the rotational angle in the frontal plane.

Application of Frequency Analysis

We calculated symmetry, reproduction, and abnormality indices, using the Fourier transform technique, for the waveforms of the frontal plane. Abnormality was defined as the difference between an individual's wave pattern and the basic waveform, which consists of only two, six, and ten frequency cycles. The three indices were calculated as follows:

1. Reproduction index $= \sum_{n=1}^{10} P_{2n-1}/P_2$

2. Symmetry index $= \sum_{n=1}^{5} P_{4n}/P_2$

3. Abnormality index $= \left(\sum_{n=1}^{20} P_n - S \right)/S$

$P_n = sqr(A^2_n + B^2_n)$
$S = P_2 + P_6 + P_{10}$

The lower values calculated for each index denote the ability of the subject to perform tasks effectively.

Results

Rotational Angle (Table 1)

Frontal Plane

There were significant differences ($P < 0.001$) between the normal subjects and the patients, the mean rotational angle for the patients being smaller than that for the normal subjects.

Sagittal Plane

There was a significant difference ($P < 0.001$) between the normal subjects and the advanced and end-stage patients. The mean rotational angle for the advanced and end-stage patients was larger than that for the normal subjects.

Horizontal Plane

There was no difference between the normal subjects and the patients.

Table 1. Mean angles of the pelvic rotation on the three planes.

	Normal subjects (n = 52)	Patients (n = 46)	
		Early stage (n = 17)	Advanced and end-stage (n = 29)
Rotational angle on frontal plane (degrees)	6.6 ± 1.7	4.9 ± 1.5***	3.8 ± 1.2***
Rotational angle on sagittal plane (degrees)	3.6 ± 1.1	4.5 ± 2.0	7.0 ± 2.5***
Rotational angle on horizontal plane (degrees)	7.5 ± 3.3	7.5 ± 3.1	7.2 ± 2.9

mean ± SD, *** $P < 0.001$, Significant difference between normal subjects and patients
The mean rotational angle of the patients was smaller than that of the normal subjects on the frontal plane, and larger on the sagittal plane

Observations of the Waveform of the Rotational Angle

Frontal Plane

There were three typical patterns in the waveform of the normal subjects on the frontal plane:

(1) Symmetrical pattern
(2) Highly reproducible pattern on repeated gait cycles
(3) Two-peak pattern almost synchronous with the perpendicular component of the floor reaction force

In the normal subjects, the pelvis dropped on the swing phase side during free walking, with the Trendelenburg phenomenon observed in all cases. In contrast, the waveforms of the advanced and end-stage patients were irregular, asymmetrical, and poorly reproducible on repeated gait cycles, compared with those of the normal subjects. The waveforms in these patients did not show a two-peak pattern, but rather numerous small peaks (Fig. 2) [7].

Sagittal Plane

In the normal subjects, the pelvis was in anteversion at heel strike and had approximately two rotation cycles per gait cycle. In most of the patients, the

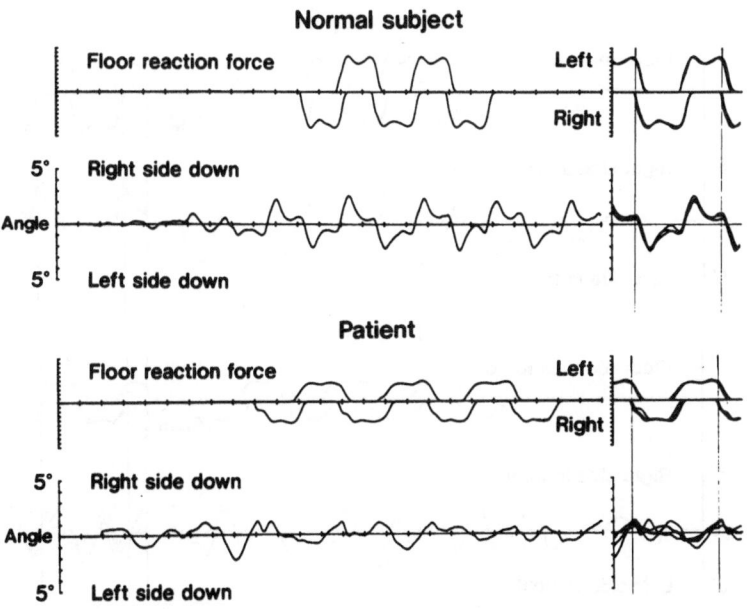

Fig. 2. Typical waveforms of a normal subject (*above*) and a patient with right advanced coxarthrosis (*below*) on the frontal plane (from [7] with permission)

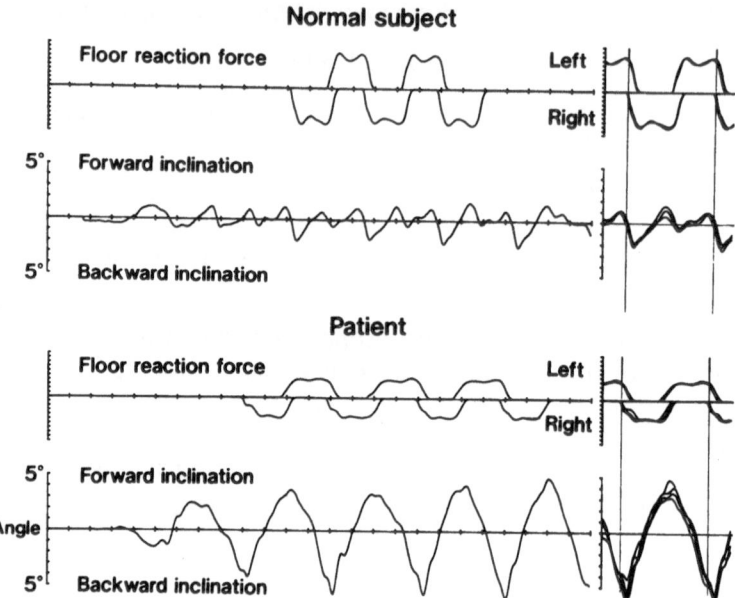

Fig. 3. Typical waveforms of a normal subject (*above*) and a patient with right advanced coxarthrosis (*below*) on the sagittal plane (from [7] with permission)

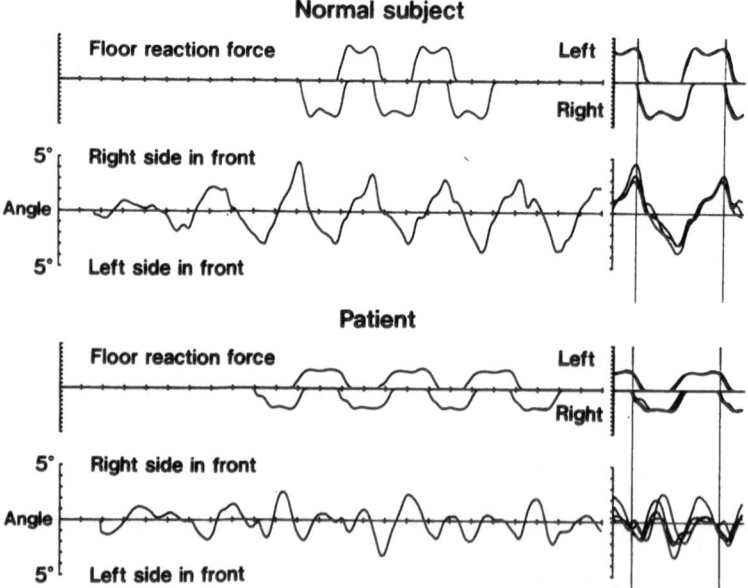

Fig. 4. Typical waveforms of a normal subject (*above*) and a patient with right advanced coxarthrosis (*below*) on the horizontal plane (from [7] with permission)

pelvis inclined backward at heel strike of the affected side, with some patients showing one rotation cycle per gait cycle (Fig. 3) [7].

Horizontal Plane

In the normal subjects, the waveforms were symmetrical and peaked at heel strike, whereas in the majority of patients, the waveforms were asymmetrical and peaked late at mid-stance on the affected side (Fig. 4) [7].

Application of Frequency Analysis (Table 2) [7]

Reproduction Index

There was a difference ($P < 0.05$) between the normal subjects and the early-stage patients and there was a significant difference ($P < 0.001$) between the normal subjects and the advanced and-end stage patients. This result objectively represents the lack of symmetry of the patients' waveforms.

Symmetry Index

There was a significant difference ($P < 0.05$) between the normal subjects and the early-stage patients and there was a significant difference ($P < 0.001$) between the normal subjects and the advanced and end-stage patients. This result objectively represents the lack of reproduction of the patients' waveforms.

Abnormality Index

There was a difference ($P < 0.01$) between the normal subjects and the early-stage patients and there was a significant difference ($P < 0.001$) between the normal subjects and the advanced and end-stage patients. This result objec-

Table 2. Results of the three indices using Fourier transform.

Index	Normal subjects (n = 40)	Patients (n = 44)	
		Early-stage (n = 15)	Advanced and end-stage (n = 29)
Reproduction	0.39 ± 0.15	0.54 ± 0.24*	0.93 ± 0.57***
Symmetry	0.21 ± 0.11	0.36 ± 0.20*	0.63 ± 0.45***
Abnormality	0.43 ± 0.12	0.64 ± 0.21**	1.10 ± 0.60***

mean ± SD, *$P < 0.05$; **$P < 0.01$; ***$P < 0.001$, Significant differences between normal subjects and patients
These results show that the waveforms of the patients are poorly reproducible, asymmetrical, and different from the normal pattern (from [7] with permission)

tively represents the difference of the patients' waveforms from the basic
waveform, which consists of only two, six, and ten frequency cycles.

Discussion

Objective analyses of human walking were not made until the mid-nineteenth
century because of the lack of devices suitable for this purpose. As pointed out
above, since the end of the nineteenth century, such analyses have developed
along with the development of photographic methods. However, there are
various difficulties and problems with accuracy involved with photographic
methods; in contrast, the gas rate-sensor is very accurate and handy. Using this
device, we can measure angular change at real time. However, we must take
care with this method, because the angle measured using this device is a
relative and not an absolute angle.

In the frontal plane, the mean rotational angle for the patients was significantly smaller than that for the normal subjects, while in the sagittal plane it
was larger. These differences are thought to be attributable, in the frontal
plane, to an attempt by the patient to minimize pain at impact and, in the sagittal plane, to pelvic compensation for the affected movement of the disabled hip
joint. However, since we were not able to distinguish a patient from a normal
subject by observation of the rotational angle in the frontal plane or that in the
sagittal plane, we plotted our the results for the 52 normal subjects and 29 ad-

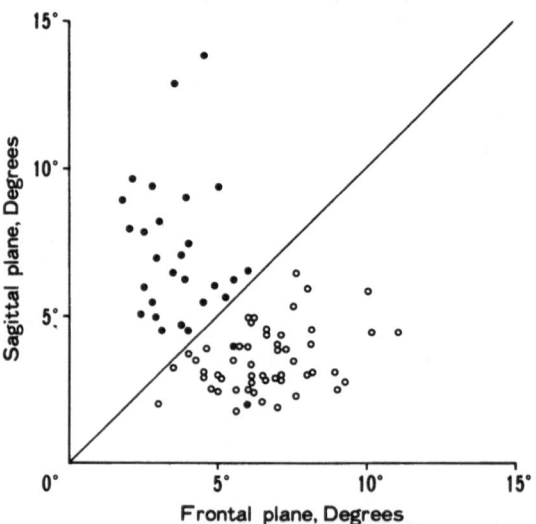

Fig. 5. The rotational angles of the 52 normal subjects and 29 advanced and end-stage
patients were plotted on this graph, with the X and Y axes representing the angles of the
frontal and sagittal planes, respectively. *Open circles*, normal subjects ($n = 52$); *solid
circles*, patients ($n = 29$) (from [7] with permission)

vanced and end-stage patients on a graph, with the X and Y axes representing the angles of the frontal and sagittal planes, respectively. As seen in Fig. 5 [7] the two groups were easily distinguished, with the advanced and end-stage patients and normal subjects found in the upper-left and lower-right of the graph, respectively. In other words, it was possible, to a large extent, to differentiate the two groups according to whether the rotational angle was larger in the frontal or the sagittal plane. Thus, we were able to distinguish a patient from a normal subject by using the ratio of the rotational angle in the sagittal plane to that in the frontal plane. A ratio larger than one was considered to be abnormal.

The waveform of the frontal plane was the most characteristic of the three waveforms drawn in this study. In the frontal plane, the waveforms of the normal subjects were regular, symmetrical, and highly reproducible, and demonstrated two peaks in one stance phase. Thurston et al. [5] reported that these two peaks were influenced by the swinging of the leg and were synchronized with toe off and heel strike. These investigators [5] also reported that the pelvis dropped in the swing phase during walking, a feature known as the Trendelenburg phenomenon. Normal subjects can prevent this drop of the pelvis by operating their abductor muscles at stance phase when still, but the pelvis drops in the swing phase side during free walking.

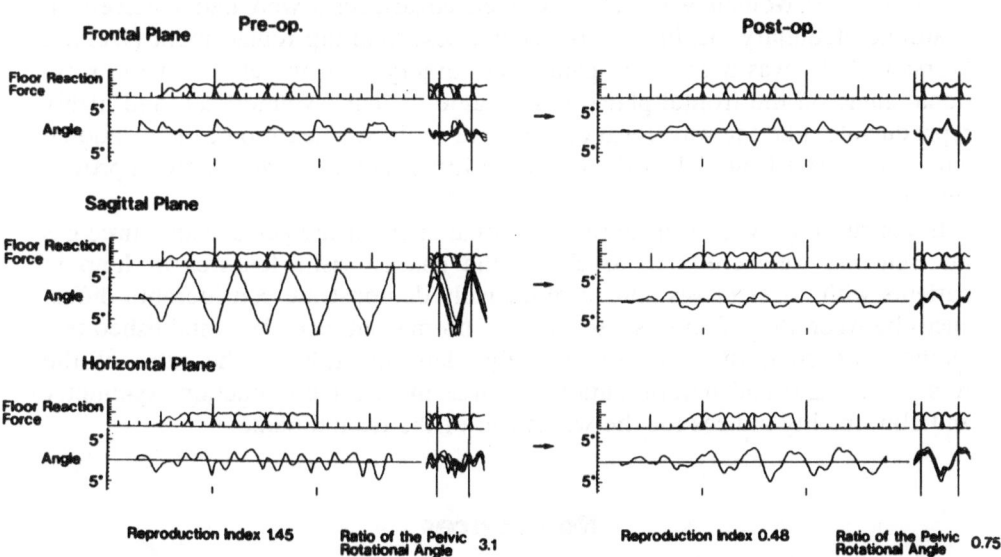

Fig. 6. Findings in a 51-year-old woman with left advanced coxarthrosis measured before (*Pre-op.*) and after surgery (*Post-op.*). The ratio of the rotational angle, and the three indices, were greatly improved. Pre-op. and Post-op. findings, respectively, were; reproduction index, 1.45 and 0.48; symmetry index, 2.07 and 0.16; abnormality index, 2.65 and 0.44; ratio of the pelvic rotational angle, 3.1 and 0.75; speed, 35 m/min and 47 m/min; JOA score, 44 and 99. *JOA*, Japanese Orthopaedic Association [8]

In the patient group, the waveforms were irregular, asymmetrical, and poorly reproducible on repeated gait cycles. Although we were able to roughly distinguish a patient from a normal subject by observation of the waveform in the frontal plane, it is desirable that to express the wave pattern with a numerical value. We therefore calculated symmetry, reproduction, and abnormality indices for the waveforms of the frontal plane, using the Fourier transform technique. When these indices were calculated for each case, significant differences ($P < 0.001$) were seen between the waveforms for the normal subjects and the advanced and end-stage patients. We were thus able to establish quantitative analysis of the limp in coxarthrosis.

Eight patients who showed unilateral involvement were measured again at 1½ years after total hip replacement. The findings after surgery were compared with those before. The Japanese Orthopaedic Association (JOA) score [8] for coxarthrosis increased from 55 before surjery to 90 after surgery, thus showing significant improvement in clinical evaluation. There was also significant improvement in the ratio of the rotational angle and the three indices. Thus, we suggest that this ratio and these three indices are useful for the quantitive analysis of improvement after surgery.

Case Report

A 51-year-old woman with left advanced coxarthrosis who had received an anatomic medullary locking (AML) cementless total hip replacement (DePuy, Warsaw, Ind.) was measured again after surgery. The waveform of the rotational angle on the frontal plane had became regular, symmetrical, and highly reproducible, like the waveform of a normal subject. The ratio of the rotational angle decreased from 3.1 to 0.75 and the three indices were greatly improved (Fig. 6).

In conclusion, by measuring the rotational angle of the pelvis using three gas rate-sensors, we were able to clarify the characteristic features of limp in patients with coxarthrosis and to objectively demonstrate considerable differences between these features and those of normal subjects. We established two methods of quantitive analysis, namely, determination of the ratio of the rotational angle and determination of three indices (reproduction, symmetry and abnormality indices for the waveform of the frontal plane.

References

1. Murray MP, Droght AB, Kory RC (1964) Walking patterns of normal men. J Bone Joint Surg [Am] 46A:335–360
2. Levens AS, Berkeley CE, Inman VT, Blosser JA (1948) Transverse rotation of the segments of the lower extremity in locomotion. J Bone Joint Surg [Am] 30A:859–872
3. Inman VT, Ralston HJ, Todd F (1981) Human walking. Williams and Wilkins, Baltimore, pp 22–61

4. Stokes VP, Andersson C, Forssberg H (1989) Rotational and translational movement features of the pelvis and thorax during adult human locomotion. J Biomech 22:43–50
5. Thurston AJ, Harris JD (1983) Normal kinematics of the lumbar spine and pelvis. Spine 8:199–205
6. Van Leeuwen JL, Vink P, Spoor CW, Deegenaars WC, Fraterman H, Verbout AJ (1988) A technique for measuring pelvic rotations during walking on a treadmill. IEEE Trans Biomed Eng 35:485–488
7. Kawate K, Ohneda Y, Kakihana T, Tamai S (1992) Assessment of limp in patient with coxarthrosis: Three dimensional measurement of the pelvic rotation. J Jpn Orthop Assoc 66:643–656
8. Shima Y, Tagawa H, Ueno R, et al (1971) Comparative study of several treatments for coxarthrosis: Creating guidelines for evaluation; and evaluation of long-term results (in Japanese). J Jpn Orthop Assoc 45:813–833

6

Tridimensional Motion Analysis of Trunk Versus Pelvis Movement in Hip Abductor Lurch

Mitsuo Suzuki[1] and Nobutoshi Yamazaki[2]

Summary. Hip abductor lurch movements were studied to determine gait features and to quantatively evaluate the effects of therapy. Spatial displacement and angulation of the pelvis and thorax were measured by a combination of a light emission diode-position sensor detector (LED-PSD) optical system and gyrosensors. These data were analyzed with the authors' original programs. One hundred and sixty-one patients (545 tests) consisting of 94 normal controls (362 tests) and 67 patients with hip abductor lurch (183 tests) were analyzed. Pre- and postoperative evaluations were performed on nine total hip arthroplasties for unilateral coxoarthrosis. Relative movements between pelvic and thoracic axes showed increases on all tridimensional planes in patients with hip abductor lurch, as compared with results in normal controls. Pre- and postoperative comparative analysis revealed marked improvement of limping motion. This analysis provides quantitative evaluation of the efficacy of therapeutic intervention and the monitoring of rehabilitation programs, and, as such, serves as a useful clinical guide.

Key words: Hip abductor lurch—Motion analysis—Compensatory motion

Introduction

Hip abductor lurch, a characteristic gait in patients with hip joint disease, is due to failure of the abductor muscle group (gluteus medius and minimus, tensor facia latae, etc.) and is accompanied by specific conspicuous body motion. This lurch is synonymous with "elastisches Sturz-hinken" proposed by Lorenz [1], a typical example of which is the Duchenne-Trendelenburg' limp

[1]Department of Orthopedic Surgery, Murayama National Hospital, 2-37-1 Gakuen, Musashimurayama, Tokyo, 208 Japan
[2]Faculty of Science and Technology, Keio University, 3-14-1 Hiyoshi, Kohoku-ku, Yokohama, 223 Japan

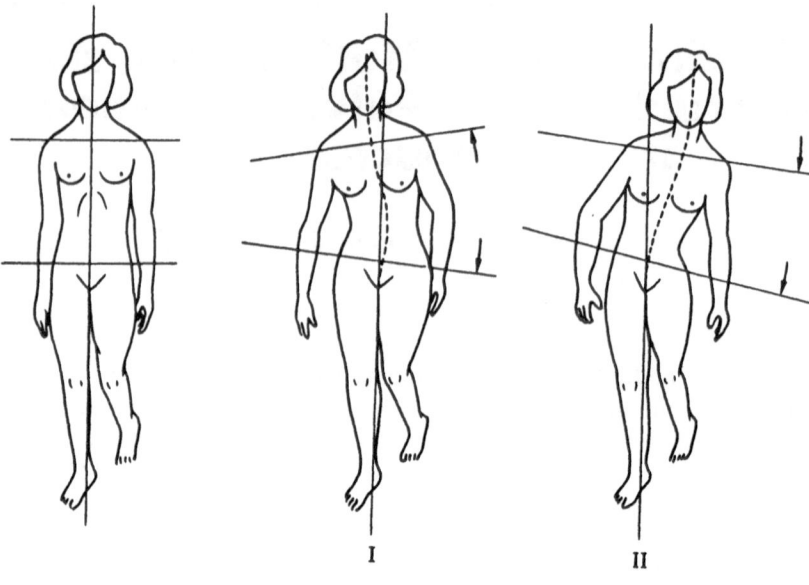

I   II

Normal Gait   Dechenne-Trendelenburg' Limping

Fig. 1. Note the tilting of the transverse axis of the pelvis and the upper trunk. In normal walking, the two axes stay almost horizontal and parallel. In hip abductor lurching, the contralateral side of the pelvis subsides in the phase of ipsilateral mono-pedal weight-bearing, and the trunk bends compensatorily towards the ipsilateral side. *I*, Sufficient compensation; *II*, insufficient compensation

(Figs. 1, 2). In this limp, when the affected side of the hip is weighted, the contralateral side of the pelvis subsides and the trunk slants toward the affected side of the hip. This compensatory motion is comprehensively described in Pauwel's [2,3] model of balance, in which model the medial lever arm is shortened by trunk inclination or displacement ipsilaterally, to compensate for abductor muscle insufficiency.

Our predecessors observed the characteristics of limping (predominantly movements within the frontal plane) by inspection alone, but were unable to record them objectively with quantitative measuring devices because of technical limitations. Recent technical progress in biomechanical measurements and the development of assessment program approaches have permitted detailed analyses of abnormal gait, which analyses are now at the stage of clinical application [4]; these will provides a basis for determining the adequacy of surgical design, the quality of design for arthroplasty devices, and the adequacy of prescriptions for braces and rehabilitation training. The purpose of this study was to apply a qualitative system of tridimensional measurement of motion to hip abductor lurch in order to determine the characteristics of the abnormal gait patterns involved and to evaluate the effects of therapy [5–9].

Fig. 2. Monopedal weight-bearing to
the side of mallum coxae luxans. Note
subsidence of the contralateral side of
the pelvis and compensatory lumbar
scoliosis

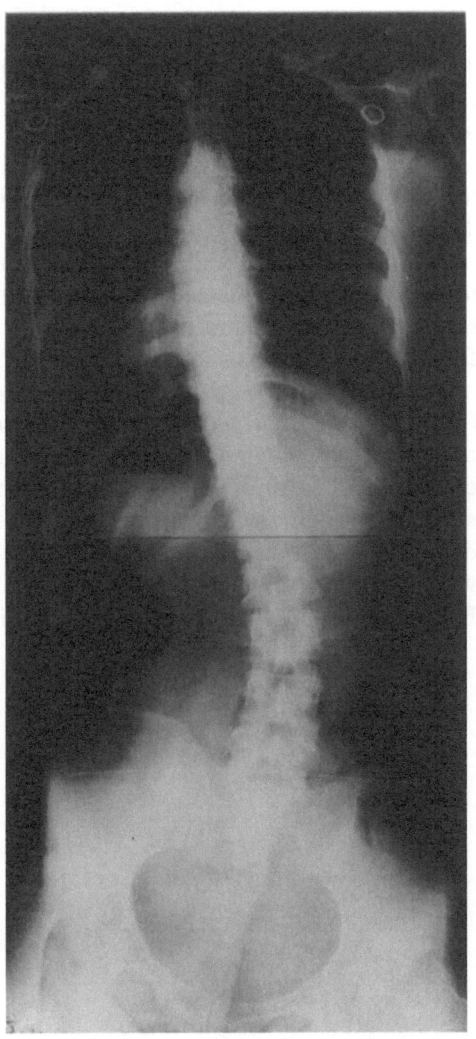

Materials and Methods

Subjects

We examined 161 subjects (545 tests), consisting of 94 normal controls (362
tests), 25 patients with osteoarthritis (OA) of the hip (68 tests), 18 with
rheumatoid arthritis (RA) (40 tests), 10 with avascular necrosis of the femoral
head (20 tests), 3 with untreated congenital dislocation of the hip (12 tests),
and 11 with other hip conditions (43 tests).

Measurement

Two jigs of mark units, each consisting of three light emission diode (LED) targets (a total of six) and a gyrosensor (a total of two), were attached to the area of the upper thorax and the pelvic area on the back of each subject. The legs of the jigs were erected on the skin of the posterior spinal processes of the second thoracic vertebra and on the bilateral posterior superior iliac spines. The spatial distance between the LED targets on the jigs and characteristic points on the body (bilateral acromial processes and bilateral anterior superior iliac spines) were measured, and the spatial positional relationships among these measuring points were determined. The motion of these six targets was followed-up and recorded by using two position sensor detector (PSD) cameras behind the subject with the angle of rotation along the prependicular axis (on the horizontal plane) being obtained from the gyrosensors. The space measured was 60 cm wide × 150 cm long × 200 cm deep. Gait samples during central 10-s duration of walking 10.8 m on long force plates were collected several times in synchronization with floor reaction force. Since gait form depends on velocity, samples were collected during free walking (the speed at which the subject walked most comfortably) and during slightly faster walking. Each measurement required about 10 min. Gait data were input, at a sampling

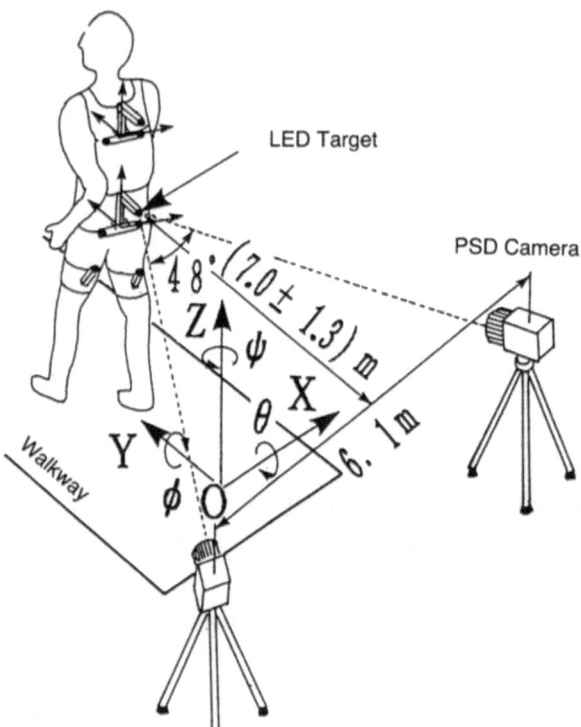

Fig. 3. Measurement set-up. *LED*, light emission diode; *PSD*, position spot detector

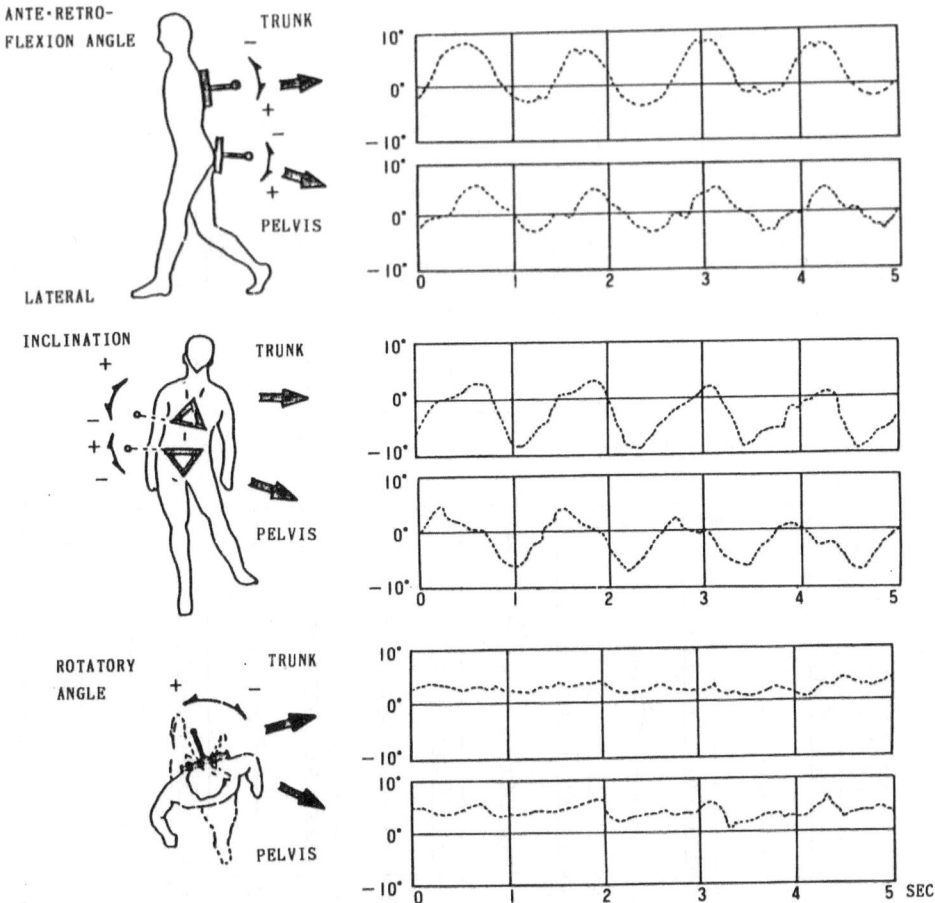

Fig. 4. Tridimensional motion measurement of trunk and pelvis

frequency of 100 Hz, for processing and analysis, using a program developed by the authors (Figs. 3, 4).

Results

Normal gait showed smaller movements of the trunk at the thoracic level and pelvic axes on all three dimensional planes than did hip abductor lurch. Figure 5 shows the mean amplitudes of positional and angular changes in the pelvic and trunk axes during free walking in a normal adult. On a Lissajous diagram of mutual movements of the two axes on each of the planes, these values produced a uniform distribution pattern in a small area near the origin of motion. Since the rotatory motion of the pelvic axis (motion in the horizontal

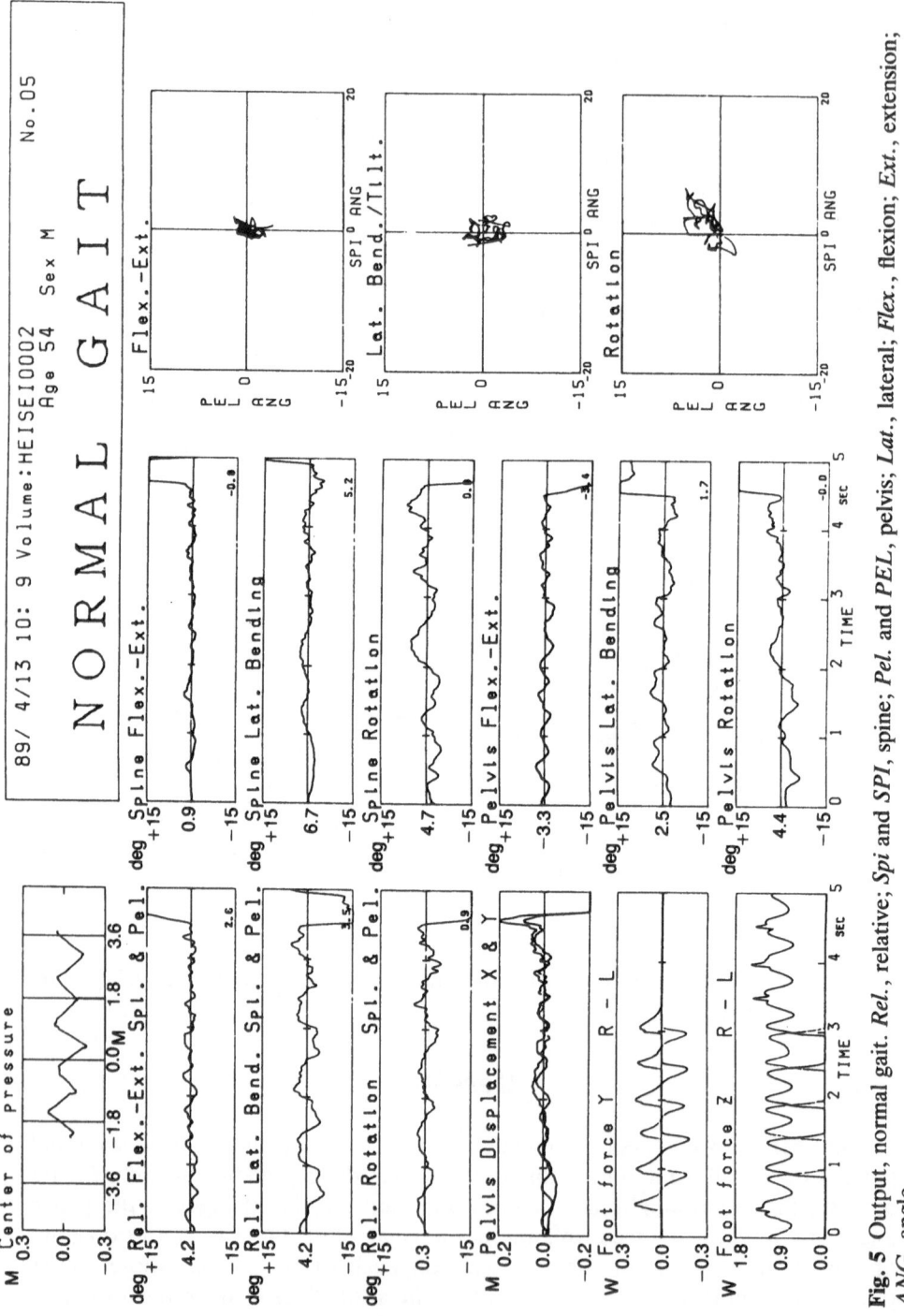

Fig. 5 Output, normal gait. *Rel.*, relative; *Spi* and *SPI*, spine; *Pel.* and *PEL*, pelvis; *Lat.*, lateral; *Flex.*, flexion; *Ext.*, extension; *ANG*, angle

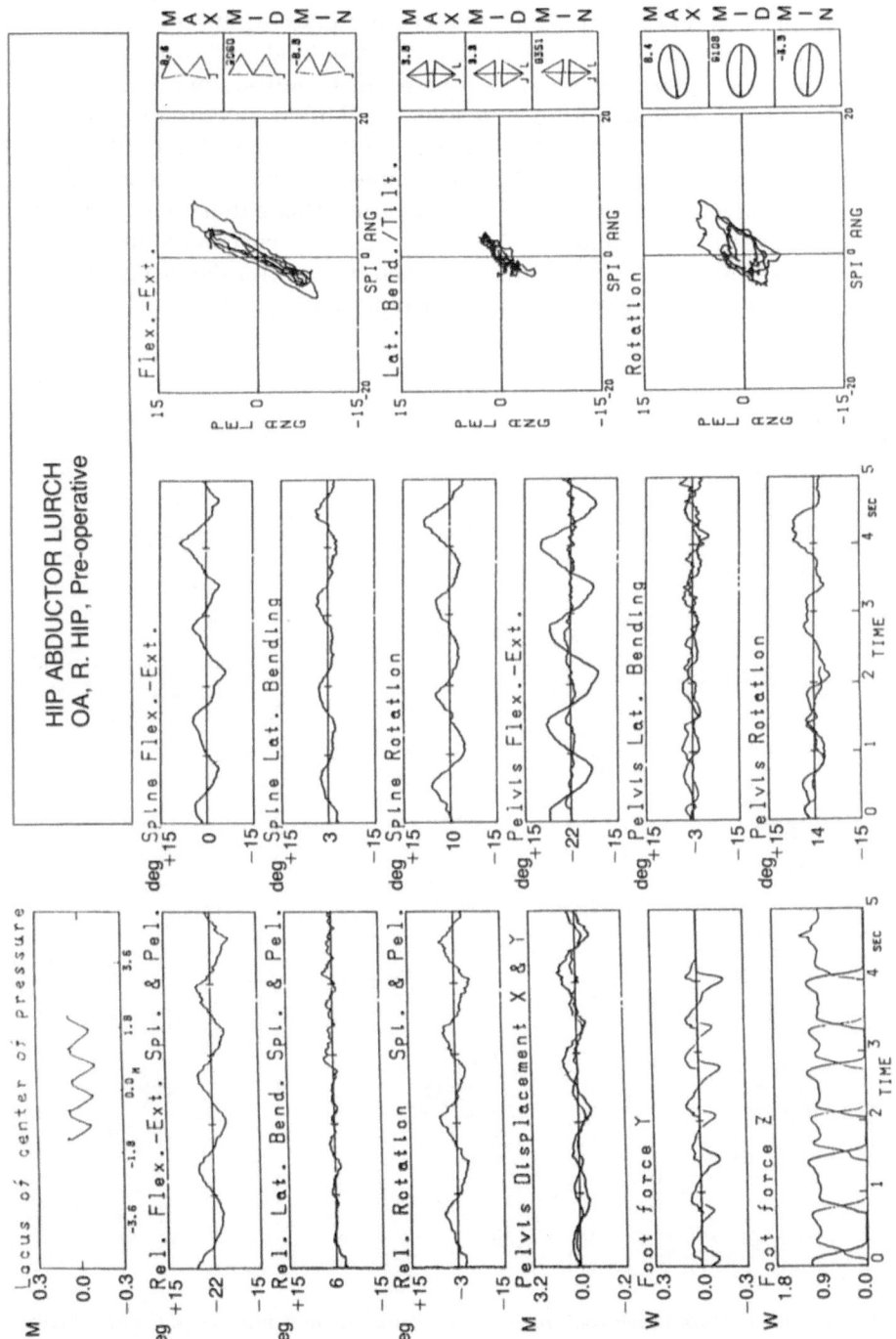

Fig. 6. Output, osteoarthritis of the right hip, preoperative

plane) shows fairly large individual differences, the area of distribution on a Lissajous diagram is relatively large, with a circular pattern, in some cases.

Those patients who had limps showed greater amplitudes of positional and angular changes in the two axes than did the normal controls. In hip abductor lurch, both types of amplitude were increased. In addition, relative movements between the two axes showed a characteristic tendency. Specifically, the Lissajous diagram demonstrated ellipsoidal distribution around the origin, rather than merely enlarged uniform circular distribution. This pattern represents a correlation between the movements of the two axes, this correlation indicating that the two axes move in compensation for each other, with the inclination of the ellipsoidal axis in the Lissajous diagram representing the degree of com-

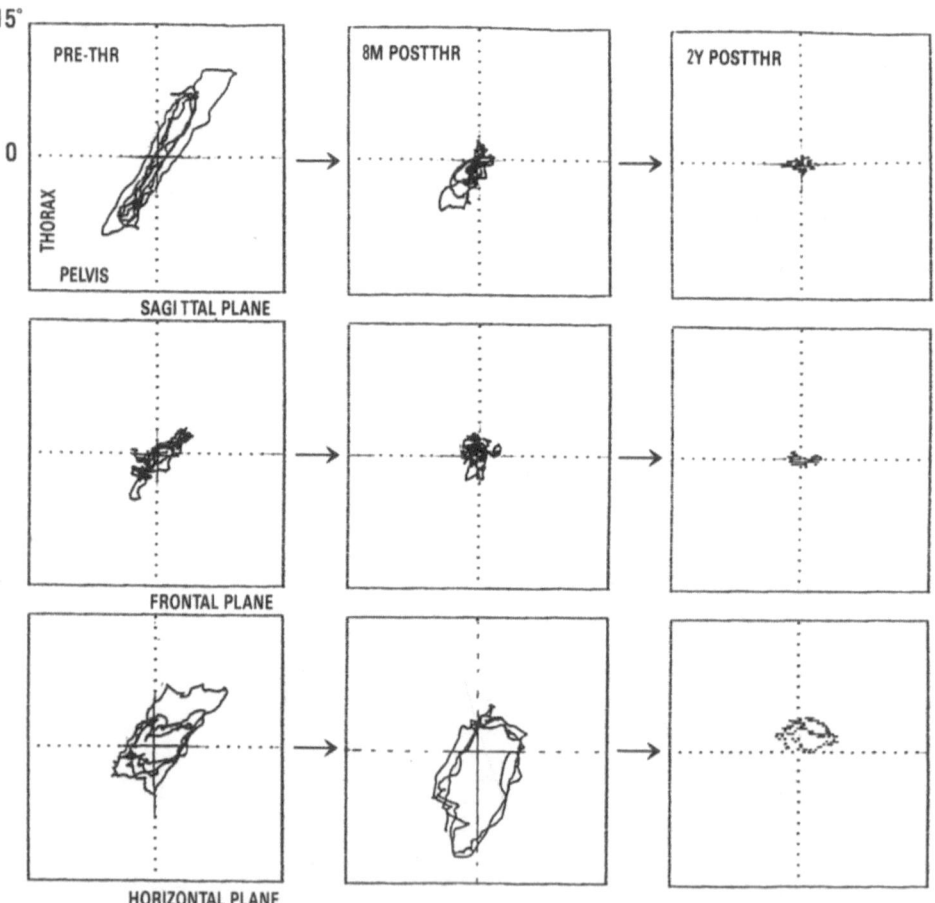

Fig. 7. Pre- and postoperative features, osteoarthritis of right hip, 57-year-old female. Improvement of limping is demonstrated. *Pre-THR*, before total hip replacement; *M*, months; *Y*, years

pensation. Clinically, each inclination corresponds to Duchenne-Trendelenburg limp in the frontal plane, to ante-retroflexion gait in the sagittal plane, and to pelvic gait in the horizontal plane. Figure 6 shows the output data for a patient showing hip abductor lurch.

When total hip arthroplasty was performed, correcting the positional relationships of the joint, followed by abductor muscle exercise and gait form training, the relative movement between the two axes improved markedly, approaching normal movement (Fig. 7).

Discussion

Measurement Systems

The following kinesimetric systems are commercially available: for video-digitization or detection of luminous spots; MOVIAS 3D (NAC, Tokyo, Japan), Ariel (Ariel, Los Angeles, Calif.), and Peak (Peak, Englewood, Colo.). Infrared strobo reflex types are Elite (BTS, Milan, Italy), Vicon (Oxford Metrics, Oxford, England), and the Lavic MAS series (Emtec, Tokyo, Japan). Infrared LED types are Selspot II (Selcom AB, Partille, Sweden), LOCUS III D (Anima, Tokyo, Japan), and the PSD system (Hamamatsu Photonics, Hamamatsu, Japan). These systems, each with its own characteristic performance, are now used for motion analysis in sports, rehabilitation, and human engineering, as well as in robot technology, automobiles, and aircraft. Compared with these commercially-available devices, a major part of our device was custom-built for use in clinical tests.

The authors developed a unique program suitable for clinical use, being convenient in scale and having a simple procedure, by combining various systems made in Japan (LED Targets and PSD camera, Hamamatsu Photonics, Hamamatsu, Japan; floor reaction force plates, the Kyowa Dengyo Tokyo, Japan; and gyrosensor, the Anima, Tokyo, Japan). Precision tests of the hardware revealed that the error of the angle in the space for measurement was less than $0.8°$, and that for displacement was less than $2.6\,mm$ in the frontal plane.

Measuring Conditions

Since gait form is dependent on velocity and cadence, walking samples were obtained from each subject during free walking (gait at which walking was easiest) and during slightly faster walking for comparison. Kinesimetric data were found to be appropriate when the measured free walking values were normalized with velocity. It was not possible to measure the degree of slippage between the skeleton and the skin and between the skin and the jig. However, the supporting points of the legs of the jig are considered to be minimal slippage spots during walking. Furthermore, even if a jig moves away slightly from a definite point, the spatial positional relationship between the mark and the characteristic point on the body is already known, so that the two points are interchangeable. Therefore, the slippage error should be smaller than that occurring when a mark is directly attached to the external surface of the acromial process or the anterior and posterior iliac spinae. These factors ensure the reliability of the measured values of abnormal gait forms in comparison with those of the normal control group.

Clinical Considerations

Hip abductor lurch occurs biomechanically when there is abductor muscle insufficiency against the affected hip moment. The hip moment, on this occasion, equals the product of the lever arm length (distance between the bearing center of the joint and the body weight gravity line) and whole body weight. Consequently, the contralateral side of the pelvis sinks, while the trunk reflexively corresponds to the subsidence with an inclination or displacement towards the affected hip side, thus attempting to shorten the lever arm and to compensate the abductor muscle weakness. These compensation mechanisms are comprehensible in terms of Pauwels' balance model [2–3]. Elongations of the lever arms are seen clinically in osteoarthritis of the hip as supralateral migration, mushroom deformity of the femoral head, subluxation, etc., whereas, abductor muscle insufficiencies are seen in the case of mechanical upward displacement of the greater trochanter, and in hypoactive muscle atrophy due to coxalgy. Typical Duchenne-Trendelenburg lurch, however, is rarely seen in OA of the hip. The osteoarthritic patient walks protectively, evading large sway motion, to avoid coxalgy. Further, as the OA stage advances, hip contracture and lumbar spondylosis develop. Such stiffness considerably modifies the limping pattern. In the measurement of limping, the authors have often observed less limping in the advanced stage than in the early stage in patients with OA of the hip. These contradictory phenomena should be taken into account when evaluating limping analysis.

Regarding the evaluation of limping, our method of, motion analysis is clinically useful in monitoring the effects of surgery, pre- and postoperative procedures, rehabilitation training prescriptions, and those of other therapeutic interventions. When total hip arthroplasty is performed for the patient with hip abductor lurch to correct the anatomical positional relationships of the joint, and this is followed by training of the abductor muscles and gait form, the gait pattern improves markedly, approaching that of the normal control pattern (Fig. 7). Regarding the relative rotatory movements of the pelvis and trunk, despite the correction of deformity, the gait pattern established in preoperative years prevails, preventing these values from approaching the normal pattern (Fig. 7).

Our measurement and analysis system will provide a useful clinical guide that should be helpful for determining indications for hip surgery, for designing hip devices, for producing appropriate prescriptions for braces, and for creating rehabilitation training.

References

1. Lorenz A (1920) Die sogenannte angeborene Hüftverrenkung. Ihre Pathogenese und Therapie. Ferdinand Erke, Stuttgart
2. Pauwels F (1965) Gesammelte Abhandlungen zur funktionellen Anatomie des Bewegungsapparates. Springer, Berlin Heidelberg New York
3. Pauwels F (1976) Biomechanics of the normal and diseased hip. Theoretical foundation: Technique and results of treatment. Springer, Berlin Heidelberg New York
4. Stokes VP, Andersson C, Forssberg H, et al (1988) Rotational and translational movement features of the pelvis and thorax during adult human locomotion. J Biomechanics 22:43–50
5. Suzuki M, Yamazaki N (1988) Motion analysis for hip abductor lurch (in Japanese). Hip Joint 14:327–331
6. Suzuki M, Yamazaki N (1992) Tridimensional motion analysis for hip diseases (in Japanese). J. Jpn Orthop. Assoc. 66(8):S1427
7. Suzuki M, Yamazaki N (1992) Tridimensional motion analysis for gait disturbances (in Japanese). Jpn J. Rehab. Med. 29(11):886
8. Tsuchiya K, Yamazaki N, Suzuki M, et al (1989) Introduction to gait analysis (in Japanese). Ishiyaku Shuppan, Tokyo
9. Yamazaki N (1982) General remarks on motion analysis (in Japanese). Sogo Rehabilitation 10(2):225–230

7

Three-Dimensional Analysis of Trunk Movements During Walking in Patients with Osteoarthritic Hip

Shirou Hirose, Kazuhiko Sawai, Tomokazu Hattori, and Shigeo Niwa[1]

Summary. Although previous authors have described the movements of body segments during normal walking, there are very few studies concerned with trunk movement in pathological gait. The aim of this study was to identify the significant features of the three-dimensional trunk motion that occurs during walking in patients with osteoarthritis of the hip, using a motion analysis system developed by our department. It was concluded that the center of gravity can be transferred by Duchenne's phenomenon in the frontal plane, and that the decreased range of hip motion can be compensated for by an increased rotational angle of the pelvis in the horizontal and sagittal planes.

Key words: Gait—Gait analysis—Limping—Motion analysis—Osteoarthritic hip—Trunk movement—Walking

Introduction

Trendelenburg's phenomenon and the so-called Duchenne's phenomenon are well known characteristic motions of the trunk in the frontal plane during limping. Several reports have been published on trunk motion during normal walking [1–6]. However, very few studies have addressed the trunk motion that occurs, especially in the horizontal and sagittal planes, in limping related to disorders of the hip joints.

Accordingly, we attempted to clarify the distinctive features of trunk and pelvis motion in patients with osteoarthritic hips, examining the features in three dimensions during walking. For this purpose, we used an on-line TV analysis system (Aichi Medical University TV system; ATV system) that was developed at our university [7].

---

[1] Department of Orthopaedic Surgery, Aichi Medical University, 21 Karimata Yazako, Nagakute, Aichi, 480-11 Japan

Material and Methods

Patients

Thirty-one patients with osteoarthritic hips (50 hips) which caused limping were investigated. All were women, ranging in age from 30 to 73 years (average, 50.0 years). Ten healthy women (20 hips) served as normal controls; their ages ranged from 54 to 74 years (average, 59.5 years). Regarding the motion of the upper trunk in the frontal plane, the patients were divided into three groups: (1) those who were positive for Duchenne's phenomenon; (2) those showing the lateral shift phenomenon; and (3) those who were negative for these phenomena. Regarding the motion of the pelvis in the frontal plane, the patients were divided into two groups: (1) those who were positive for Trendelenburg's phenomenon and (2) those who were negative for this phenonemon. Regarding the motion of the upper trunk and the pelvis in the horizontal plane, the patients were divided into three groups: (1) those with limited internal rotation in unilateral hip; (2) those with such a limitation in bilateral hips; and (3) those with no such limitations in either hip.

System

Our ATV system consists of retro-reflective markers, TV-cameras with infrared light-emitting diode (LED) stroboscopes, video processing units, force platforms, and a 16-bit personal computer (Fig. 1). Infrared rays, which are recorded and displayed on the TV monitor, are projected from LEDs synchronized with the TV camera exposure and reflected by bright markers attached to anatomical landmarks on body segments. The video processing unit can measure and record the position of the bright markers on X-Y coordinates. The personal computer calculates the angle between the line of two bright markers and the horizontal or vertical axis line during walking.

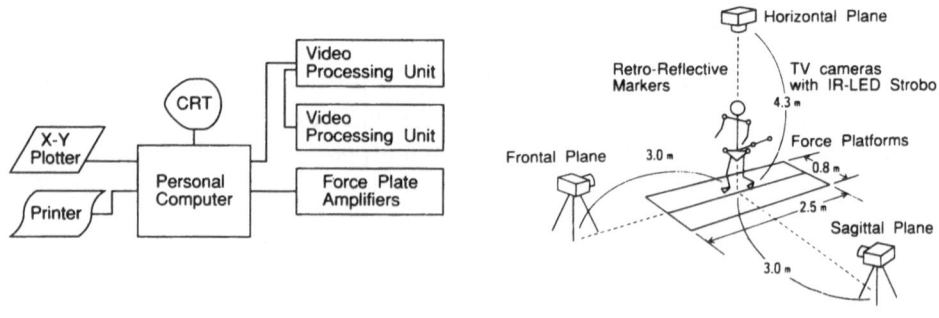

Block Diagram   **TV camera Configuration**

Fig. 1. Components and set-up of Aichi Medical University TV system

Three cameras are set up, two at a height of 0.95 m from the floor, in front and at the side of the subject, and one overhead, to estimate body motion in the frontal, horizontal, and sagittal planes.

The distance from the center of the force platform to the anterior-posterior or lateral camera is 3 m, and the distance to the overhead camera is 4.3 m. According to a preliminary accuracy test, maximum error was less than 1° in this measurement system when the subject was positioned within 1 m of frontal deviation from the light axis of the camera [8].

Procedure

The subjects walked on the force platforms at free speed with no restrictions. The angle between the line of two markers on both acromions and a frontal line parallel to the floor was measured to represent the rotational movement of the upper trunk in the frontal and/or horizontal planes. Likewise, the angle between the line of two markers on both anterior superior iliac spines and a frontal line parallel to the floor was measured to represent the rotational movement of the pelvis in the frontal plane. Also, the angle between the line of two markers on a glass fiber rod that was affixed to the sacrum and a sagittal line parallel to the floor was measured to represent the rotational movement of the pelvis in the horizontal and/or sagittal planes. In addition to the calculation of the rotational angle at each turning point during the gait cycle, the displacement angle, i.e., the difference of the angle between each turning point, was calculated.

The following pathologic movements of the trunk were identified in frontal view by inspection prior to measurement. In this study, Duchenne's phenomenon was defined as a bending of the upper trunk toward the side of the supporting limb. Additionally, the lateral shift phenomenon was categorized as a horizontal displacement of the upper trunk instead of marked lateral bending. Trendelenburg's phenomenon was defined as a drop of the pelvis to the side opposite the supporting limb.

Results

In the Figures, the vertical axis represents the rotational and displacement angle of the trunk, and a positive angle value indicates upward rotation on the swinging limb side in the frontal plane and forward rotation on the swinging limb side in the horizontal and/or sagittal planes.

Frontal Plane

Motion of the Upper Trunk (Fig. 2)

In the normal control group, a slight upward rotation occurred on the swinging limb side just after the beginning of the single support, the rotation being

76   S. Hirose et al.

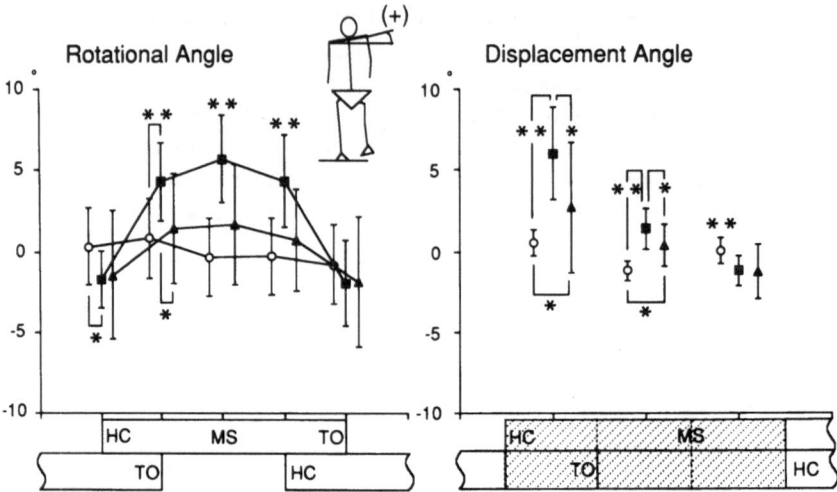

Fig. 2. Upper trunk motion in the groups who were positive for Duchenne's phenomenon (*closed squares*; $n = 10$) and those who showed the lateral shift phenomenon (*closed triangles*; $n = 19$) in the frontal plane. HC, Heel contact; MS, mid-stance; TO, toe-off. *Open circles*, normal group ($n = 20$). *$P < 0.05$; **$P < 0.01$

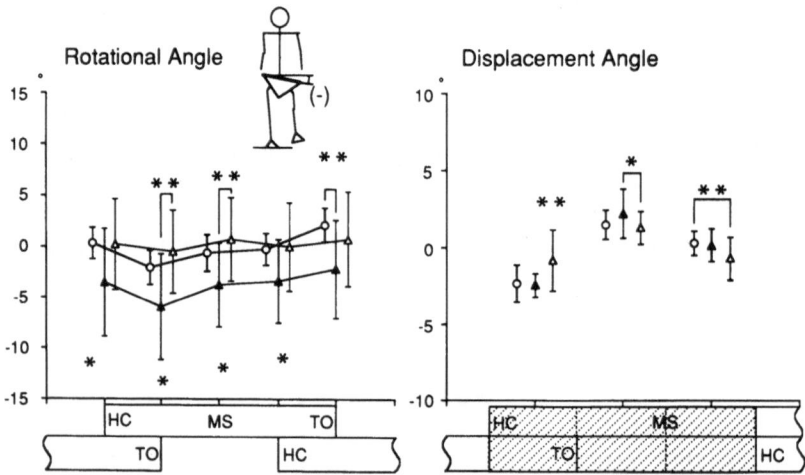

Fig. 3. Pelvic motion in the groups who were positive for Trendenenburg's phenomenon in the frontal plane (*closed triangles* (+); $n = 12$) and those who were negative (*open triangles* (−); $n = 38$). *Open circles*, normal group ($n = 20$). *$P < 0.05$; **$P < 0.01$

reduced at mid-stance. A marked upward rotation was maintained on the swinging limb side during the early part of the stance phase in the group who were positive for Duchenne's phenomenon (D(+) group). The group showing the lateral shift phenomenon (lateral shift group), in whom the rotational angle was smaller than in the D(+) group, showed a motion pattern identical to the D(+) group.

Motion of the Pelvis (Fig. 3)

Downward rotation was observed on the swinging limb side in normal controls just after the beginning of the single support.

In the group who were positive for Trendelenburg's phenomenon (T(+) group), pelvic tilting continued throughought the whole gait cycle on the swinging limb side. Furthermore, downward rotation from the tilting position in the pelvis occurred just after the beginning of the single support. The downward rotational angle of the pelvis was smaller in the group who were negative for Trendelenburg's phenomenon (T(−) group) than in the normal control group and the T(+) group.

Horizontal Plane

Motion of the Pelvis (Fig. 4)

A symmetrical forward rotation was observed on the swinging limb side in normal controls. An oblique position of the pelvis, in which the affected side was anterior, was maintained during the entire gait cycle in the group with limitation of internal rotation of the unilateral hip (unilateral limitation group). The amplitude of forward rotation was greater in the group with limitation of internal rotation of the bilateral hips (bilateral limitation group) than in the

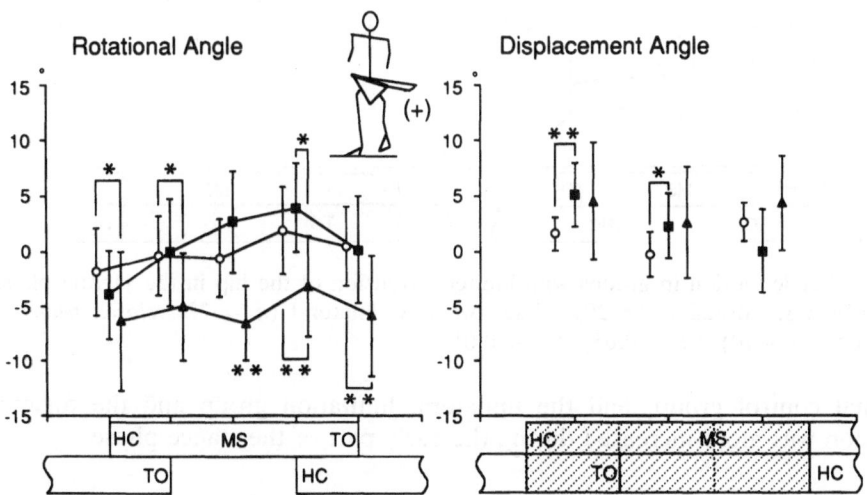

Fig. 4. Pelvic motion in groups with limited internal rotation of the hip in the horizontal plane. *Open circles*, Normal ($n = 20$); *closed squares*, bilateral ($n = 6$); *closed triangles*, unilateral ($n = 8$). $^{*}P < 0.05$; $^{**}P < 0.01$

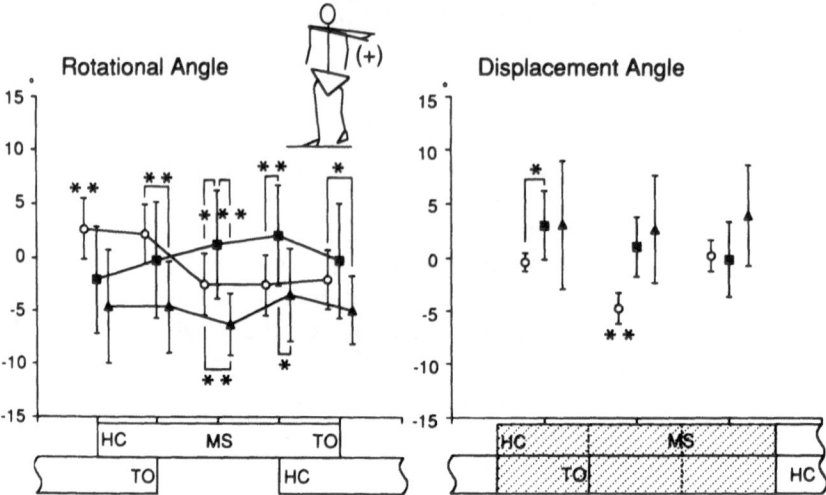

Fig. 5. Upper trunk motion in groups with limited internal rotation of the hip in the horizontal plane. *Symbols* and *numbers in groups* as in Fig. 4

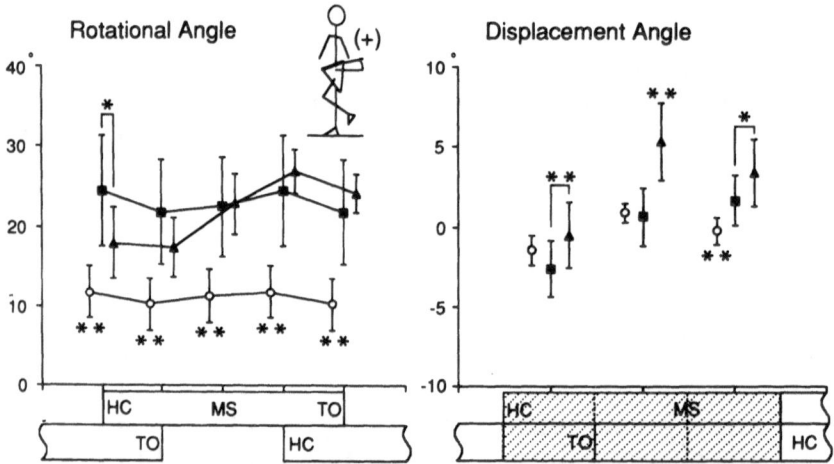

Fig. 6. Pelvic motion in groups with limited extension of the hip in the sagittal plane. *Open circles*, normal (*n* = 20); *closed squares*, bilateral (*n* = 22); *closed triangles*, urilateral (*n* = 10). *P < 0.05; **P < 0.01

normal control group, and the unilateral limitation group and the forward rotation was more increased during the early part of the stance phase.

Motion of the Upper Trunk (Fig. 5)

Symmetrical backward rotation of the upper trunk was demonstrated on the swinging limb side in normal controls. In the bilateral limitation group, in contrast, forward rotation was observed. Additionally, rotation of the upper trunk was not synchronized with that of the pelvis in normal controls, whereas these motions were synchronized in the unilateral and bilateral limitation groups.

Sagittal Plane

Motion of the Pelvis (Fig. 6)

With regard to the motion of pelvis in the sagittal plane, the patients were divided into three groups: (1) those with limited extension of unilateral hip; (2) those with this limitation in the bilateral hips; and (3) those with no such limitations in either hip. The forward tilting angle of the pelvis was greater in the group with limited extension of the bilateral hips than in the normal controls. The forward rotation was markedly increased in the group with limited extension of the unilateral hip during single limb support of the affected side, and this rotation was reduced after toe-off.

Discussion

Several studies have reported that the pelvis drops to the side opposite the supporting limb during normal gait [1,5,6]. This study confirms this phenomenon. The horizontal displacement of the upper trunk, which occurred toward

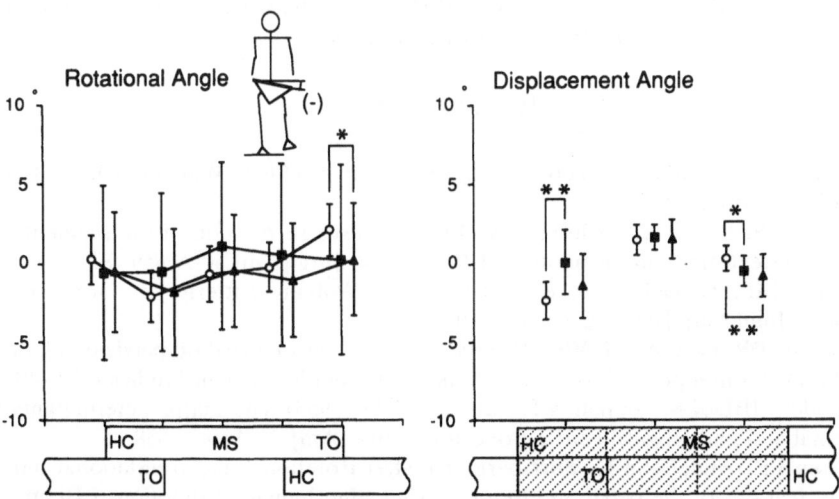

Fig. 7. Pelvic motion in groups who were positive for Duchenne's phenomenon and who showed the lateral shift phenomenon in the frontal plane. *Open circles*, normal ($n = 20$); *closed squares*, Duchenne (+) ($n = 10$); *closed triangles*, lateral shift ($n = 19$). *$P < 0.05$; **$P < 0.01$

the side opposite the supporting limb during walking, we called the lateral shift phenomenon. This phenomenon is distinguished from Duchenne's phenomenon in appearance by lack of the marked lateral bending of the upper trunk. However, motion patterns similar to those of the D(+) group were also revealed in the lateral shift group with the ATV system.

It is generally assumed that Duchenne's phenomenon originates in compensation for a drop of the pelvis on swinging limb side in patients with dysfunction of the hip abductor muscles [9]. Regarding pelvic motion in the D(+) group, the pelvis on the swinging limb side rotated upward rather than downward, and we observed a downward rotation during the latter part of the stance phase (Fig. 7). The authors consider that Duchenne's phenomenon, which reduces the resultant force on the hip by transferring the center of gravity toward the supporting limb, can occur not only in patients with hip abductor weakness but also in those with severe pain in the hip or a laterally migrated femoral head [10].

Patients with osteoarthritic hips usually exhibit a position of flexion, adduction and/or external rotation. According to previous reports, during walking, normal hips extend gradually during stance phase (mean extension angle 10° [3]–15° [11]), and rotate internally from just before heel-contact to late stance phase (mean internal rotation angle 4° [11]–7° [2]).

Regarding motion in the horizontal plane, the patients were selected in terms of the presence and absence of limited internal rotation of the hip; regarding motion in the sagittal plane, the patients were selected in terms of the presence or absence of limited extension of the hip. In these two groups, i.e., those with limited internal rotation, and those with limited extension, it was revealed that the rotational angle of the pelvis was increased in each plane to compensate for the reduced range of hip motion.

References

1. Inman VT, Ralston HJ, Todd F (1981) Human walking. Williams and Wilkins, Baltimore, pp 22–61
2. Levens AS, Inman VT, Blosser JA (1948) Transverse rotation of the segments of the lower extremity in locomotion. J Bone Joint Surg [Am] 30-A:859–872
3. Murray MP, Drought AB, Kory RC (1964) Walking patterns of normal men. J Bone Joint Surg [Am] 46-A:335–360
4. Nottrodt JW, Charteris J, Wall JC (1982) The effects of speed on pelvic oscillations in the horizontal plane during level walking. J Hum Movement Studies 8:27–40
5. Saunders JBDeCM, Inman VT, Eberhart HD (1953) The major determinants in normal and pathological gait. J Bone Joint Surg [Am] 35-A:543–558
6. Stokes VP, Andersson C, Forssberg H (1989) Rotational and translational movement features of the pelvis and thorax during adult human locomotion. J Biomechanics 22:43–50
7. Hattori T, Niwa S, Sawai K, Mitsui T (1986) The development of the on-line motion analysis system using television techniques. Proceedings of ann meeting of the Japanese society for orthopaedic biomechanics 8:207–212

8. Hirose S, Sawai K, Hattori T, Yamamoto T, Niwa S (1990) Three-dimensional analysis of segmental movements of human body during gait. Part 1. Application of our ATV-system. Cent Jpn J Orthop Surg Traumatol 33:1358–1362
9. Debrunner HU (1982) Orthopaedic diagnosis, 2nd revised edn. Georg Thieme, Stuttgart, pp 189–195
10. Whittle MW (1991) Gait analysis: An introduction. Butterworth-Heinemann, Oxford, pp 91–129
11. Johnston RC, Smidt GL (1969) Measurement of hip-joint motion during walking. Evaluation of an electrogoniometric method. J Bone Joint Surg [Am] 51-A: 1083–1094

8

Gait Analysis Following Pelvic Osteotomy

Toshiaki Hamada[1], Takashi Aoyama, Fumio Kasahara[2],
Atsumu Tokura[3], Takayuki Miura, and Hisashi Iwata[4]

Summary. Gait analysis was conducted in 53 patients who had undergone pelvic osteotomy: 39 with rotational acetabular osteotomy (RAO), 11 with Salter innominate osteotomy, and 3 with Chiari pelvic osteotomy. A 16-mm cinecamera and a force-plate were used to analyze gait features and evaluate postoperative recovery of ambulation. Despite the fact that RAO is, when compared with the Salter and Chiari methods, a more invasive technique, postoperative recovery of ambulatory function was in no way deficient, and patients achieved close to a normal gait. The most effective parameters for assessing the speed of postoperative gait recovery were stride length, stance duration, change in the angle of the hip joint, time for appearance of the first peak in the vertical component of the force-plate, and the acceleration of the anterior component of the anteroposterior ground reaction force, all of which improved in the 6th–8th postoperative month. Parameters which remained unimproved for a longer postoperative period were: inclination of the trunk, decreased double knee action, shallowness of the force-plate vertical component curve, and deceleration of the posterior component of the anteroposterior ground reaction force.

Key words: Hip—Osteoarthritis of the hip—Gait analysis—Pelvic osteotomy—Salter innominate osteotomy—Chiari pelvic osteotomy—Rotational acetabular osteotomy

[1] Department of Orthopaedic Surgery, Nishio Municipal Hospital, 6 Kamiawara, Kumami-cho, Nishio, Aichi, Japan
[2] Rehabilitation Engineering Center for Employment Injuries, Komei 1-10-6, Minato-ku, Nagoya, Japan
[3] Department of Orthopaedic Surgery, Chubu Rosai Hospital, Komei 1-10-6, Minato-ku, Nagoya, Japan
[4] Department of Orthopaedic Surgery, Nagoya University School of Medicine, Turuma-cho 65, Syowa-ku, Nagoya, Japan

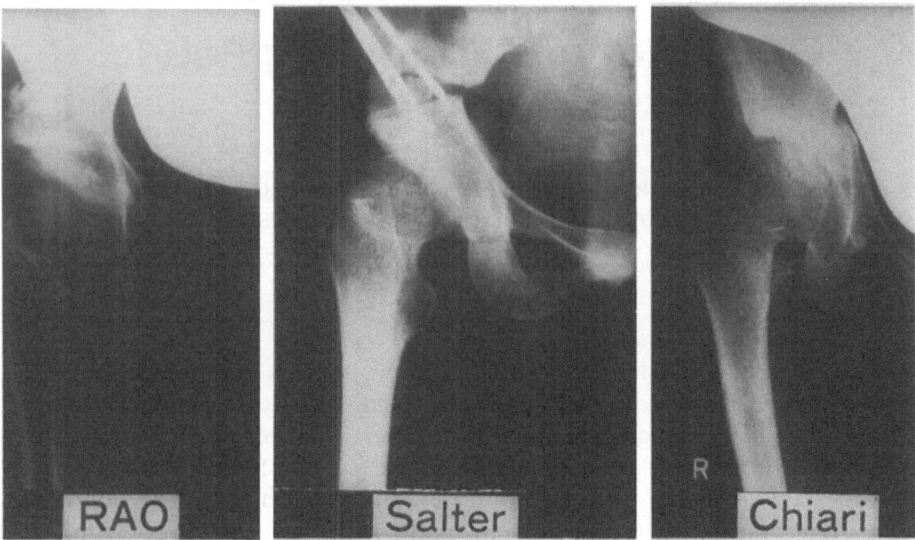

Fig. 1. Postoperative radiography of three types of pelvic osteotomy. *RAO*, Rotational acetabular osteotomy

Introduction

There have been many studies of normal and abnormal gait. The majority of reports on the load-bearing joints of the lower limbs have focused on disorders of the hip joint, since many kinds of hip joint disease cause severe disturbance of ambulatory function. For example, Gore and Murray [1] and Murray et al. [2–4] have done extensive research on patients who have had artificial hip replacement surgery, and in Japan there is Iida's [5] detailed report. To date, however, to our knowledge, there have been no studies of gait analysis in patients who have undergone rotational acetabular osteotomy (RAO).

Accordingly, to determine the effects of this procedure on postoperative gait, we analyzed the gait of RAO patients in comparison with that of normal subjects and patients treated by the Salter or Chiari methods. (Fig. 1)

Materials and Methods

Subjects

A total of 39 RAO patients were our subjects. Patients who underwent bilateral osteotomy, in which analysis would have been difficult, and those in whom other operations, such as femoral osteotomy, were performed were excluded from the study. All 39 patients, except one, were women. Age at the time of operation ranged from 12 to 55 years (mean, 29.5 years).

Table 1. Subject profiles.

	Number of cases (n)	Sex (female)	Age in years (average)	Radiological classification			Length of follow-up
				Pre arthrosis	Early arthrosis	Advanced arthrosis	
RAO	39	38	12–55 (29.5)	32	4	3	1–4 Years
Salter	11	11	20–36 (29.5)	10	1		1 Years
Chiari	3	3	24–38 (32.3)		2	1	1 Year

For *RAO*, rotational acetabular osteotomy, data were obtained preoperatively in 27 patients and postoperatively in 24

Preoperative X-rays revealed pre-arthrosis in the great majority of the patients (32); 4 patients were in early stage and 3 in the progressive stage of arthrosis. Surgical procedures in all patients followed the technique described by Ninomiya and Tagawa [6].

Postoperative treatment consisted of setting the quadriceps immediately after the operation. Range of motion (ROM) exercises for the hip joint commenced in the 4th postoperative week and included rising without load and walking with a cane. From the 6th postoperative week walking under a partial load and joint resistance exercises were introduced. Permission for full load was given in the 14th week.

Gait analysis in the RAO group, first conducted 4 months after surgery, when ambulation under a load became possible, was repeated monthly thereafter for at least a year, if possible. The longest period of such observations was 4 years.

Similar analyses were conducted for 1 year in a control group consisting of 11 patients who underwent operation by the Salter method and 3 who underwent operation by the Chiari method. The age of the Salter group was 20–36 years (mean, 29.5); the group consisted of ten patients who had shown pre-arthrosis and one in the early stage. The age of the Chiari group was 24–38 years (mean, 32.3) and this group consisted of two patients in the early stage and one in the advanced stage of arthrosis (Table 1).

Gait analysis was also conducted in a control group of ten healthy female subjects, all 20 years of age.

Method of Analysis

Ground reaction force measurements were taken with a force-plate, and kinematic data were obtained with a 16-mm Bolex cinecamera (Bolex H16; Bolex International S.A., Trendon, Switzerland). Two 400 × 1000 mm force-plates, made by Kyowa Dengyo (Tokyo, Japan), were used to measure vertical, longitudinal, and lateral motion components. For kinematic measurements, markers were attached to various parts of the body and the 16 mm cinecamera was used to observe movement in the frontal and sagittal planes; coordinates were then plotted from film measurements.

Kinematic data were collected at 62 Hz and 32 frames/s. Force-plate data were passed through a strain amplifier and converted from an analog to a digital signal. The results were then input to a minicomputer (Eclipes 5/130; Japan Data General, Tokyo, Japan) and plotted on a graph.

The 16-mm film was also analyzed, using a motion analyzer (NAC MOVIAS model 200 A; NAC, Tokyo, Japan) and graph pen (Graf/pen GP-6; NAC). The results were fed into the minicomputer and plotted on a graph. All data described in this paper were stored on floppy disks [7].

In the frontal plane, to measure the inclination of the pelvis, markers were attached to both anterior superior iliac spines, while other markers were attached to the superior border of the sternum and the pubic symphysis to measure the inclination of the trunk. In the sagittal plane, spots were affixed at the acromion, the greater trochanter, the lateral joint line of the knee, the lateral malleolus(ankle), the capitis of the fifth metatarsal, and the heel to measure the angle and speed of change in the angle of the hip joint and knee joint during locomotion.

Subjects were allowed to walk freely with their normal gait, and one cadence was recorded after the subject had walked 5 m. The parameters thus obtained

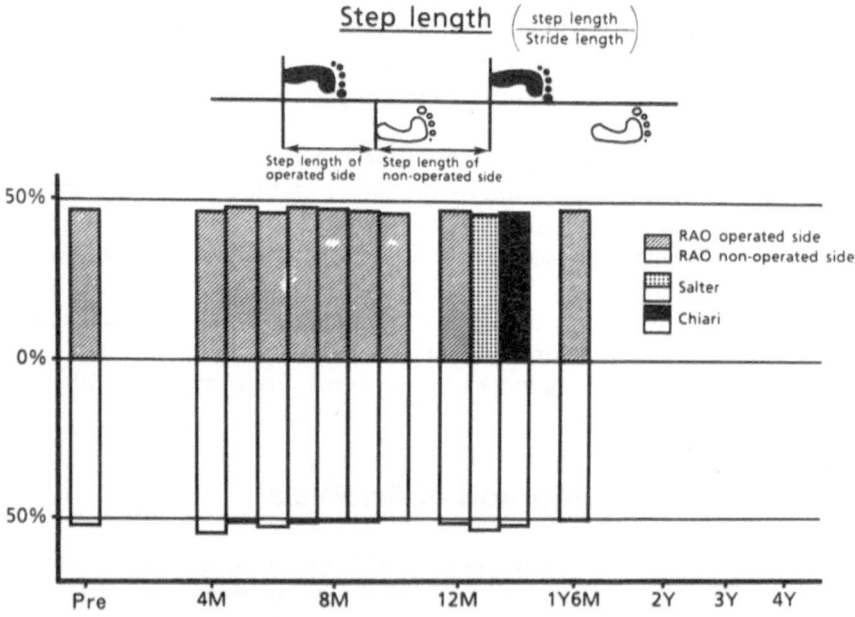

Fig. 2. Step length. *Y*, Years; *M*, months; *diagonally hatched columns* indicate RAO operated side; *dotted columns*, Salter operated side; *closed cloumns*, Chiari operated side; *open columns*, non-operated sides

were: in the sagittal plane, changes in angle of the lower limbs, speed of change in angle, step length, and stride length; in the frontal plane, angle of inclination of the trunk and pelvis, and the three ground reaction force components.

Results

Parameters that did not change pre- and postoperatively were step length, cadence, and pelvic tilt when standing on the affected side.

Step Length (Fig. 2)

There was almost a 1 : 1 correlation between step length in the healthy controls and in the osteotomy patients, showing that the osteotomy patients had little pain and had good ambulatory function.

Pelvic Inclination with Weight on the Affected Side

When patients rested weight on the affected side, the pelvis tilted about 4° toward the normal side both before and after surgery, but the difference was not significant in comparison with the controls.

Cadence

Cadence was stable. Regular cadence is a prerequisite for making comparisons with force-plate data. In this study, however, walking paced with a metronome was judged as a load condition, and all tests were conducted with free pacing. Even so, cadence for osteotomy patients remained within the range of 100–110 min, thereby meeting the prerequisite. From this we ascertained that cadence would not affect comparisons with the force-plate data.

Parameters that improved in the early postoperative period were: stride length, stance duration, range of motion of the hip joint during locomotion, time for appearance of the first peak in the vertical component of the ground reaction force, and the acceleration of the anterior component of the antero-posterior (A–P) ground reaction force.

Stride Length (Fig. 3)

Stride length shortened once in the 4th postoperative month, but at 6 months was back to preoperative levels, and after 7 months had stabilized. When the mean and standard deviation (SD) was expressed as a proportion of height, if height was taken to be 100, the mean ± SD for the healthy controls was 74.42% + 6.69%, showing no significant difference from the results for the osteotomy patients.

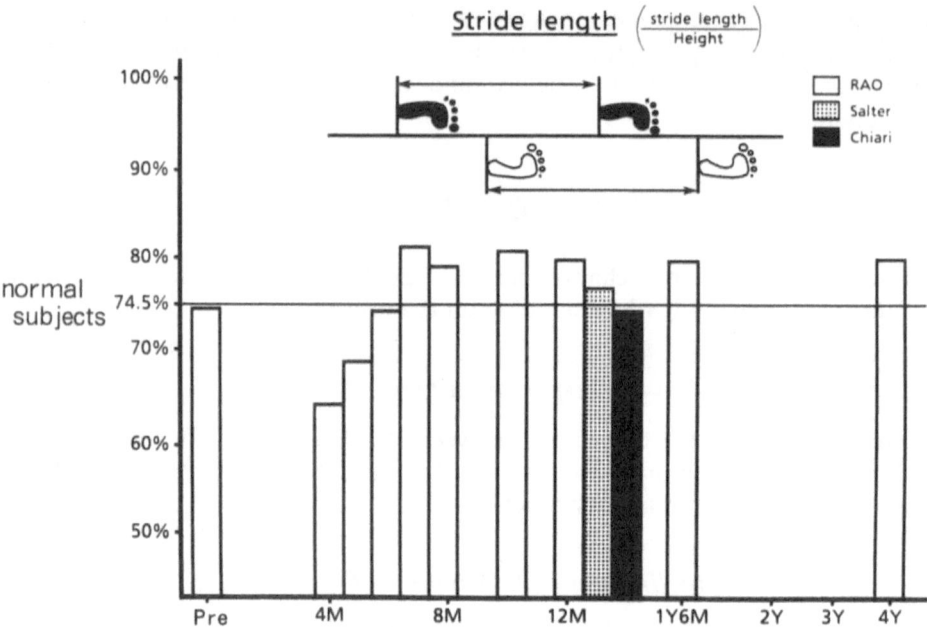

Fig. 3. Stride length. *Open columns*, RAO; *dotted columns*, Salter; *closed columns*, Chiari

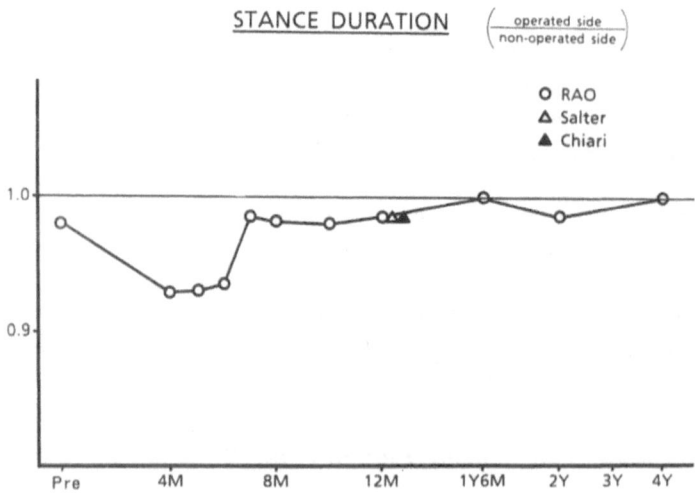

Fig. 4. Stance duration. *Open circles*, RAO; *open triangles*, Salter; *closed triangles*, Chiari

Stance Duration (Fig. 4)

The time spent with weight on the affected side versus the normal side also showed improvements 7 months after surgery. This period, in which stride length and stance duration improved, corresponded to the time it took for patients to relinquish reliance on the cane for walking indoors.

Range of Motion of the Hip Joint During Locomotion

The range of motion of the hip joint during locomotion also improved at 8 months, with range exceeding that in the preoperative period.

Time for Appearance of First Peak in the Vertical Component of the Ground Reaction Force (Fig. 5)

At 6 months, the time prior to this appearance had improved.

Acceleration of Ground Reaction Force

The acceleration equalled that of the normal side 6 months after surgery, declined slightly after that time, but then returned to normal and stabilized at 1 year.

Parameters that did not improve after surgery were: inclination of the trunk toward the affected side when placing weight on the affected leg, decreased double knee action, shallowness of the vertical component curve, and a decline in the deceleration of the posterior component of the A–P ground reaction force.

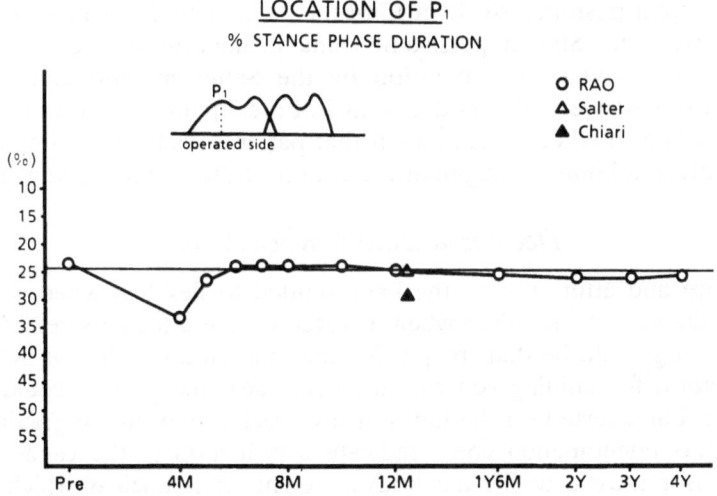

Fig. 5. Time for appearance of the first peak (*P1*) in the vertical component of the ground reaction force. *Symbols*, As in Fig. 4

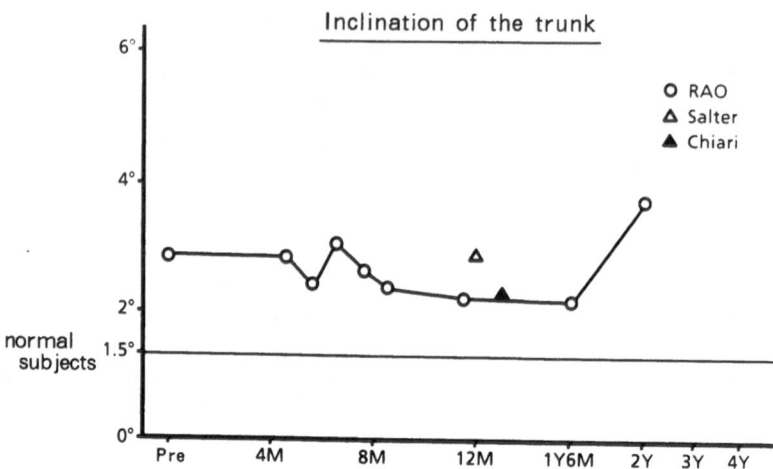

Fig. 6. Inclination of the trunk toward the affected side with weight on the affected leg. *Symbols*, As in Fig. 4

Inclination of the Trunk Toward the Affected Side with Weight on the Affected Leg (Fig. 6)

In osteotomy patients, both before and after surgery, the trunk inclined 2°–3° toward the affected side when weight was placed on the affected leg. In contrast, inclination of the trunk in normal subjects was less than 1.5°. Of the 39 patients, we obtained preoperative data in 21. In 16 of these 21 patients, preoperative trunk inclination was 2° or greater, and inclination continued for as long as 1 year postoperatively in 13 of the 23 patients for whom we obtained postoperative data. Similar persistent trunk inclination was seen in 6 of the 8 patients who underwent operation by the Salter method and for whom postoperative data were obtained, and in 2 of the 3 patients who were operated on by the Chiari method. It is thought that patients probably incline the trunk over the affected femur to augment the action of the gluteus medius muscle.

Decreased Double Knee Action

Both before and after surgery the knee tended to flex less when weight was rested on the normal side than when it rested on the affected side. The reason for this finding could be that, by putting the knee in a position of light flexion in preparation for landing (on the affected side), the patient could perhaps reduce the load exerted on the hip joint by acceleration due to gravity. Alternatively, this phenomenon could indicate a reduction in the capacity of the knee extensor muscle to lift the body's weight. It is unclear which of these explanations is correct, but, regardless of the reason, the loss of double knee action persisted for a long time.

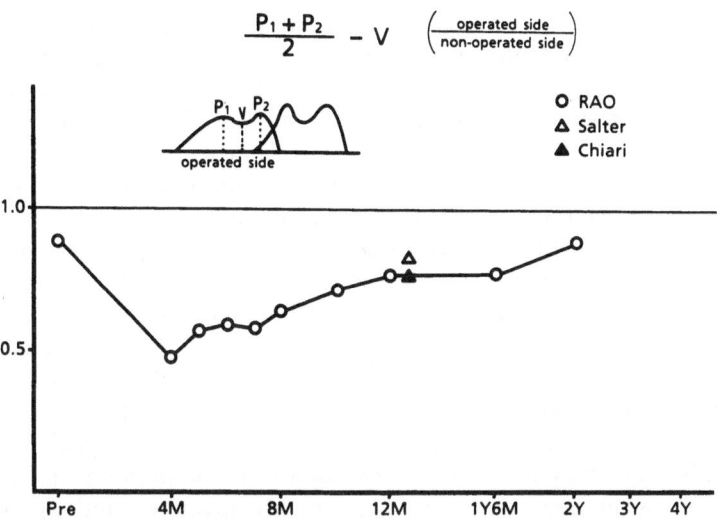

Fig. 7. Shallow vertical component curve. *P1*, First peak; *P2*, second peak; *V*, vertical component. *Open circles*, RAO; *open triangles*, Salter; *closed triangles*, Chiari

Shallow Vertical Component Curve (Fig. 7)

Of the force-plate measurements, the dip in the vertical component curve remained shallow for a prolonged time. This finding is directly related to the loss of double knee action, as described above, and shows the effect of reduced acceleration due to gravity. The same tendency was seen in the Salter and Chiari patients.

Deceleration of the Posterior Component of the A–P Ground Reaction Force

Deceleration showed the worst postoperative improvement, with recovery taking anywhere from 1.5 to 2 years.

Discussion

Past studies of gait analysis of subjects with hip joint diseases date as far back as Ahlback and Lindahl [8], who studied the excessive motion of other joints in patients who had undergone operations for joint fixation. Murray, with different collaborators, used multiple-exposure cameras and ground reaction force to analyze gait in hip disorder patients. In their series of studies, Murray et al. [2] analyzed gait accompanied by pain; Murray and Gore [1] analyzed gait after hip-fixation operations; Murray et al. [3] analyzed changes in gait before and after artificial hip replacement; and Murray et al. [4] analyzed variations in gait depending on the type of artificial joint used. Wall et al. [9] used foot switches to study cycles and ground reaction force in patients who had received artificial joints, and they measured muscle electrical potential by surface electromyography. Jacobs et al. [10] analyzed the cycle of vertical component force in force-plate studies of patients with degenerative hip joint disease. Here in Japan, Tokura [11] used a force-plate to study patients with degenerative hip joint disease, and Iida [5] published a detailed report on postoperative gait in hip joint replacement patients. However, to date, as stated above, there have been no reports of gait analysis in pelvic osteotomy patients.

Many studies have addressed the roentgenographic and clinical assessment of the RAO, Salter and Chiari pelvic osteotomy methods used in this study, all of which are all widely known osteotomy techniques [6,12,13]. Designed to correct dysplastic hips, these osteotomy procedures require exposure of the pelvis, and, although they are superior techniques for improving the centrifugality of the femoral head, they are also highly invasive, and the risk of postoperative loss of muscle tone is high. Kawakita et al. [14] used a Cybex II (Lumex; Bay Shore, N.Y.) to make sequential measurements of the strength of the abductor muscles after RAO surgery, and found that muscle strength 3 months after surgery had declined to a mean 67.8% of preoperative measurements. At 6 months, however, abductor muscle strength was restored to close to preoperative levels, and it continued to improve thereafter. These findings correlate closely with the parameters used in this study for recovery in the early postoperative period—stride length, stance duration and time of the first peak on the vertical component of the ground reaction force, range of motion of the hip joint during ambulation, and acceleration of the anterior component of the A–P ground reaction force—and this correlation would indicate that there has been a restoration of abductor muscle strength.

In the clinical setting, the major problem is how to correctly evaluate the gait of the patient who presents for examination: What parameters of patient gait should be measured, and how should these be assessed? Often, what is required is only a relativistic comparison; that is, a comparison of patient gait with that of other patients, or a comparison of the same patient's gait from past to present.

Naturally, when making comparisons between patients, it is necessary that we know beforehand the type of disease. For example, there is no point in making comparisons between a person with an artificial leg and a post-cerebulovascular accident (CVA) hemiplegic, nor is there any call to do so. A comparison with other patients of necessity means: Is the progress of postoperative gait recovery better or worse than the mean recovery parameters in other patients with the same disorder? Thus, the parameters extracted from this study of gait analysis are parameters for assessing the postoperative recovery of pelvic osteotomy patients, and should be useful for evaluating the individual patient's progress in rehabilitation.

The force-plate and the 16-mm cinecamera used in this study involve complex instrumentation and a great effort in collecting data. Therefore, these methods have the disadvantage of being impractical for use in the clinical setting. Nevertheless, the parameters extracted by this method (i.e., stride length and stance duration) are factors that can be measured easily, even on an outpatient basis, through the use of such basic materials as shoes with a foot switch, and painting or drawing materials. Furthermore, the development in recent years of semiconductor cameras and affordable personal computers has made it possible to analyze gait measurements without much effort or inconvenience as part of daily clinical practice.

The results of this study show that even the comparatively invasive RAO technique permits postoperative recovery of gait that is comparable to that obtained by employing the Salter or Chiari methods, in that all the patients examined were able to return to almost normal walking patterns.

References

1. Gore DR, Murray MP (1975) Walking patterns of men with unilateral surgical hip fusion. J Bone Joint Surg [Am] 57A:759–765
2. Murray MP, Gore DR, Clarkson BH (1971) Walking patterns of patients with unilateral hip pain due to osteoarthritis and avascular necrosis. J Bone Joint Surg [Am] 53A:259–274
3. Murray MP, Wisconsin W, Brewer BJ, Zuege RC (1972) Kinesiologic measurements of functional performance before and after McKee-Farrar total hip replacement. J Bone Joint Surg [Am] 54A:237–256
4. Murray MP, Gore DR, Brewer BJ (1979) A comparison of the functional performance of patients with Charnley and Muller total hip replacement. Acta Orthop Scand 50:563–569
5. Iida H, Masuda K, Itou T (1983) Gait analysis before and after Charnley total hip replacement (in Japanese). Cent Jpn J Orthop Surg Traumatol 26:544–548
6. Ninomiya S, Tagawa H (1984) Rotational acetabular osteotomy for the dysplastic hip. J Bone Joint Surg [Am] 66A:430–437
7. Kasahara F, Aoyama T (1977) Gait analysis in the single leg support phase using simultaneous measurements of ground reaction force and kinematic cinephotography (in Japanese). Proceedings of the 18th annual conference of the Japan Human Engineering Association. Jpn J Ergonomics [Suppl] 13:9–10
8. Ahlback SO, Lindahl O (1966) Hip arthrodesis: The connection between function and position. Acta Orthop Scand 37:77–87
9. Wall JC, Ashburn A, Klenerman L (1981) Gait analysis in the assessment of functional performance before and after total hip replacement. J Biomed Eng 3:121–127
10. Jacobs NA, Skorecki J, Charnley J (1972) Analysis of vertical component force in normal and pathological gait. J Biomech 5:11–34
11. Tokura A (1974) Kinesiologic and kinematic analysis of walking pattern in osteoarthritis of the hip (in Japanese). J Jpn Orthop Assoc 48:1–11

12. Chiari K (1974) Medial displacement osteotomy of the pelvis. Clin Orthop 98:55–73
13. Salter RB (1961) Innominate osteotomy in the treatment of congenital dislocation and subluxation of the hip. J Bone Joint Surg [Br] 43B:518–539
14. Kawakita T, Matsumoto T, Miyamori K (1989) A study of hip abduction strength in patients treated by rotational acetabular osteotomy (in Japanese). Cent Jpn J Orthop Surg Traumatol 32:872–874

9

Gait Analysis of Unilateral Total Hip Replacement: Quantitative Analysis of the Vertical Floor Reaction Force

Makoto Okuno, Kichizo Yamamoto, Ryota Teshima, Tetsuya Otsuka, and Noriyuki Takasu[1]

Summary. The subjects of this study were 31 patients, all female, with an average age of 60.1 years, who had had unilateral total hip replacement due to osteoarthrosis of the hip joint. Ten Charnley's low friction arthroplasties and 21 Omnifit systems were implanted from 1984 to 1990, and the mean follow-up period was 5.0 years. Gait studies were conducted on a walkway to measure the floor reaction force at free speed without the use of canes or crutches. Multiple regression analysis was applied to the recorded data and various parameters were obtained for the quantitative evaluation of gait, both pre- and postoperatively, in patients who had received total hip replacements (THR). We found that the similarity index for the "weighing-off" effect reached the highest value in the 3 years after surgery, and then decreased gradually during the follow-up period. The gain in the similarity index of acceleration effect was also significant in the 3 years after operation. The recovery of the index value was more rapid in patients with the Omnifit (Osteonics Corp., Allendale, N.J.) type than in these with the Charnley (Charnley's Low Friction Arthroplasties, Thackray Ltd., London, England) type THR. For the kinesiological and quantitative evaluation of THR patients pre- and postoperatively, gait analysis, using measurements of the floor reaction force, appears to be a suitable technique for long-term follow-up.

Key words: Gait analysis—Floor reaction force—Total hip replacement—Multiple regression analysis—Similarity index—Charnley type THR—Omnifit type THR

Introduction

Since 1984, to carry out kinesiological assessments of patients with total hip replacements (THR), both pre- and postoperatively, we have employed a gait

---

[1] Department of Orthopaedic Surgery, Tottori University Faculty of Medicine, 36-1 Nishimachi, Yonago, Tottori, 683 Japan

analysis system involving the use of a floor reaction force (FRF) plate. However, since obtaining reproducible quantitative results and making correct evaluations are difficult with this method, due to problems with measuring the parameters and interpreting them, we carried out this study to determine which of these parameters were of value for such assessments.

Materials and Methods

Patients

Our study population consisted of 31 patients (all female) with unilateral osteoarthrosis of the hip joint; they had no degenerative disorders in the other weight-bearing joints (e.g., knees and/or ankles). The patients had an average age of 60.1 years (range, 47–77 years), and the mean follow-up period was 5.0 years (range, 2–8 years).

Implants

Ten patients were fitted with Charnley's low friction arthroplasties (Thackray Ltd., London, England) with cement fixation (C-type group) and 21 patients received Omnifit hip systems (Osteonics Corp., Allendale, N.J.) without cement fixation (O-type group). Patients in the former group were 56.4 years of age (average) and those in the latter group had an average age of 61.8 years. The follow-up period of the C-type group was 6.9 years (range, 6–8 years) and that of the O-type group was 3.8 years (range, 2–7 years).

Gait Examination

Examinations of the gait were conducted before surgery and 3 and 6 months afterward; following which they were carried out annually. The routine method used was as follows: Each patient was examined as she walked back and forth on a 6.15-m walkway, in the center of which a force plate (60-cm long and 40-cm wide, Kistler Instrumente AG, Winterthur, Switzerland), of the single-step type, was embedded. The patient walked with bare feet at a pace that was comfortable, placing her foot on the force plate; five to ten trials were performed for each foot. The FRF data were calculated with a personal computer (PC-9801 RF, NEC, Tokyo, Japan), and the normalized FRF was plotted and printed out with a plotter (MP 4300; Graphtec, Tokyo, Japan) and a printer (PC-PR101 G2, NEC).

The recorded vertical (Fz), fore-aft (Fy), and medio-lateral (Fx) components of the FRF during the stance phase of gait are shown in Fig. 1. The characteristic points of the Fz component were subjected to multivariate analysis. The "weighing-off" effect (i.e., the acceleration of the center of gravity of the body during ambulation), deceleration effect, acceleration effect, and impulse of the vertical reaction force (VRF) were obtained by using Eqs. 1, 2, 3, and 4:

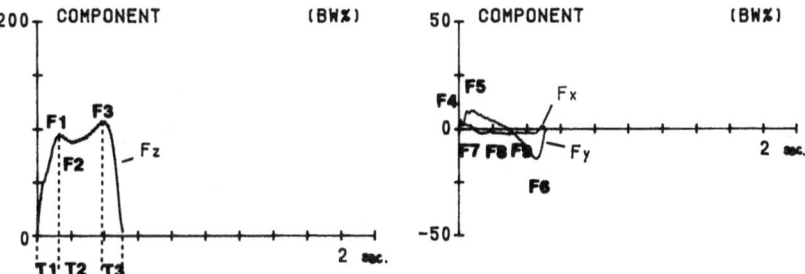

Fig. 1. Normalized floor reaction forces and the characteristic points required to obtain the parameters. *BW*, Body weight; *Fz*, vertical component of the floor reaction force; *Fx*, medio-lateral component; *Fy*, fore-aft component; (see text for detailed definitions for $F_1 \sim F_9$, $T_1 \sim T_3$)

"Weighing-off" effect; $\dfrac{(F_1 + F_3)}{2} - F_2$ (BW%) (1)

Deceleration effect; F_1/T_1 (BW%/s) (2)

Acceleration effect; F_3/T_3 (BW%/s) (3)

Impulse of VRF; $Iz = Fz\ dt$ (kgf·s) (4)

where F_1 indicates heel-strike peak of the VRF in early stance phase, F_2 indicates the minimum value during midstance, and F_3 indicates heel-off peak in late stance phase. T_1 and T_2 are the times from zero to F_1 and F_3 to zero, respectively. Iz is the integrated value of the Fz wave by time, and indicates the total impulse force during stance phase. BW is body weight. The similarity index, i.e., the ratio of the value for the affected limb to that for the unaffected limb, was obtained by using Eq. 5.

Similarity index; $\dfrac{\text{Value for affected limb}}{\text{Value for unaffected limb}} \times a \times b \times c$ (5)

where a is the supplementary coefficient by cadence, b is the supplementary coefficient by Fx component, and c is the supplementary coefficient by locus.

These three supplementary coefficients were obtained from multiple regression analyses performed individually for each subject.

Results

"Weighing-Off" Effect and Impulse of Vertical Reaction Force

Changes in the similarity indexes of the "weighing-off" effect and the impulse of VRF are shown in Fig. 2. The similarity index of the "weighing-off" effect was 0.75 (±0.15, SD) preoperatively; its value was lowest in the 3 months after surgery, but increased gradually, up to 0.90 (±0.13) the 3 years after THR

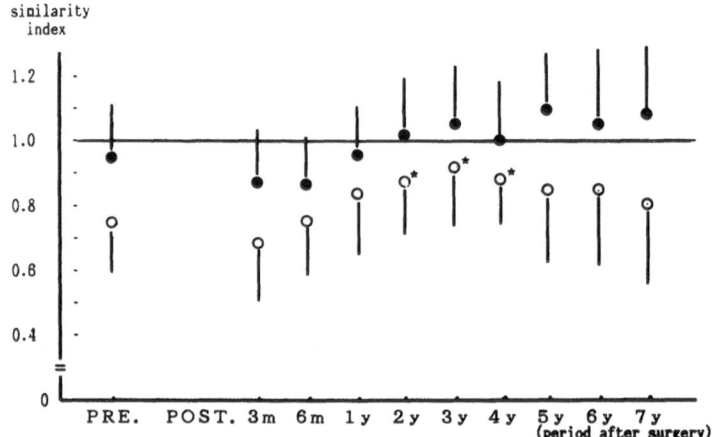

Fig. 2. Changes in the similarity indexes of the "weighing-off" effect (*open circles*) and the impulse of the vertical reaction force (*closed circles*). y, Years; m, months. Mean ± SD, *$P < 0.05$

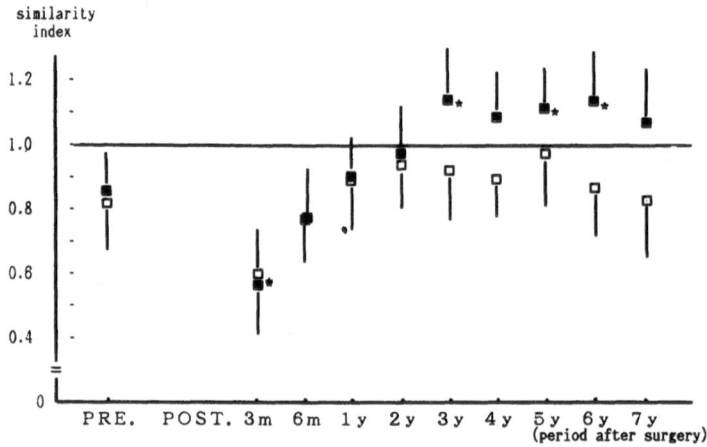

Fig. 3. Changes in the similarity indexes of the deceleration effect (*open squares*) and acceleration effect (*closed squares*). Mean ± SD, *$P < 0.05$

($P < 0.05$). Thereafter, this index gradually decreased. In contrast, the similarity indexes of the impulse of VRF did not vary as much as the "weighing-off" effect indexes within the follow-up period, and maintained a value of more than 1.0 between 2 and 7 years after surgery.

Deceleration and Acceleration Effects

Changes in the similarity indexes of the deceleration and acceleration effects are shown in Fig. 3. Although the two indexes were almost equal preoper-

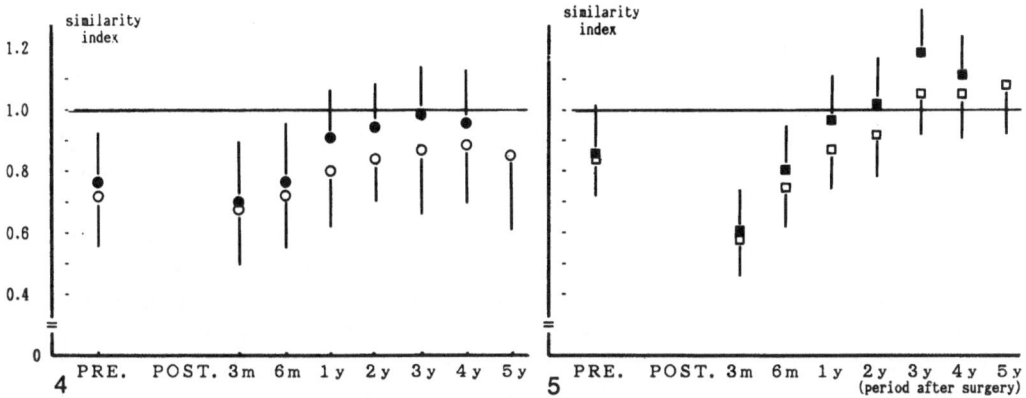

Fig. 4. Comparison of "weighing-off" effect similarity indexes between Charnley type (*C-type, open circles*) total hip replacement (*THR*) and Omnifit type (*O-type, closed circles*). Mean ± SD

Fig. 5. Comparison of acceleration effect similarity indexes between Charnley type (*C-type, open squares*) THR and Omnifit type (*O-type, closed squares*). Mean ± SD

atively and 2 years after the operation, the values of the acceleration effect indexes were markedly higher than those of the deceleration indexes 3 years after surgery ($P < 0.05$ for preoperative value). This latter tendency remained for several years in the follow-up period.

Comparison between C-type and O-type THR

Changes in the indexes were not statistically significant as a whole. The similarity indexes of the "weighing-off" and acceleration effects are shown in Figs. 4 and 5, respectively. The "weighing-off" effect indexes of C-type THR patients did not reach the value of 0.9. In the O-type patients, the similarity index of the "weighing-off" effect increased more and at a faster rate than in the C-type. However, the differences between the two groups in this index were not significant by the inter-group comparison test. The indexes of the acceleration effect in the O-type THR increased more rapidly than in the C-type group, although this difference was also not significant. These findings are discussed below.

Discussion

There have been some studies of gait analysis in THR patients [1–7]. Stauffer et al. [1] emphasized the significance of the biomechanical analysis of gait following THR, and Hattori et al. [2] established the recognition of FRF patterns in THR patients. Fujita et al. [3] analyzed the gait of THR patients by using measurements of the joints, e.g., the shoulder, hip, knee, and ankle and so on. Rydell [4] measured the forces on femoral prostheses directly, by using a transducer. Yano [5], Ouchi [6], and Okuno [7] used multivariate analysis to analyze the gait in pre- and post THR patients.

However, clinical evaluations of patients with THR are generally obtained only by physical examination (e.g., range of motion), and by the use of scoring systems, such as that of the Japanese Orthopaedic Association [8], roentgeno-graphs, and so on. As the number of patients with arthroplasties is increasing (in own Department we have had more than 500 in the last 19 years), we believe that kinesiological assessments of patients who have had implant sur-gery should be carried out in the out-patient department. We have already begun using various evaluation systems, e.g., measurement of the bone mineral content of the prosthetic hip and/or knee by single photon absorptiometry and dual X-ray absorptiometry (DEXA), but these methods are static assessments. Therefore, to obtain a kinesiological evaluation, we employ gait analysis, using the floor reaction force plate, and we measure muscle strength with the "hydromusculator", a device for this purpose (O.G. Giken Co. Ltd., Okayama, Japan). This is done in accordance with the concept of hip abductor muscle function described by Inman [9].

We believe it is necessary to simplify these examination systems. With the idea of establishing a method of gait analysis assessment, we used multivariate analysis of the vertical reaction force (VRF, Fz component). However, com-parisons between affected patients and normal subjects were found to be difficult. We therefore introduced the concept of the similarity index, described here.

According to Stauffer et al. [1], the "weighing-off" and acceleration effects indicate smooth high-speed ambulation. In this study, we found it remarkable

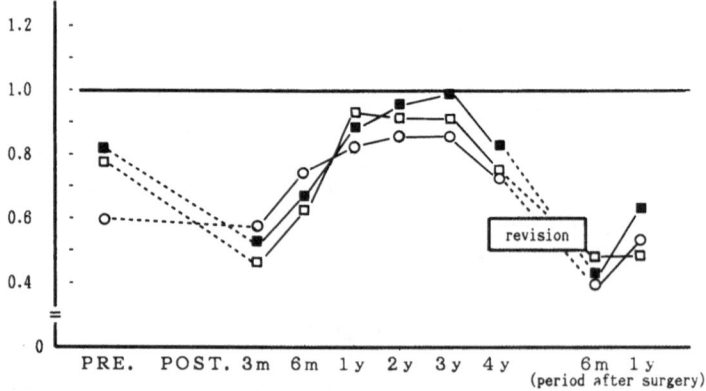

Fig. 6. Changes in similarity indexes in a patient who had revision surgery (female; 58 years of age at primary THR). *Open circles*, "weighing-off" effect; *open squares*, deceleration effect; *closed squares*, acceleration effect

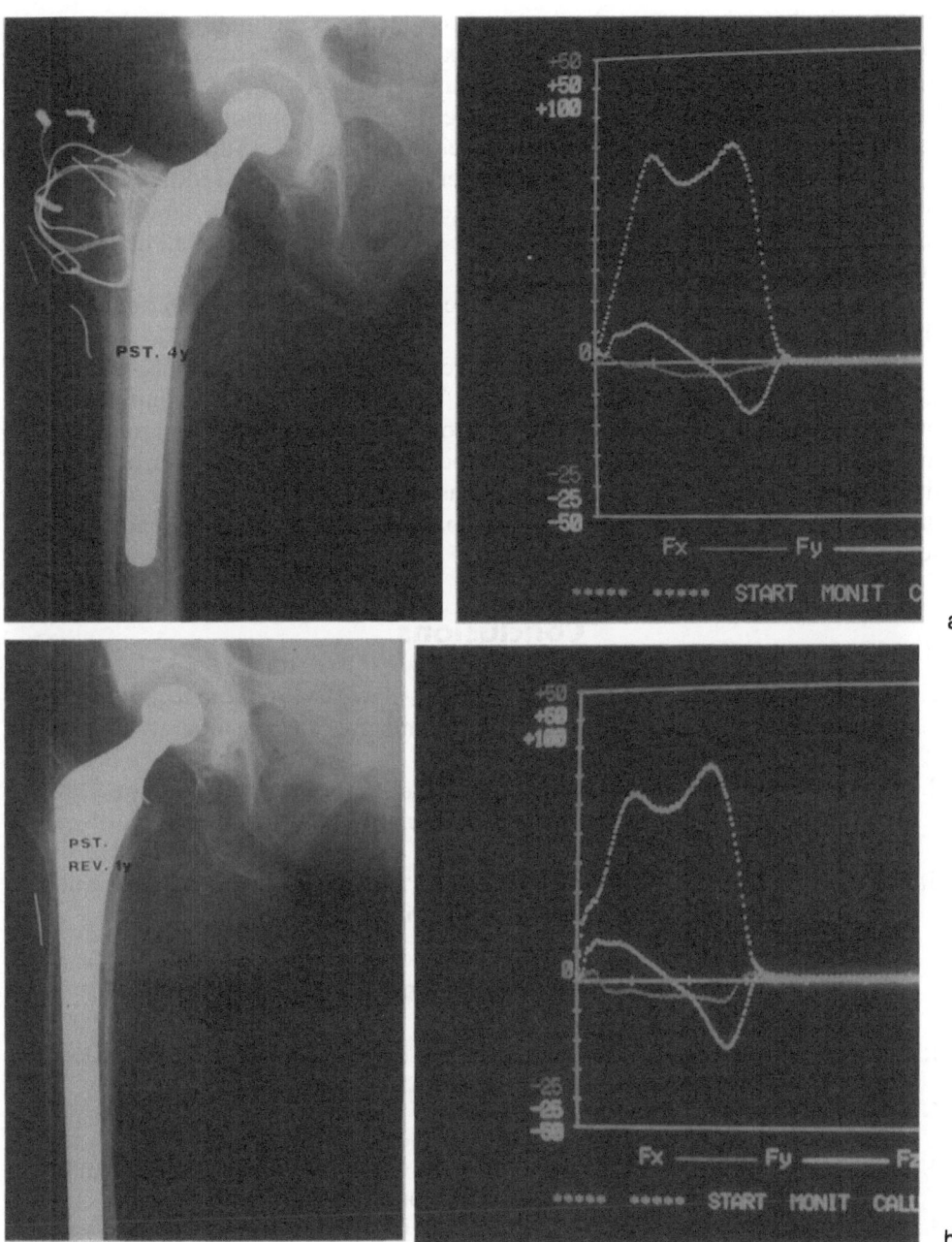

Fig. 7a,b. X-ray and walking pattern in the patient whose indexes are described in Fig. 6, **a** 4 years after primary operation (*PST 4y*) **b** 1 year after femoral stem revision surgery (*PST REV 1y*)

that the similarity indexes of these two effects increased gradually after surgery and that their values remained at a peak for 3 years postoperatively. This finding seems to indicate that it takes about 3 years for the muscle strength of hip abductors and extensors to be recovered. However, it is necessary to take careful note of the finding that these indexes slowly decreased after this period.

Actually, we experienced a revised surgery patient who had reductions in all similarity index values 4 years after surgery. We secured the femoral stem revision without cement 5 years after the primary operation, and she is now able to walk without crutches (Figs. 6, 7).

Regarding the deceleration effect and the impulse of VRF, we did not obtain significant results in this study.

There was no significant difference in the change of similarity indexes in the Charnley's (C-type) and Omnifit (O-type) THRs during the follow-up period, but the values of the indexes did change more rapidly in the O-type group than in the C-type patients. Because greater trochanteric osteotomy is performed in the Charnley's procedure, more time seems to be required for the recovery of hip abductor muscle power in the C-type patients. None of the patients in our series complained of so-called thigh pain.

Conclusions

1. Gait analysis, using the floor reaction force plate, is a suitable technique for the kinesiological assessment of THR patients pre- and postoperatively, and is of value for long-term follow-up.
2. The use of similarity indexes obtained from the vertical reaction force (Fz component) enables the clear discrimination and quantification of gait differences.
3. Recovery of the similarity indexes of the "weighing-off" and acceleration effects is a good indicator of ambulation ability.

References

1. Stauffer RN, Smidt GL, Wadsworth JB (1974) Clinical and biomechanical analysis of gait following Charnley total hip replacement. Clin Orthop 99:70–77
2. Hattori T, Shimono T, Hasegawa H, et al (1982) Gait analysis after total hip replacement for osteoarthritis (in Japanese). Orthop Biomechanics 4:55–60
3. Fujita M, Norimatsu T, Chiba G, et al (1980) Gait analysis after Charnley-Müller total hip replacement (in Japanese). Orthop Biomechanics 2:33–37
4. Rydell N (1965) Forces in the hip joint. Part II, intravital measurements. In: Kenedi RM (ed) Biomechanics and related bio-engineering topics. Pergamon, Oxford, pp 351–357
5. Yano H (1979) Gait analysis after total hip replacement (in Japanese). Orthop Mook 7:206–223
6. Ouchi E, Terayama K, Watanabe S, et al (1980) Comparison between the floor reaction force and clinical evaluation in patients with total hip replacement (in Japanese). Orthop Biomechanics 2:119–124

7. Okuno M (1989) Clinical study on total hip replacement with gait analysis (in Japanese). J Yonago Med Assoc 40:137–154
8. Imura S, Matsunaga T (1990) Handbook of the hip joint—diagnosis and treatment (in Japanese). Nankodo, Tokyo, p 161
9. Inman VT (1947) Functional aspects of the abductor muscles of the hip. J Bone Joint Surg 29:607–619

10

—Overview—
Clinical Gait Analysis in Hip Patients

Tomokazu Hattori, Shirou Hirose, Kazuhiko Sawai, and Shigeo Niwa[1]

Summary. The relationship between ascending and descending trunk movements and the biphasic pattern of the vertical reaction force was demonstrated by practical accelerometry and reaction force measurement. The
vertical reaction force directly represents the kinematic acceleration of vertical body movement. The abnormal trunk movement of hip patients was
investigated with a TV motion analysis system. Patients with lateral trunk
bending and shifting showed considerable lateral trunk displacement, shortening the lever arm for the dynamic weight due to body mass, gravitation,
and vertical acceleration. This bending and shifting may reduce the required
abductor force and the hip resultant force. In clinical gait analysis, it is
important to correctly interpret gait parameters and abnormal gait patterns
in terms of biomechanical considerations, and to establish criteria for gait
disorders.

Key words: Gait analysis—Hip patients—Biomechanics

Introduction

Gait analysis in hip patients is generally performed for two different reasons:
(1) the estimation of hip resultant force as a mechanical stress and (2) clinical
gait assessment in kinesiology. In the former case, resultant force estimation
requires accurate measurements of body segmental motion, and ground reaction
force and center of pressure, as well as certain assumptions regarding the mass
and center of gravity of each body segment. The resultant force can then be
estimated by solving the force moment equation with several force components
and their lever arm length around the hip joint. This research-oriented gait
analysis is difficult to perform clinically.

---

[1] Department of Orthopedic Surgery, Aichi Medical University, 21 Yazako Karimata,
Nagakute, Aichi, 480-11 Japan

Clinical gait assessment, i.e., clinical gait analysis, deals with abnormalities of segmental movements and ground reaction forces themselves, as manifested by patients who shows various types of constrained walking due to severe pain, muscle weakness, or joint contracture. Accordingly, gait analysis is a quantitative assessment, made by employing various measurement systems, rather than merely noting observations. These abnormal movements are definitely related to the joint resultant force because of the gain or loss of segmental acceleration and lever arm length and their effects on force moment equilibrium.

If we consider the biomechanics of the hip joint in bipedal human walking, it appears that gait analysis could be essential in the assessment of joint function.

Ground Reaction Force

Ground reaction force measurement is usually made in terms of both simple and diverse gait assessments, using a single or twin-plate system. Gait parameters can be derived from measured force components, peak force values, force vector angles, and stance time in the single-plate system, while in a twin-plate system, double support time, as a bipedal stepping transition, and step length can be derived. Further, the gait cycle and its composition can be determined in a twin long-plate system where ground reaction force can be measured for several steps on both sides all gait events can be observed in terms of the loading status.

In many clinical gait studies of hip patients, an average value for each category is compared with the normal average for various gait parameters, and conclusive judgements are made following statistical evaluation and the determination of standard deviation. Scattergrams of clinical evaluation and gait parameters may also be described, along with correlation coefficients. However, the results of these general approaches are sometimes unsatisfactory in that superficial classifications of different gait patterns are made and possibly confused, even when the clinical diagnosis or the treatment is the same. We believe, therefore, that individual gait abnormalities should be distinguished as far as possible for further investigation. Of course, there is no clear definition of normal gait, due to wide variations among different age groups, and according to sex or personal habits, but, if sufficient data are obtained for normal persons, then statistical tests can be applied for the judgement of abnormality [1,2].

We performed practical measurements of 25 joints in 15 patients with hip osteoarthritis, using a computerized twin long-plate system consisting of two long force plates ($2.5\,m \times 0.4\,m$; Kistler Type Z12888; Switzerland) and a 16-bit personal computer (NEC PC-9801VX, Tokyo). Typical gait parameters in individual patients were examined and the abnormalities were compared to previous results in 190 joints of 95 normal subjects [2], with the differences in the statistical probability being 5% ($\bar{X} \pm 1.96\sigma$) and 1% ($\bar{X} \pm 2.58\sigma$). Con-

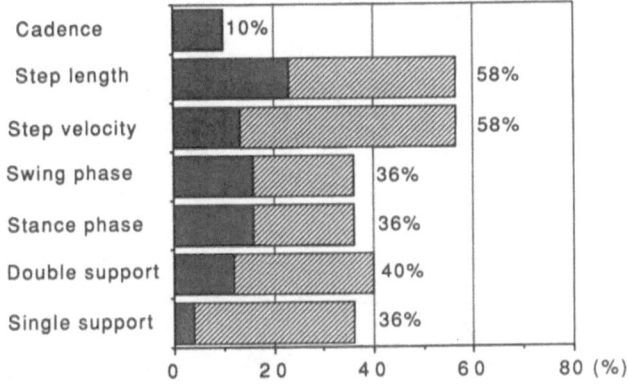

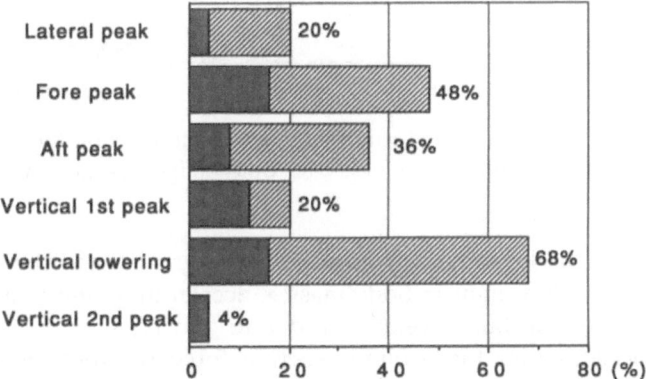

Fig. 1a,b. Rates of occurrence of abnormalities. **a** Cadence, step length, and gait cycle composition; **b** ground reaction force parameters. *Cross-hatched area*, $P < 0.05$ ($\bar{X} \pm 1.96\sigma$); *diagonally-striped area*, $P < 0.01$ ($\bar{X} \pm 2.58\sigma$)

siderable abnormalities were observed in the lowering of the vertical force component, and in step length, step velocity, and peak value of the fore force component. The rate of occurrence these abnormalities may retrospectively suggest the significance of each gait parameter in hip osteoarthritis (Fig. 1a,b).

Regarding the lowering of the vertical force component, normal subjects show a characteristic biphasic waveform, with the peak values in each phase exceeding the static body weight of the subject, and considerable lowering below the body weight occurring in between the two peaks [3–6] (Fig. 2a). Subjects with gait disorders, however, lose this characteristic biphasic pattern, and the vertical component remains at around the static body weight throughout the stance period. This is a useful sign in clinical gait analysis that the subject has some dysfunction in locomotion as demonstrated in Fig. 2a.

However, the physical implications of this biphasic waveform are often insufficiently understood. In Newtonian physics, force is produced by mass and

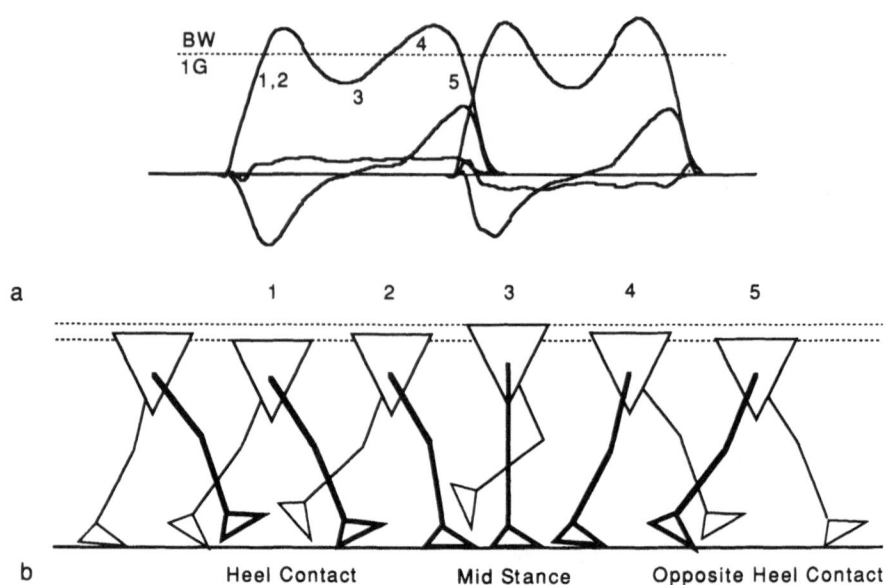

Fig. 2a,b. Ground reaction force and gait model. **a** Ground reaction force components and **b** compass gait model with hinged legs. *BW*, Body weight; *G*, gravity

acceleration. Thus, body weight is produced by body mass and the gravitation of the earth, i.e., body weight = body mass × acceleration due to gravity; this is a static force with constant acceleration due to gravity.

In a dynamic situation, the ground reaction force is expressed as a reaction of the stepping force during walking; this is produced by body mass, acceleration due to gravity, and extra kinematic acceleration of ascending or descending body movements, i.e., dynamic vertical force = body mass × (acceleration due to gravity ± kinematic acceleration). The concept of kinematic acceleration introduced here can be expressed by using a compass gait model with hinged legs (Fig. 2b).

In normal free walking, heel contact (1) is made in a descending body movement due to the astride position of hip flexion and knee extension where the body mass center remains behind the stepping foot, with the opposite hip and knee being extended. Then a little knee flexion (2) absorbs the stepping impact and creates smooth load transfer in the early part of the stance phase. The acceleration of the descending body movement cooperates with the acceleration due to gravity and increases the overall vertical force onto the ground. Therefore, the first peak of the biphasic waveform is observed as part of an increase in a single measurement of the reaction force.

Extension of the knee and hip (2) to (3) produce an ascending body movement, applying counter force onto the ground, where the body obtains upward acceleration according to the integration of the applied force, and the body mass center is situated right above the supporting foot, reaching the highest

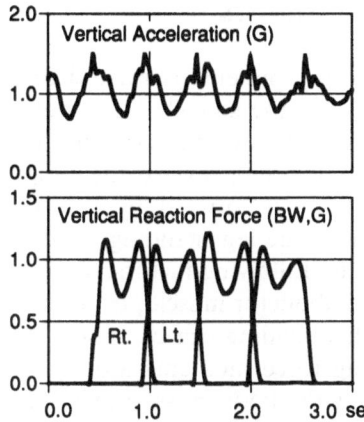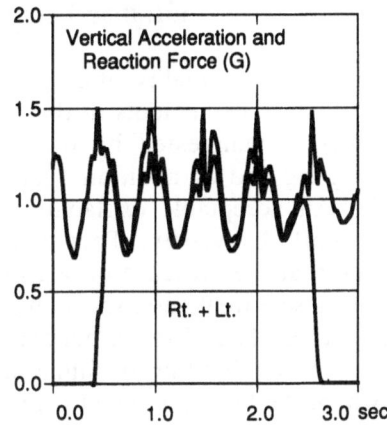

Fig. 3. Vertical acceleration and ground reaction force; finite impulse response digital filter with cut-off frequency of 10 Hz was applied for noise reduction. *Rt.*, Right; *Lt.*, left

position with a single straight leg (3). This acceleration acts against the acceleration due to gravity, and therefore a considerable lessening of the vertical force can be observed as the lower part of the biphasic waveform.

Again the descending body movement begins with further hip and knee extension; asthe opposite leg is swung forward with hip flexion and knee extension, the body mass center is situated ahead of the supporting foot (4), and it changes to the next astride position with the opposite heel contact (5). Thus the downward acceleration due to the descending body movement again cooperates with the acceleration due to gravity and increases the overall vertical force onto the ground. The second peak is also a part of the increase in the single measurement of the reaction force. Therefore the vertical component directly represents kinematic acceleration of the body, and the biphasic waveform is a consequence of alternating downward and upward acceleration. This was demonstrated by practical measurements using a twin long force plate system with a strain gauge accelerometer (Kyowa Electronic Instruments AS-10HB, Tokyo, Japan) attached to the upper trunk. The vertical acceleration is overlaid onto the vertical force diagram, which is well matched with the vertical reaction force (Fig. 3).

The reduction or disappearance of the biphasic sign suggests gentle walking with less descending and ascending body movement and kinematic acceleration, thus this reduction or disappearance may reduce joint moments and joint resultant forces in the lower limbs, and produce what is called poor loading capacity.

Upper Trunk Movements

A typical gait abnormality often observed in hip patients is related to upper
trunk and pelvis movements in the frontal plane, the so-called Trendelenburg
sign, which is manifested by the pelvis dropping on the side of the non-
supporting leg and by additional lateral trunk bending toward the supporting
leg side. An acceptable explanation has been put forward by investigators
at New York University. They used a static one-leg standing model, and
concluded that the presence of one or more the following factors: severe pain,
abductor muscle weakness, short lever arm of abductor muscle, or joint instability, may cause lateral trunk bending [5]. This bending can be simply demonstrated by an equilibrium of falling and rising force moments around the hip
joint, where the falling moment is produced by the body mass, gravity, and the
lever arm for the center of the body mass, and the rising moment is produced
by the muscle contraction force of the hip abductor and the lever arm for the
hip abductor (Fig. 4a). Hip resultant force can then be calculated from the
force components of the equilibrium.

Thus, the dropping of the pelvis is the ultimate result of insufficient rising
moments, due to abductor muscle weakness or to a short lever arm for the hip
abductor muscles, and the lateral trunk bending is the compensation required
to reduce the falling moment and the hip resultant force, which it does by
shortening the lever arm, bringing the body mass center close to the supporting leg side. Proceeding to dynamic biomechanics, two more kinematic
components should be involved in the one-leg standing model. As mentioned
above, upward or downward acceleration due to ascending or descending body
movements naturally participates in the equilibrium. Lateral trunk bending

Fig. 4a,b. Gait models of single support. **a** Static model, **b** dynamic model. F, Abductor
muscle force; m, body mass; g, acceleration due to gravity; a, kinematic vertical
acceleration; m_t upper trunk mass; b, Kinematic lateral acceleration

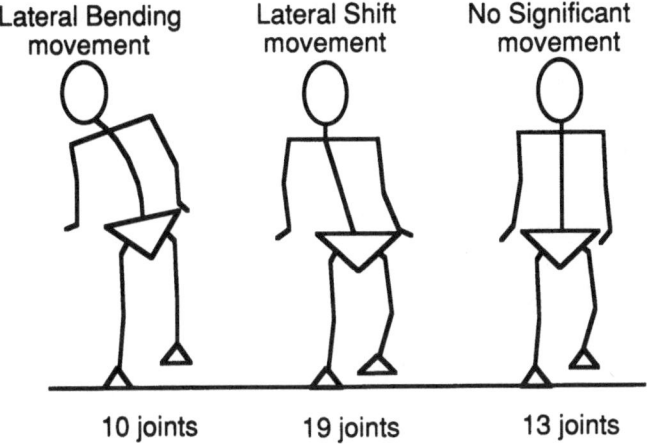

Fig. 5. Classification of lateral trunk movement

may also generate a horizontal acceleration and may thus affect the equilibrium (Fig. 4b). However there are, to date, to our knowledge, few biomechanical studies of upper trunk movement.

Accordingly, we measured lateral trunk movement in 21 patients with osteoarthritis of the hip and in 10 healthy subjects as normal controls, using our original TV motion analysis system [2,7,8]. Optical-reflective markers were attached bilaterally on both acromions and on the anterior superior iliac spine. The middle point of both shoulder markers was calculated; this was used to measure the lateral trunk movement. The linear velocity and acceleration of the movement were calculated from the displacement. Before the kinematic measurement was carried out, the patients were divided into three groups by an experienced clinician, who made careful observations of trunk bending. The groups were: trunk lateral bending, trunk lateral shifting, and no significant movement (Fig. 5).

The lateral bending group showed more than 60 mm lateral displacement in one entire step, and almost 40 mm during the double support period. The lateral shifting group showed almost 60 mm lateral displacement, but less than 30 mm during the double support period (Fig. 6) [9,10].

Both normal subjects and the lateral bending and shifting patients showed maximum lateral displacement at single support, with the displacement being considerable in the patients (Fig. 7). The maximum velocity of lateral movement was observed during the early single support period in normal subjects, while the lateral bending subjects had a forward phase shift showing great velocity at heel contact, where the body mass center quickly moved toward the stepping leg side to shorten the lever arm before the single support period.

Regarding lateral acceleration, different types of complicated waveform were observed in the normal subjects and in the patients. Although the lateral acceleration in normal subjects faded away at the end of the single support

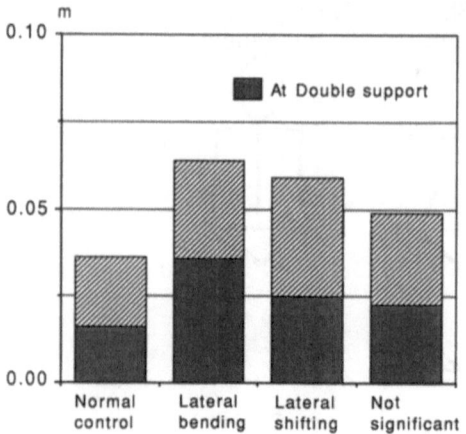

Fig. 6. Lateral displacement of upper trunk

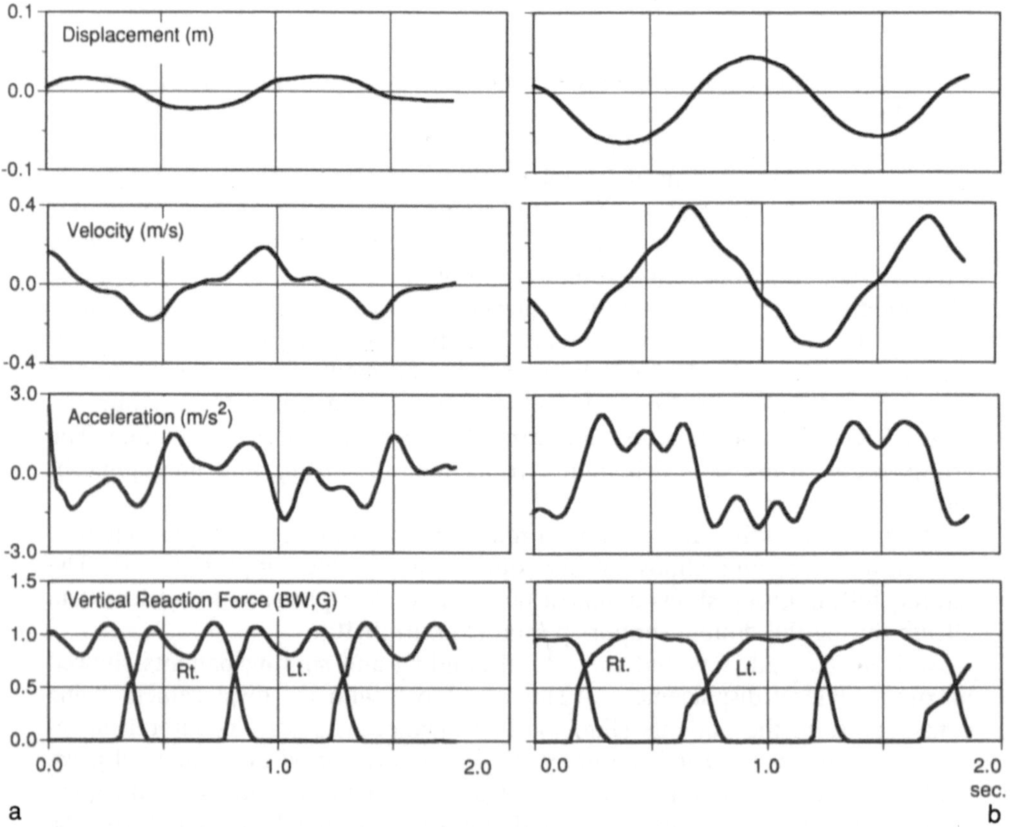

Fig. 7a,b. Lateral trunk movement and vertical reaction force. **a** normal female, aged 57; **b** bilateral osteoarthritis of the hip; female, aged 52. *Rt.*, right; *Lt.*, left. Finite impulse response digital filter with cut-off frequency of 10 Hz was applied for noise reduction

period, lateral bending subjects showed continuous and nonuniform acceleration toward the non-supporting side throughout the single support period, and rapid turnover in breaking the excessive movement during the following double support period. Accordingly, it appears that the acceleration always generates the falling moment, rather than the rising moment, in the single support period, and this negatively affects the force moment equilibrium.

It is suggested that lateral trunk bending and shifting movements produce lever arm shortening during the single support period by means of quick trunk movements made during the double support period, and that falling moments are produced by acceleration of the trunk movement rather than by the expected rising moments in the single support period. However, the falling moment seems to be a favorable factor in the occurrence of the subsequent large trunk displacement toward the opposite side.

Discussion

Gait analysis with advanced system developments has been successful in human science and research-oriented studies, and clinical gait analysis has been confirmed as a valid functional evaluation or quantitative assessment of gait performance, to be used in stead of descriptive records. However, some surgeons still have major questions, namely, what benefit can be obtained? Is such analysis worthwhile for practical treatment, considering the cost and time involved? Answers to these questions depend on the application and understanding of biomechanics.

Current gait analysis systems are computerized and technologically advanced; their cost is lower than that of previous systems and less knowledge is required for their practical use. Gait analysis provides a great deal of information on gait performance and can exclude some gait parameters that are only theoretical or are insignificant in particular gait disorders. Patient gait often involves such compensatory phenomena, induced by the original dysfunction, as pain, muscle weakness, and joint contracture. It is suggested that the key points for successful clinical gait analysis are the correct interpretation of gait parameters and abnormal gait patterns, and the establishment of criteria for gait abnormalities. Thus, it is necessary to introduce stable and reliable gait analysis systems to sites where they can be used clinically, both for reassessing fundamental knowledge and for obtaining new information.

References

1. Murray MP, Drought AB, Kory RC (1964) Walking patterns of normal man. J Bone Joint Surg [Am] 46:335–360
2. Hattori T (1992) Current trends in gait analysis: Advanced techniques for data acquisition and analysis. In: Niwa S, Perren SM, Hattori T (eds) Biomechanics in orthopedics, Springer, Tokyo, pp 102–120

3. Inman VT, Ralston H, Tod F (1981) Human walking. Willams and Wilkins, Baltimore
4. Paul JP (1992) Fundamentals of gait analysis. In: Goh JCH, Nather A (eds) Proceedings of 7th international congress on biomedical engineering. National University of Singapore Press, Singapore, pp 579–582
5. Whittle M (1991) Gait analysis: An introduction. Butterworth-Heinemann, Oxford
6. Pedotti A, Ghista DN (1981) Human locomotion analysis. In: Ghista DN, Roaf R (eds) orthopaedic mechanics: Procedures and devices, vol 2. Academic, London, pp 111–174
7. Hattori T, Niwa S, Sawai K, Furukawa R (1986) The development of on-line motion analysis system using television techniques (in Japanese). Orthop Biomechanics 8:207–212
8. Hattori T, Hirose S, Kuwahara T, Sawai K, Niwa S (1991) Direct linear transformation method for three-dimensional kinematic measurement (in Japanese). Orthop Biomechanics 13:411–417
9. Hirose S, Sawai K, Hattori T, Niwa S (1990) Three-dimensional analysis of segmental movements of the human body during gait: Third report (in Japanese). Orthop Biomechanics 12:303–306
10. Hattori T, Hirose S, Sawai K, Niwa S (1992) Gait analysis of upper trunk and pelvic movements in normals and hip osteoarthritis. In: Goh JCH, Nather A (eds) Proceedings of 7th international congress on biomedical engineering. National University of Singapore Press, Singapore, pp 454–455

III

Hip Joint Surgery and Simulation

A. Total Hip Arthroplasty

11

Simple Solid Models for Surgical Planning in Hip Arthroplasties

Hiromi Ohtsuka, Kazuhiko Sawai, and Shigeo Niwa[1]

Summary. Our simple solid model, made of foam polystyrol, is helpful in understanding the complex shape of bony geometry (such as that found in hip joints) and is useful for carrying out surgical planning and rehearsal operations.

Key words: Hip joint—Total hip arthroplasty—Solid model—Computerized tomography—Surgical planning—Rehearsal operation—Three-dimensional image

Introduction

This simple solid model is easily made and is very economical compared with a three-dimensional surface reconstruction image or other milled solid models.

It is essential when planning THA to grasp the detailed shape of the hip. Prior to surgical treatment for the hip, especially total hip arthroplasty (THA), the three-dimensional shape of the bone is difficult to grasp from preoperative X-ray pictures alone. Accordingly, several studies [1,2] have recently been published on three-dimensional surface reconstruction images, produced by using a television monitor, and on solid models [1,3] made with a milling system, based on computed tomographic (CT) data for the hips. Since 1985, it has been possible to discuss operative procedures in detail, using solid models produced by life-size enlargements of contiguous CT slice images [4]. This type of solid model is useful for gaining information about the size and the shape of an object; it also helps in surgical planning and in rehearsal operations.

---
[1] Department of Orthopaedic Surgery, Aichi Medical University, 21 Karimata, Yazako, Nagakute, Aichi, 480-11 Japan

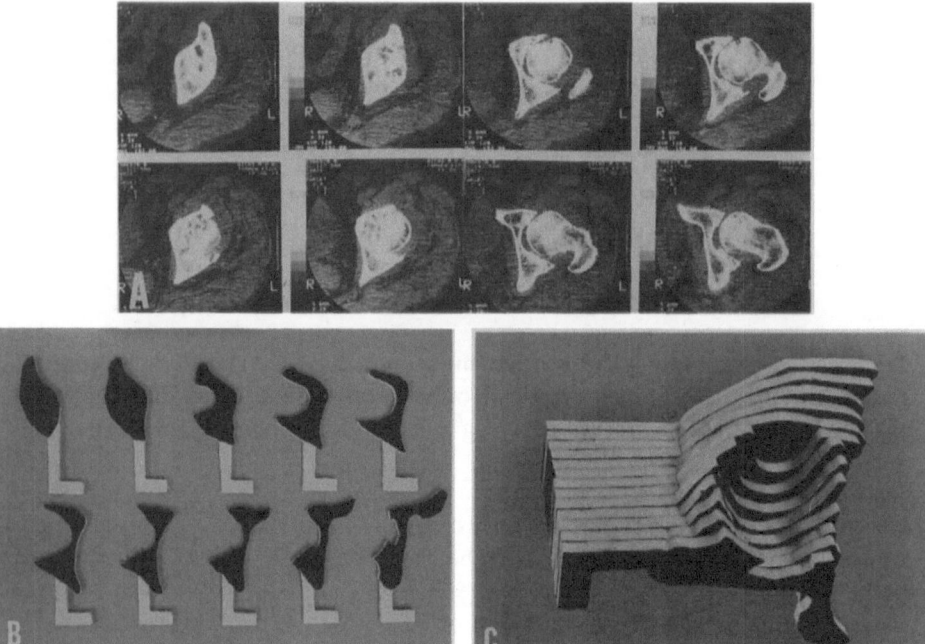

Fig. 1A–C. Method of producing a simple solid model. **A** Life-size enlargement of contiguous CT images, **B** cut-off images, **C** simple solid model

Methods

Contiguous slices were obtained from the hip joint by CT scans that included a region from 2 cm above the acetabulum to just above the obturator fossa, taking care to have an accurate position. Contiguous slices were also obtained from the femur (from the top of the femoral head to various optional levels) and, finally, from the femoral condyle, to indicate the baseline in the frontal plane. The CT scans were usually done in 5-mm-thick slices in ordinal sequence. The load area of the hips was scanned in 1.5-mm-thick slices with 3-mm separation, in order to distinguish the femoral head from the acetabulum. Contiguous slice images were developed to enable target reconstruction of the bone with distinct contours (Fig. 1A). The CT slice images were traced on a foam polystyrol plate, 3- or 5-mm-thick, from which the traced images were cut out (Fig. 1B). The solid models were produced by piling up these CT slices according to the baseline in the frontal dimension (Fig. 1C).

Comparison of the solid models and the external dimensions of cadaveric hips is shown in Table 1. The dimensions of the acetabulum, femoral head, and femoral shaft in the solid models were compared with those in an actual cadaveric specimen. This comparative examination demonstrated that the acet-

Table 1. Comparison of cadaver and solid model.

Site	Axis	Cadaver (mm)	Solid model (mm)	Difference (mm)
Acetabulum	X	26.6	27.6	1.0
	Y	53.9	53.3–55.5	−0.6–1.6
	Z	56.8	56.9–58.0	0.1–1.2
Femur	X	44.1	42.5–45.9	−1.6–1.8
	Y	44.5	43.1–45.5	−1.6–1.0
	Z	44.2	43.6–54.0	−0.6–9.8
Shaft	X	29.4	29.1	0.3
	Y	24.8	24.5	0.3

X, Sagittal axis; Y, frontal axis; Z, coronal axis

abular models were accurate in size to within 1 mm in the frontal axis, 1.6 mm in the sagittal axis, and 1.2 mm in the coronal axis. Femoral head models were accurate in size to within 1.8 mm and 1.6 mm in the frontal and sagittal axes, respectively. However, in the coronal axis, this model was accurate in size only to within 9.8 mm, this being the greatest error in this series. The femoral shaft models were accurate in size to within 0.3 mm in both the frontal and sagittal axes.

Case Reports

From 1985 to 1991, simple sold models were made for 350 hip joints; 143 of these were utilized for total hip replacements (THRs) and revision surgery.

Case 1

This patient was a 22-year-old male with a fracture of the right femoral neck due to a traffic accident. He had had a bipolar-type femoral head replacement performed at an emergency hospital. Two years after surgery, thigh pain occurred due to aseptic loosening of the component (Fig. 2A). Revision surgery, carried out at our university hospital, provided a cementless bipolar-type femoral head replacement (Fig. 2C). Preoperatively, the simple solid model (Fig. 2B) of the femur (made prior to surgery) clearly demonstrated the location, extent, and volume of the bone loss in the medullary canal. The surgical planning and rehearsal operation (Fig. 2B) helped us to avoid some pitfalls in the operation.

Case 2

This patient was a 32-year-old female with a bone tumor in the left proximal femur; prosthetic replacement was performed in another hospital, using the tumor femoral prosthesis. Fifteen years after surgery, hip pain occurred, due to proximal migration of the component (Fig. 3A). Revision surgery, done with a

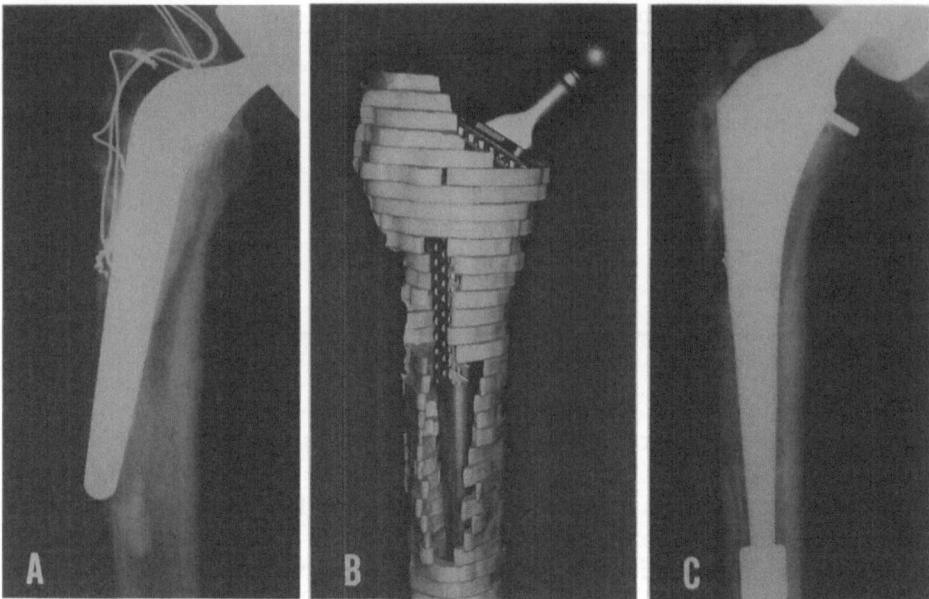

Fig. 2A–C. A Preoperative plain X-ray picture, **B** solid model of the femur and rehearsal operation, **C** postoperative plain X-ray picture

cementless THA (Fig. 3B), was performed at our university hospital. Before revision surgery, a simple solid model (Fig. 3C) of the acetabulum was made to gain information about its size and shape. Preoperatively, surgical planning and a rehearsal operation (Fig. 3D) was done by using the solid model technique. The socket type and size, as well as the bone volume and the method for grafting were determined, which, again, allowed us to avoid common pitfalls in the operation.

Discussion

It is difficult to grasp three-dimensional shapes, such as the craniofacial complex, from plain X-ray pictures. In the light of this difficulty, it is not surprising that three-dimensional surface reconstruction on a television monitor produced with computed tomographic data, was first developed in the fields of craniofacial [5,6], oral [7] and plastic [3,8] surgery. Three-dimensional surface reconstruction has been employed by some orthopedic surgeons [1,2] for surgical planning and rehearsal operations in hip surgery. Further, solid models have recently been produced from CT data by using a computer aided design and manufacturing (CAD, CAM) system [3]. These solid models are now used widely in surgical planning and rehearsal operations, especially in craniofacial, oral, and plastic surgery.

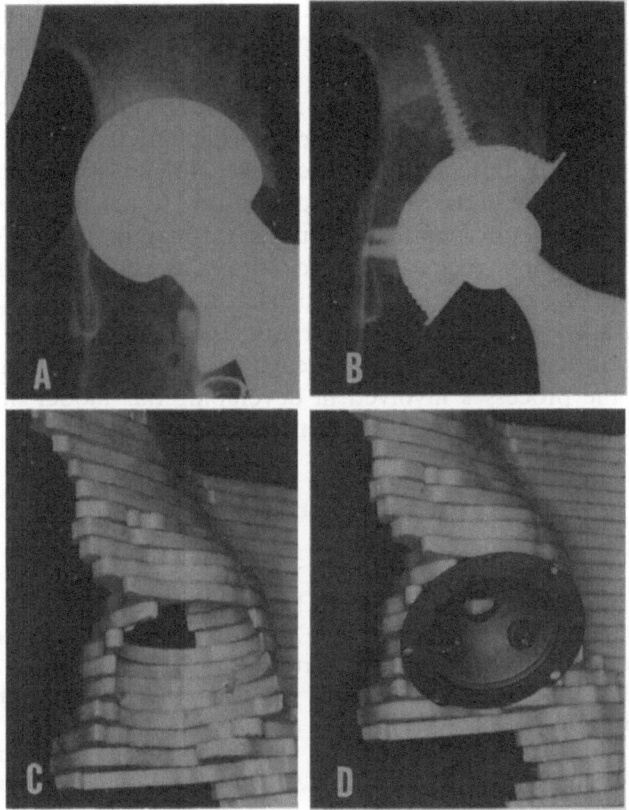

Fig. 3A–D. Case 2. **A** Preoperative plain X-ray picture, **B** postoperative plain X-ray picture, **C** solid model of the acetabulum, **D** rehearsal operation

Since 1985, simple polystyrol solid models have been developed in our department to assist in the multilateral understanding of complex shapes such as hip joints. This system has proven to be extremely beneficial in surgical planning and in rehearsal operations. These simple solid models can be made from foam polystyrol plates that are readily available on the market. These models are easier to produce and are more economical than both three-dimensional reconstruction surface images produced on a television monitor and other solid models produced by using a milling system (both of these models being based on CT data). These latter two models are both exorbitantly expensive compared to the simple solid model. Further, three-dimensional reconstruction surface image on a television monitor has significant limitations.

A simple solid model was well simulated from the cadaveric acetabulum and femur within a measuring error of 1.8 mm in the frontal and sagittal axes. However, a significant problem was the large error in the coronal axis of the convex surface, i.e., the 9.8-mm error in a femoral 44.2 mm in diameter. This simple solid model, which is made by piling up foam polystyrol plates, 5-mm-thick, to simulate the sum of contiguous CT slices, unfortunately has an inherent problem in that segments which are, in fact, less than 5-mm-thick are not represented realistically. This non-negligible error in the coronal axis depends upon the thickness of the CT slice, particularly on the convex surface. A possible solution to this problem is for the contiguous CT scanning to be done in thinner slices with narrow separation at the top of the convex surface.

The most advantageous point about using this simple solid model is that actual prosthetic components can be used in conjunction with the model to carry out the rehearsal operation, and to finalize the planning for surgery. In difficult operations, such as revision surgery, the solid model is particularly useful in that it helps the surgerical team avoid technical pifalls. Further refinement of the processes involved in developing this model should permit even more precise and effective results in the future.

References

1. Murphy SB, Kijewski PK, et al (1986) Computer-aided simulation, analysis, and design in orthopedic surgery. Orthop Clin North Am 17:637–649
2. Woolson ST, Fellingham LL, et al (1985) Three-dimensional imaging of bone from analysis of computed tomography data. Orthopedics 8:1269–1273
3. Lambrecht JT, Brix F (1990) Individual skull model fabrication for craniofacial surgery. J Cleft Palate 2:382–387
4. Ohtsuka H, Sawai K, et al (1989) Clinical application of solid models reconstructed from bony acetabulum by CT (in Japanese). Cent Jpn J Orthop Traumat 32:1443–1448
5. Artzy E, Frieder G, et al (1981) The theory, design, implementation, and evaluation of a three-dimensional surface detection algorithm. Comp Graphic and Image Proc 15:1–24
6. Hemmy DC, David DJ, et al (1983) Three-dimensional reconstruction of craniofacial deformity using computed tomography. Neurosurgery 13:534–541
7. Vannier MW, Marsh JL, et al (1984) Three-dimensional CT reconstruction image evaluation. Radiology 150:179–184
8. Marsh JL, Vannier MW (1983) The "Third" dimension in craniofacial surgery. Plast Reconstr Surg 71:759–767

12

Application of Three-Dimensional Milling System in Hip Disorders

Takeo Matsuno, Kiyoshi Kaneda[1], Kaoru Shimakage,
and Shigeo Matsuno[2]

Summary. Surgery of the hip and pelvis should be performed in the light of three-dimensional concepts. We recently applied a three-dimensional 3-D milling system (MCP-70-SE; Hek, Germany), with which we obtained hands-on solid models before surgery, to the preoperative planning of hip surgery. This method was used for patients with hip dysplasia who were candidates for acetabuloplasties, including shelf operation, Chiari pelvic osteotomy, and total hip replacement. Utilizing this method, it was possible to make preoperative decisions on the level and size of the shelf in the shelf operation, on the height and size of the socket, and on the size of the acetabular bone grafting in the total hip replacement. This method was also useful as a teaching tool for residents and medical students.

Key words: Three-dimensional CT scanning—Hip acetabuloplasty—Shelf operation—Total hip replacement—Simulation surgery—Three-dimensional milling system—Hip osteoarthrosis

Introduction

In Japan, we have many patients with osteoarthrosis of the hip secondary to congenital hip dislocations and hip dysplasia; these patients need acetabuloplasties, including shelf operation, Chiari pelvic osteotomy, rotational acetabular osteotomy, and total hip replacement (THR). Various radiological imaging techniques have been used for such patient since the discovery of radiologic images by Roentgen in 1895. Most of these images are in a two-dimensional (2-D) form and even the most current methods utilizing computerized tomography (CT) scanning and magnetic resonance (MR) images are mostly 2-D. These 2-D images are of little help in the field of hip surgery; such

---

[1] Department of Orthopaedics, Hokkaido University School of Medicine, North 14 West 7, Kita-ku, Sapporo, 060 Japan
[2] Bibai Rosai Hospital, Higashi 5-jo, Minami 2-chome, Bibai, Hokkai-do, 072 Japan

surgery should be performed in the light of 3-D concepts. The reconstruction of 3-D radiograms has previously been accomplished in the form of 3-D CT scanning and has been applied in the field of plastic surgery as well as in orthopedics [1–3]. However, there are some difficulties involved in obtaining solid 3-D hands-on models, even with the 3-D CT scanning method [4]. Regarding surgery for patients with hip dysplasia, we have recently applied a 3-D milling system for preoperative surgical planning [5]. Using this system, we were able to obtain solid hands-on models before surgery; these were extremely useful as both diagnostic and teaching tools for junior orthopedic surgeons. Further, simulation surgery before the operation thus became possible.

Materials and Methods

We used a Somatom-DGR CT scanner (Siemens, Erlangen, Germany) for CT scanning. The following scanning parameters were used: 4 mm slice with scanning time of 3–4 s at 150 mAs and 125 kV. Contiguous slices with minor overlapping (1 mm) from the iliac crest of the pelvis to the femoral shaft, usually 2 cm below the lessor trochanter, were made. After all slices were prepared, using the image processing system, we reconstructed a 3-D image of the model on the monitor. These models could be separated from each other at the joint space, into acetabular and femoral models. The set range of CT numbers was from 100 to 300 for analyzing bone density only. After these data were computed for controlling the milling process, the model was cut out of Obmodulan 200 (Hek; Lubeck, Germany) (high density styrofoam could be also used). The MCP-70-SE (Hek, Germany) 3-D milling system was used; with this a 3-D mold was cut. The accuracy of these models was within 0.5 mm.

We used this 3-D CT scanning method on ten with hip osteoarthrosis secondary to congenital dislocation of the hip (CDH) or hip dysplasia; six patients had early stage osteoarthrosis and four were in an advanced stage of hip osteoarthrosis. We made several analyses on the TV monitor before producing hands-on solid models at 0°, +45°, +90°, +135°, and 180°. Four such models utilizing this 3-D milling system were made for further examination.

Case Reports

Case 1

This patient, a 47-year-old female, was admitted to our hospital complaining of severe right hip pain as well as limitation of the range of motion. Radiograms revealed severe deformity of the right femoral head, which was dislocated, and severe dysplasia of the acetabulum (Fig. 1a). We planned to perform total hip replacement and several radiographical examinations were performed, including 3-D CT; a hands-on solid model of her right hip was made. This model clearly showed deformity of the femoral head as well as of the acetabu-

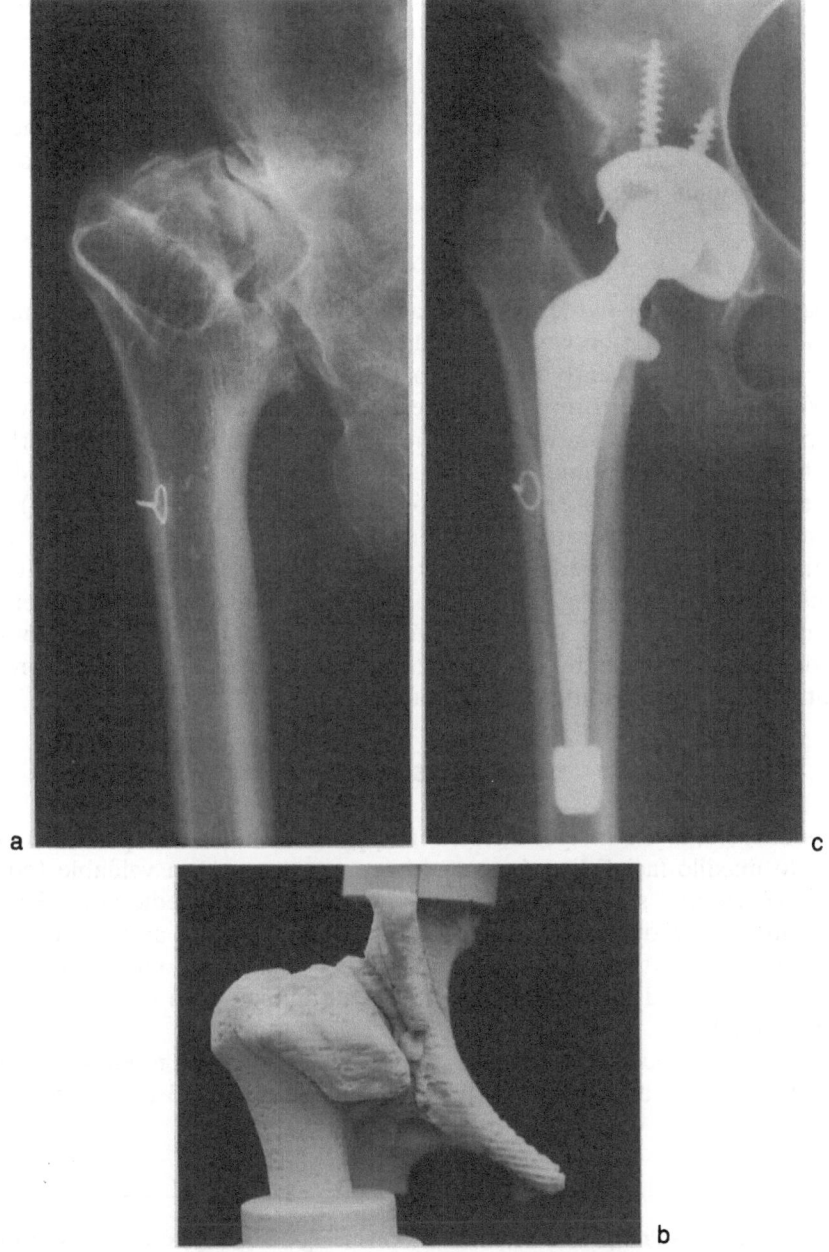

Fig. 1a–c. Case 1. **a** Preoperative radiogram of right hip. The right femoral head is dislocated, displaying severe deformity; severe acetabular dysplasia was also observed. **b** Hands-on solid model of the right hip, showing deformity of the femoral head as well as of the acetabulum. **c** Postoperative radiogram of the right hip, showing total hip replacement with a non-cemented socket (Omnifit) and a non-cemented stem (Omniflex), with no bone grafting on the acetabulum

126 T. Matsuno et al.

lum. By measuring the depth of the acetabulum on this model, we confirmed that no bone grafting of the acetabulum was required, and this was of help in determining the resection level of the femoral neck (Fig. 1b). Total hip replacement of the right hip was performed, along the lines indicated, using a non-cemented socket (Omnifit; Osteonics, Allendale, USA) and non-cemented stem (Omniflex; Osteonics, Allendale, USA), with no bone grafting required on the acetabulum (Fig. 1c).

Case 2

This patient, a 29-year-old female, was admitted to our hospital complaining of moderate left hip pain on walking. Radiograms revealed mild dysplasia of the left hip. A diagnosis of early stage osteoarthrosis of the left hip was made (Fig. 2a). We planned to perform a shelf operation. A hands-on solid model of the left hip was made, this showed deficiency of the acetabulum, particularly on the anterior (Fig. 2b). Simulation surgery was performed; this confirmed that we needed to make a shelf 1.5 cm long and 2 cm in width (Fig. 2c). We performed the shelf operation according to these measurements. The postoperative radiogram showed good bony coverage of the femoral head with the formation of the shelf (Fig. 2d). A postoperative hands-on solid model also confirmed sufficient bony coverage due to the formation of a shelf on the anterior part of the acetabulum (Fig. 2e). Two years after the operation, the patient is doing well, without any symptoms.

Discussions

Three-dimensional CT images have been applied in the field of plastic surgery, mainly to maxillo-facial disorders, and have proven to be a valuable tool for both diagnosis and surgical treatment [2,4]. In the orthopedic field, 3-D CT images are now widely used as a diagnostic tool in spinal and hip disorders [1,3]. However, all these methods utilizing such images are mostly performed on TV monitors and there have been technical difficulties in producing hands-on solid models.

It has recently become possible to produce hands-on solid models by using a 3-D milling system, applying several computing methods to 3-D CT images [5].

Fig. 2a–e. Case 2. a Preoperative radiogram, revealing early stage osteoarthrosis of the left hip, with mild acetabular dysplasia. b Preoperative hands-on solid model of the left hip, showing deficiency of the acetabulum, particularly on the anterior. c Hands-on solid model with simulation surgery, with the formation of a shelf 1.5 cm long and 2 cm in width. d Postoperative radiogram, showing good bony coverage of the femoral head with the formation of a shelf. e Postoperative hands-on solid model confirming sufficient bony coverage due to the formation of a shelf on the anterior part of the acetabulum

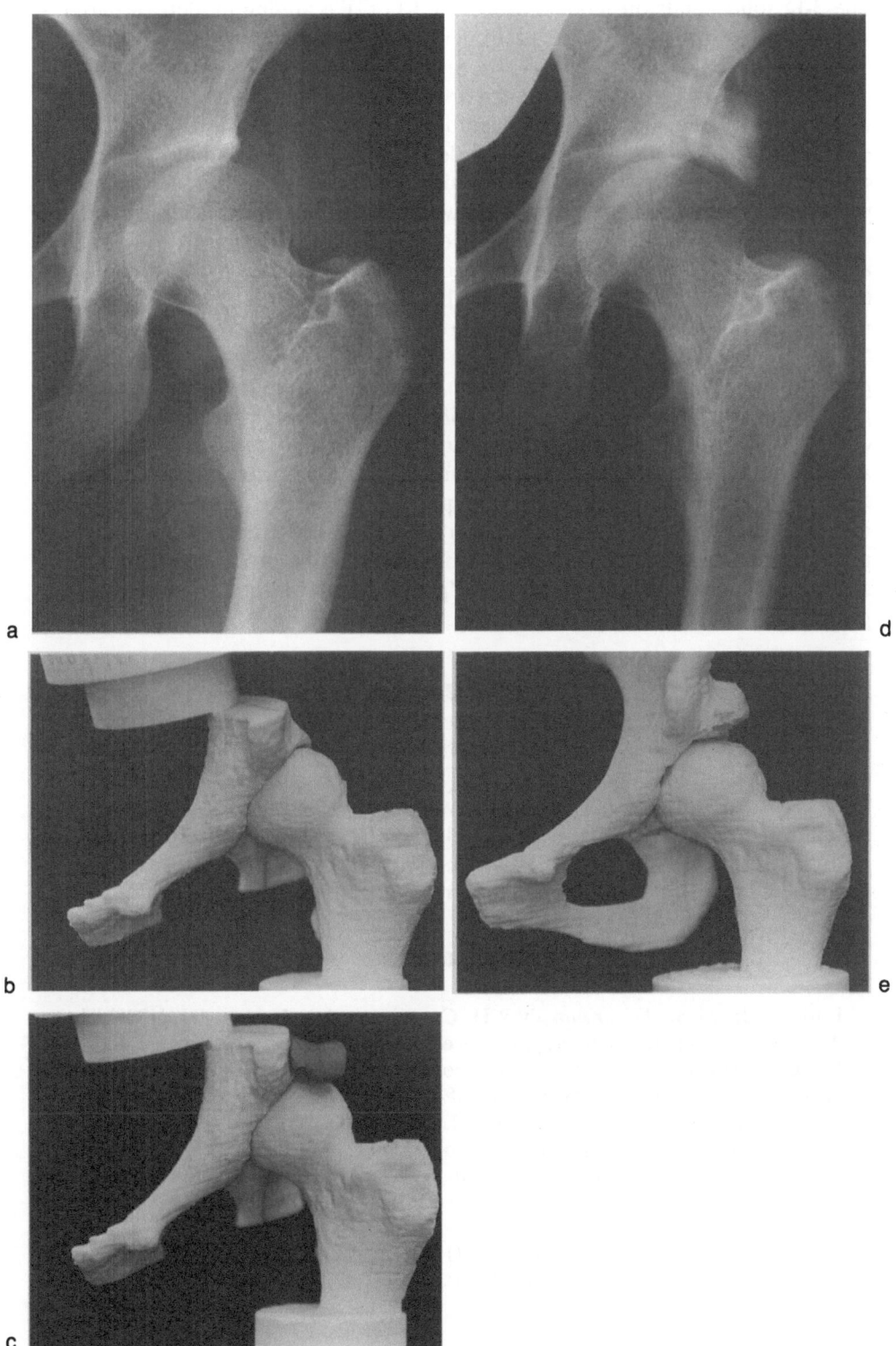

The 3-D solid hands-on models based on 3-D CT scanning models of patients with hip osteoarthrosis provide great benefits. For patients who are candidates for total hip replacement (THR), we are thus able to plan preoperatively the proper size of socket to use, whether or not we need acetabular bone grafting, where to place the socket on the acetabulum, and the level at which to resect the femoral neck. Other benefits include improved understanding of the bony structure before surgery and information which aids in the selection of a prosthesis. Further, for those patients who are candidates for shelf operation, we were able to decide on the indications for this procedure before surgery, and it was easy to predetermine the place, height, and size of the formation of the shelf, as well as its proper slant, before surgery. Preoperative decision-making on the level and angle of osteotomy in pelvic osteotomies, including Chiari pelvic osteotomy and rotational acetabular osteotomy, would also be possible with such models. Simulation surgery before operation was made possible by using these solid models. Information on postoperative results was obtained by making postoperative solid models. This method could also be applied in the future to preoperative surgical planning of operations for pelvic fracture [1], slipping of the femoral epiphysis, and other femoral osteotomies, including aseptic necrosis of the femoral head. Although this method could provide great benefits in the field of hip surgery, some problems remain to be overcome, particularly in regard to the price of this solid model, which is now prohibitively expensive.

Conclusions

We visualized the bony surface structure of the hip joint by using a 3-D computerized milling system via which we were able to make hands-on solid models based on 3-D CT scans. It was possible to perform preoperative simulation hip surgery using these models, and they were also useful as teaching tools for residents and medical students. In the very near future, this system could be adapted to use with magnetic resonance images.

References

1. Burk DL Jr, Mears DC, Kennedy WH, Cooperstein LA, Herbert DL (1985) Three-dimensional computed tomography of acetabular fractures. Radiology 155:183–186
2. Lambert PM (1989) Three-dimensional computed tomography and anatomic replicas in surgical treatment planning. Report of a case. Oral Surg 68:782–786
3. Woolson ST, Parvati D, Fellingham LL, Vassiliadis A (1986) Three-dimensional imaging of bone from computerized tomography. Clin Orthop 202:239–248
4. Fujioka M, Ouyama N, Honda T, Tujiuti J, Suzuki M, Hashimoto S, Ikeda S (1988) Holography of 3-D surface reconstructed CT images. J Comput Assist Tomogr 12:175–178
5. Matsuno T, Matsuno S, Shimakage K (1991) Application of 3-D milling system in hip arthroplasty (in Japanese). Hip Joint 17:301–303

13

Three-Dimensional Fitting of Stem by Computer Analysis

Satoru Yano, Shinji Kimura, and Tadashi Hashimoto[1]

Summary. Software was developed to check the fitting of the stem in the medullary canal before operation. This study describes the method of quantifying canal-fill in three dimensions, and of simulating stem implantation. The method cleary showed a relationship between stem and medullary canal that is not easily detected with conventional roentgenograms. Simulation is demonstrated in one case report. This method will help the surgeon in planning cementless total hip replacement.

Key words: Three-dimensional—Computer—Stem of femur—Hip—Fitting of stem—Canal-fill—Osteoarthritis

Introduction

The characteristic of press-fit total hip replacement is that maximun stem contact on the internal cortical bone surface produces more normal strain values and reduces loosening. Good canal-fill improve mechanical stability in response to various forces and moments applied to the head of the stem. Therefore the selection of stem size and shape is very important. Usually stem fit is dependent upon biplanar roentgenograms, and the percentage of the canal-fill at several levels as judged by postoperative roentgenograms is used to determine appropriate fit. Conventional roetgenograms are two-dimensional projections that may have magnification and rotation errors. The routine use of roetgenograms is inadequate for accessing the internal configuration of the femur, and is error-prone. Three-dimensional imaging is now widely accepted as one of the most modern techniques in preoperative planning. We have developed a simulation system that provides an exact three-dimensional fit between stem and femoral canal. Clinical applications of the methods are demonstrated in a case report.

[1] Department of Orthopaedic Surgery, Kasai Municipal Hospital, 1-13 Yokoo, Hojo-cho, Kasai, Hyogo, 675-23 Japan

129

S. Yano et al.

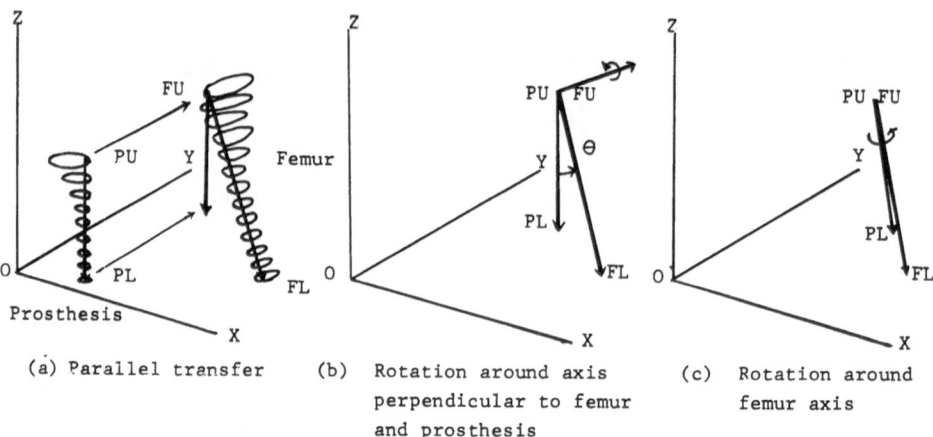

Fig. 1a–c. Fitting simulation. **a** Parallel transfer, **b** rotation around axis perpendicular to femur and prosthesis, **c** rotation around femur axis. *FU-FL*, axis of the femur; *PU-PL*, axis of the stem

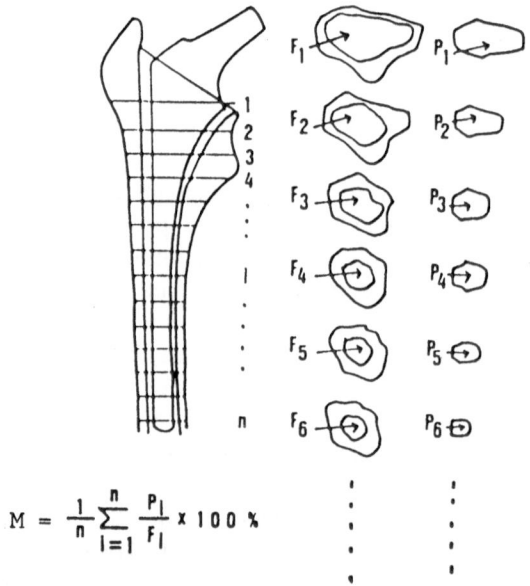

$$M = \frac{1}{n} \sum_{i=1}^{n} \frac{P_i}{F_i} \times 100 \%$$

Fig. 2. Three-dimensional canal-fill; this is defined as the percentage ratio of surface area F to surface area P. Average canal-fill is expressed as M

Materials and Methods

The patient was placed in the supine position in a gantry with legs in neutral rotation. Computerized tomographic (CT) imaging was done on a Hitachi W

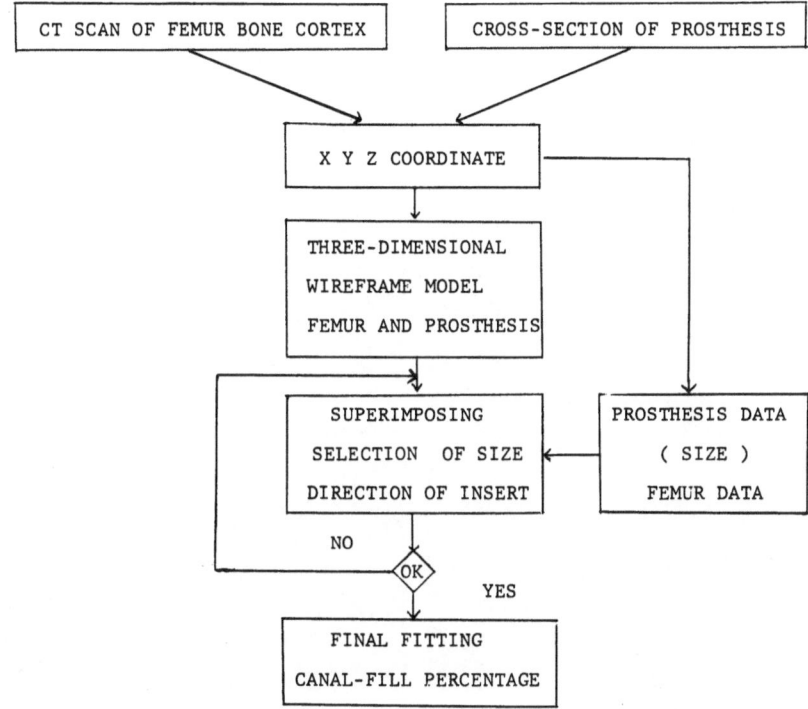

Fig. 3. Flowchart of simulation

1000 scanner (Hitachi, Tokyo, Japan). Axial scans 1.5-mm-thick were obtained every 10 mm starting at the trochanter major and continuing up to one-third of the femur shaft. CT images were magnified full-scale. The inner and outer cortical contours from individual CT images were manually digitized with a Graphtec Mitablet-II, and the x and y coordinates of each point were transferred to a NEC PC-9801 personal computer. Z axis data were incrementally transfered. The stem was marked off with pencil every centimeter perpendicular to its axis, and was molded with wire. The wire was printed on paper. The same process was then repeated, as in the CT images. These x, y, and z coordinates were then all stored as files in the disc. Computer software was developed that generated wireframe three-dimensional femoral and stem models.

Calculation

An orthogonal coordinate system was constructed, so that the axis of the stem was perpendicular to the x-y plane. The origin of the coordinate system was 0, with x representing the horizontal axis in the frontal plane, y the horizontal axis in the sagittal plane, and z the vertical axis, which was parallel to the long

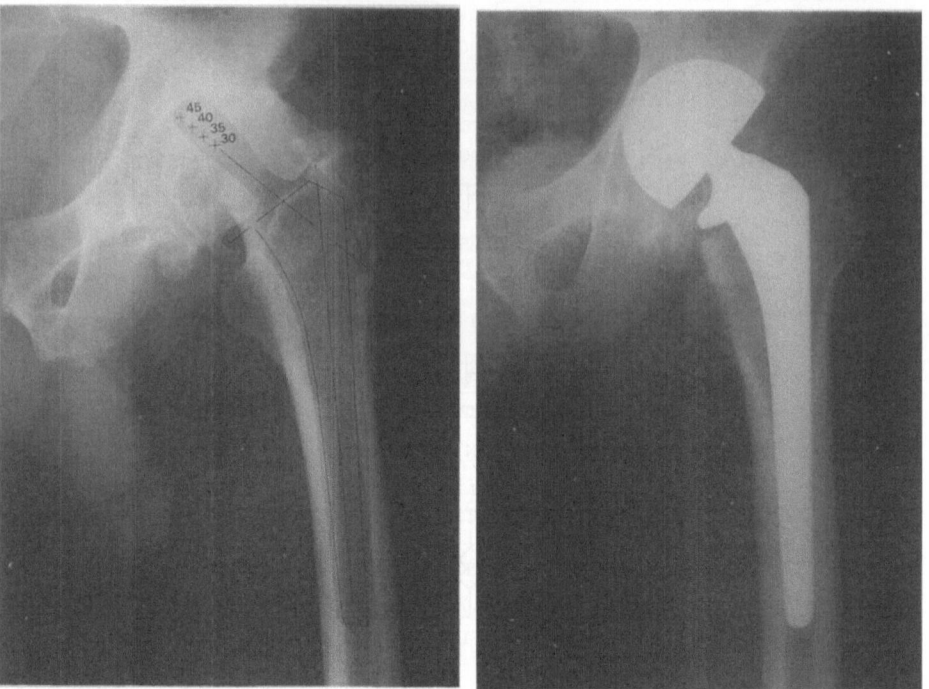

a   b

Fig. 4a,b. Roentgenograms of 44-year-old male with osteoarthritis of the left hip. **a** Preoperative anteroposterior (*a-p*) film (with overlay template in place) has a sizing error related to rotational contracture. **b** Postoperative a-p film shows the gap between stem and calcar

axis of the body and directed cranially. As the axis of the femur (FU-FL) was not parallel to that of the stem (PU-PL), PU-PL had to be translated to FU and then rotated two times using vector algebra, so that the two models could be superimposed in a cartesian coordinate system (Fig. 1).

The surface area of each cross-section P and F can be measured automatically every centimeter, and the ratio P to F expressed as a percentage in the canal-fill (Fig. 2).

Simulation of Implantation

Three-dimensional models were produced from the data in the files. The stem model was inserted into the femoral model at the desired level and positioned so that there was no intersection between the stem and cortical bone. The stem model was changed as necessary to modify the shape and size. Direct visualization of fitting in three-dimensions affords the surgeon an overall view; the process is repeated until the desired fit can be achieved. With the implant

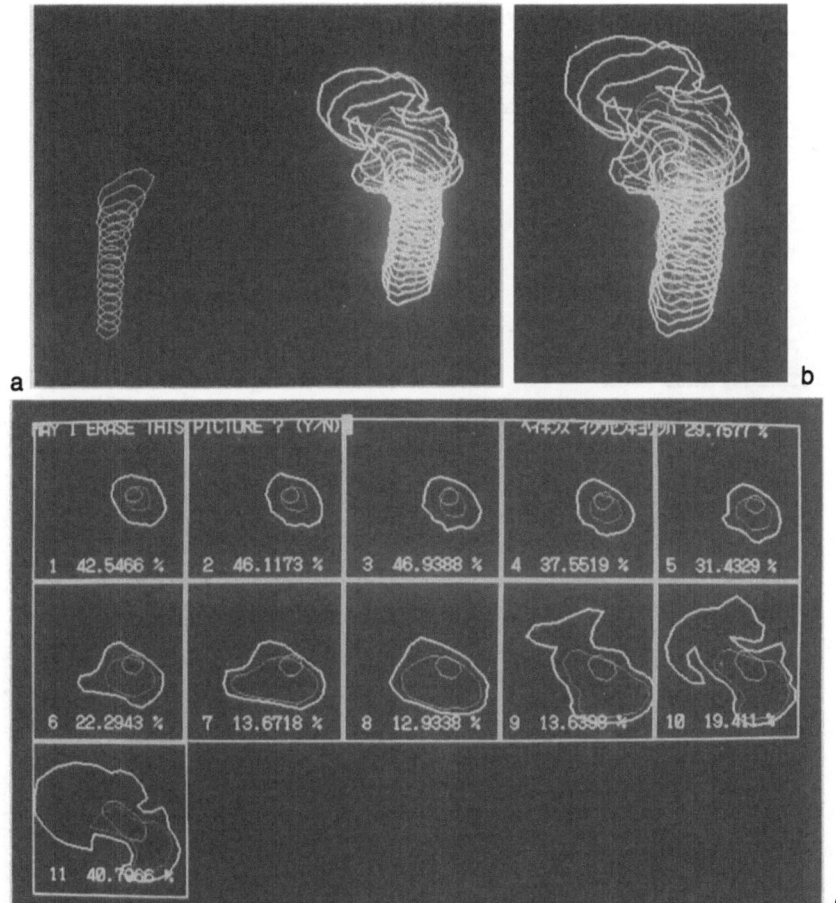

Fig. 5a–c. Simulator of internal fitting by stem (no 5). **a** Computer graphic view of stem (no 5) and femur, **b** stem is in the medullary canal. **c** Continuous axial sections demonstrate the fit of the stem. Note the contact between the stem and the cortical bone along the anterior wall. Average canal-fill percentage is 29.7%

in its final position, the percentage canal-fill may be calculated and displayed. The software was written in basic. A flowchart of the simulation is shown in Fig. 3.

Results

The three-dimensional method was useful for visualizing certain features that are often poorly defined by conventional roentgenogram. The features visualized were: (1) coronal canal-fill, (2) the location and volume of the gap between stem and bone, and (3) three-dimensional congruity between stem

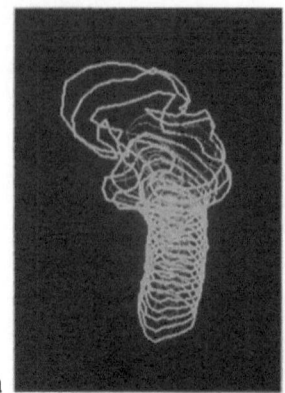

Fig. 6a–b. Simulation of final fitting by stem (no 5). **a** Stem was 1 cm distally translated. **b** Note that the contact between the stem and the cortical bone was at the distal medial wall. Average canal-fill is 41%

a

b

and femur. A decision on size is difficult when only conventional roentgenograms are available. Through three-dimensional simulation, it was possible to calculate accurate canal-fill.

Case Report

The patient described here is a 44-year-old male with osteoarthritis of the left hip, who presented with progressive pain and limitation of motion. Noncemented total hip joint replacement was selected as a treatment. Before the operation, stem fitting was done, based on the anteroposterior (a-p) roentgenogram, and the post-operative roentgenogram showed that the stem did not match the contour of the calcar (Fig. 4). Using computer simulation techniques, a three-dimensional femoral model and stem model was produced

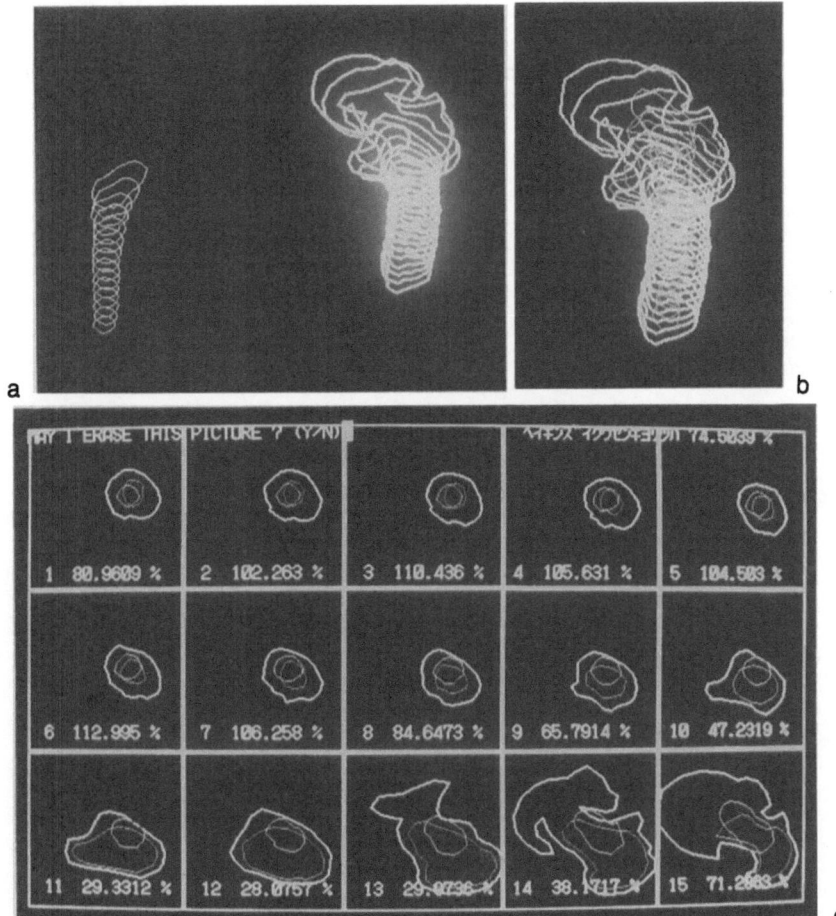

Fig. 7a–c. Simulation of initial fitting by stem (no 9). **a** Stem (no 9) and femur. **b** Stem is in the medullary canal. **c** Note the contact between the stem and the cortical bone along the anterior wall. Average canal-fill percentage is 74%

from the data in the files. Simulation was repeated until optimal fit was achieved (Figs. 5–8).

Discussion

The position of a stem to the femur is important for biomechanical stability and the need for preoperative fitting of cementless prosthesis demands accurate preoperative planning. In general, the orthopedic surgeon uses planar roent-genograms to make measurements and to decide on stem position. However, the complicated bony structures of the femur do not permit accurate evaluation

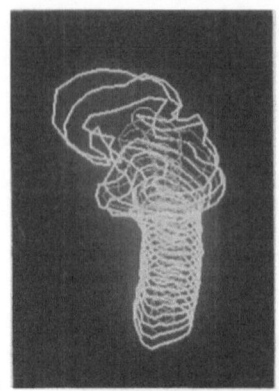

Fig. 8a,b. Simulation of final fitting by stem (no 9). **a** Stem was 3 cm translated distally. **b** Note the excellent fit at the proximal and distal area. Average canal-fill percentage is 93.2%

a

b

of the stem fit in space by means of conventional roentgenograms. Several methods have been described [1–4] for determining the stem fit in two planes, but in such two-dimensional illustrations of actual three-dimensional relationships the coronal parameters in the assessment of fit may be overlooked.

In our patient, the preoperative roentgenogram was not a true a-p projection, but rather, an oblique projection that was caused by rotational contracture in the hip joint. Figure 4 shows that the femoral template matches the contour of the femoral canal before operation; however, the stem does not match the medial wall of the femur after the operation. Thus, it would appear that none of the two-dimensional methods give a true and complete description of stem fit in three-dimensional space.

With the advent of CT, it became possible to produce cross-sectional images very easily at any point of the body, and three-dimensional images based on CT data have now been produced in computer graphics. We developed simulation software that provides an exact three-dimensional fit between the stem and the femoral canal. Utilizing this software, the prosthesis and femur can be rotated or translated to determine the best fit in any direction; the canal-fill can also be calculated and displayed at centimeter intervals. We believe our method is useful for selecting the size and shape of the stem.

References

1. Amstutz HC (1989) The anthropometric total hip femoral prosthesis. Clin Orthop Relat Res 242:105–119
2. Gorski JM, Schwarz L (1986) A device to measure X-ray magnification in pre-operative planning for cementless arthroplasty. Clin Orthop Relat Res 202:302–306
3. Müller ME (1992) Lessons of 30 years of total hip arthroplasty. Clin Orthop Relat Res 274:12–21
4. William NC (1986) Preoperative planning of total hip arthroplasty. The American Academy of Orthopedic Surgeons instructional course lectures. 36:249–251

14

Computer Simulation for the Planning of Total Hip Replacement: Improvement of the Simulation System

Hirokazu Iida, Takao Yamamuro, Ryuichi Kasai, Yoshitaka Matsusue, and Yasutaka Matsuda[1]

Summary. We have already developed a system using computerized tomography (CT) and computer simulation for estimating the optimal size and position of the socket in total hip replacement [1]. We have now improved the system, in that CT images are input not by a digitizer but by a TV camera. It takes only 10–15 min to complete the simulation in the new system. Another benefit is that the socket images are overlapped not on the contours of the acetabulum but on its real CT images. Therefore, we can easily distinguish whether sclerotic subchondral bone or soft cancellous bone contacts the socket after reaming. The new system is time-saving and provides us with more accurate preoperative information; it can easily be used routinely, supporting the surgeon's intraoperative visual assessments.

Key words: Total hip arthroplasty—Computer simulation—Bone graft—Acetabular component—Image analysis—Computed tomography—Acetabular dysplasia

Introduction

Bone grafting has been used with increasing frequency both to reconstruct the acetabulum in dysplastic hips during primary total hip replacement (THR) and to rectify the bone loss associated with the loosening process during revision THR. Solid bony support of the acetabular components is essential for successful long-term fixation. In those cases requiring acetabular bone grafts, it may be difficult to identify the optimal size and position for the acetabular component.

Although three-dimensional reconstruction of CT images is now possible and the real image of the hip joint or the pelvis can be easily revealed by this method, these images are not useful for estimating the optimal size and position

---

[1] Department of Orthopaedics, Faculty of Medicine, Kyoto University, 54 Kawara-cho, Shogoin, Sakyo-ku, Kyoto, 606 Japan

139

of the socket. For this purpose, we developed a system using computerized tomography and computer simulation [1]. We have recently improved the system, in that CT images are input not by a digitizer but by a TV camera.

Materials and Methods

The new equipment (integrated by Tectron; Kyoto, Japan) used for analysis consisted of a TV camera (NEC Corp; Tokyo, Japan), an image processor combined with a personal computer (Sumitomo Metal Industries Ltd; Osaka, Japan), and a digital color printer (Mitsubishi Electric Corp; Tokyo, Japan) (Fig. 1). We jointly developed the simulation software with Sumitomo Metal Industries.

Hip joint CT was performed with a slice interval of 5 mm. Ten to 12 slices, including tear drop line and acetabular edge, were made in such a way that each slice had the same magnification and the same coordinates. Images of the acetabulum on each slice were entered into the personal computer with the TV camera, and they were displayed on the cathode ray tube (CRT) screen. The first step of the simulation was to decide at what level the component center should be located. We usually located the probable center of the component as being in the original acetabulum. The diameter and the location of the component were chosen in relation to the antero-posterior diameter of the acetabulum and the thickness of its anterior and posterior lips. The location of the component center was decided by using a cursor on the CRT screen. The

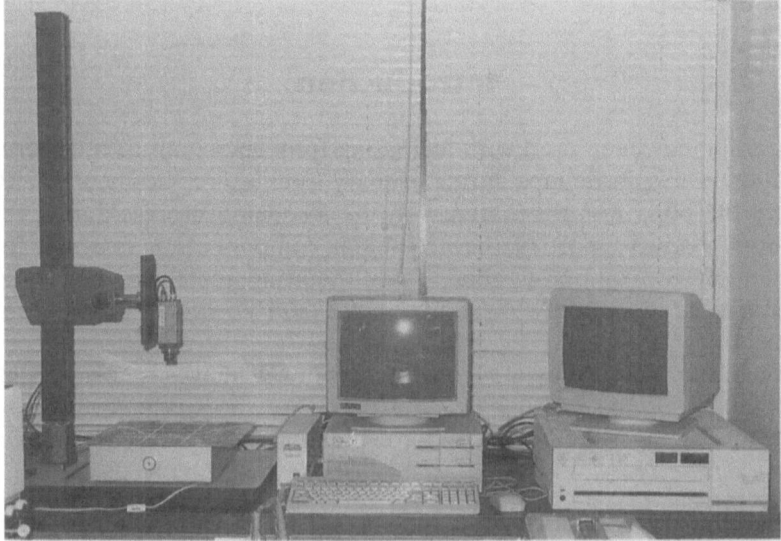

Fig. 1. The new simulation system equipment, including a TV camera, an image processor, a personal computer, and a digital color printer

shape of the acetabular component on each slice was then computed from
the location of the component center, the diameter of the component, the
inclination angle of the component, and the anteversion angle. The calculated
sections of the component were then drawn with the CT images of the acetabulum on the CRT screen. The area in need of bone graft was also shown by
a line which represented the lateral half contour of the projected image of the
component in the horizontal plane. If an adequate fit cannot be achieved in
this fashion, the component can be moved to a better location, or another
component size can be selected. The final image was then copied by the digital
color printer.

We have already performed these simulations in 110 hip joints in 88 patients,
using the old system. While it took about 1 h or more to complete the simulation in the old system, it takes only 10–15 min in the new one. Another

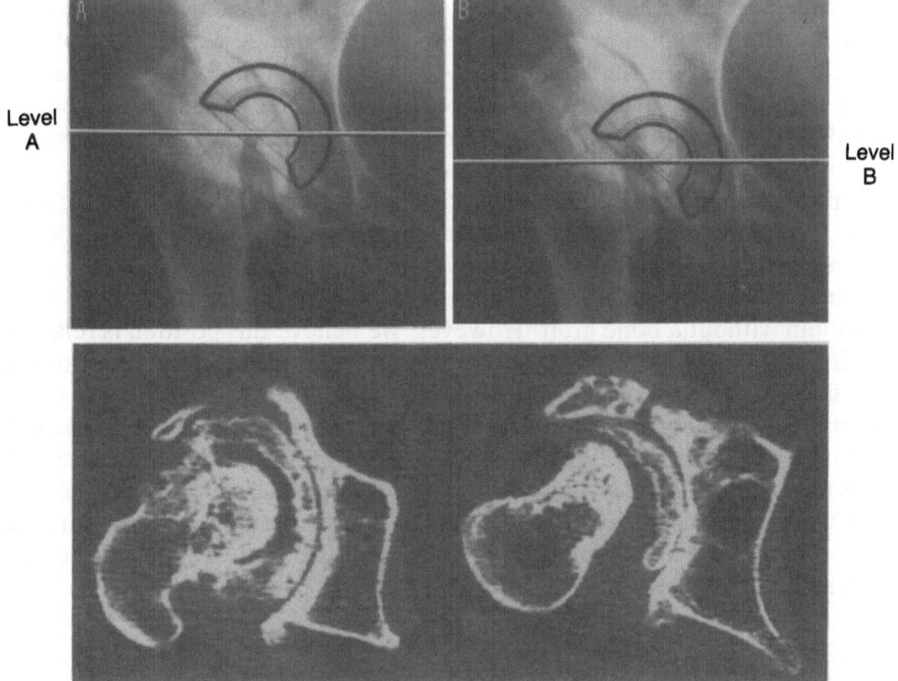

Fig. 2A,B. Case 1. *Upper panel,* preoperative planning anteroposterior (*AP*) roentgenogram of secondary osteoarthritis due to congenital dislocation of the right hip. **A**
The socket is located at a relatively high position. **B** The socket is located in the original
acetabulum. A bone graft could be necessary. *Lower panel,* each computerized tomographic (*CT*) image is a slice at the center level of the socket shown in the roentgenogram above. Although the AP diameter of the acetabulum at level A (*left*) is relatively
large, the anterior part of the acetabular floor, which is a reactive osteophyte secondary
to the anteverted femoral head, is not thick enough to ream deeply. CT image on the
right is at level B

benefit is that socket images are overlapped not on the contours of the acetabulum but on its real CT images, thus enabling us to distinguish easily whether sclerotic subchondral bone or soft cancellous bone contacts the socket after reaming. The new system is time-saving and provides us with more accurate information, and this system can easily be used routinely.

Acetabular Dysplasia

Estimating the optimal size of the socket by anteroposterior (AP) roentgenogram alone is inaccurate in this condition, since the AP diameter of the acetabulum is usually the prime factor in deciding an acceptable maximal socket diameter. Complex situations mean that decision-making is difficult in some cases. The diameter of the original acetabulum is sometimes different from that of the pseudoacetabulum, and the acetabular thickness may vary greatly depending on the level of the acetabulum. Our simulation system is quite useful for selecting the size and location of the socket in such cases. In mild acetabular dysplasia, it may also be difficult to decide whether or not bone grafts are necessary on the basis of an AP roentgenogram. Although planning on this basis can be done and might depend on the surgeon's basic concept, CT and this simulation technique offer more accurate information regarding whether bone grafting would be necessary.

A typical example is shown in Figs. 2, 3, and 4. This patient is a 65-year-old female. Her right hip joint is not severely dislocated, but there is a severe anteversion of the femoral neck and an anterior acetabular osteophyte has developed. By X-ray planning, it appeared possible to set the socket in the acetabulum without a bone graft. However, we believe this position is relatively high. If we reamed the acetabulum as usual in this position, its anterior part would have become thin or could have been penetrated. In this case, we should ream the acetabulum more inferiorly and posteriorly; a bone graft would be necessary as a result. If a bone graft were not performed, the space would be filled with bone cement, which would not be well pressurized, even if a flanged socket were to be used. Further, the inclination of the bone cement interface could be steep. We believe that these conditions are unfavorable for the long-term stability of sockets.

Fig. 3. If reaming was done up to 42-mm and was deep enough to contain a 42 mm socket at level A, the acetabular floor would be too thin and could be partially penetrated, as shown in this simulation

Fig. 4. The AP diameter of the original acetabulum at level B in Fig. 2 is relatively small. If we used a 42-mm socket to fit the original acetabulum, the acetabular floor would not be too thin, but the space between the acetabulum and the socket would be so large that a bone graft would be necessary

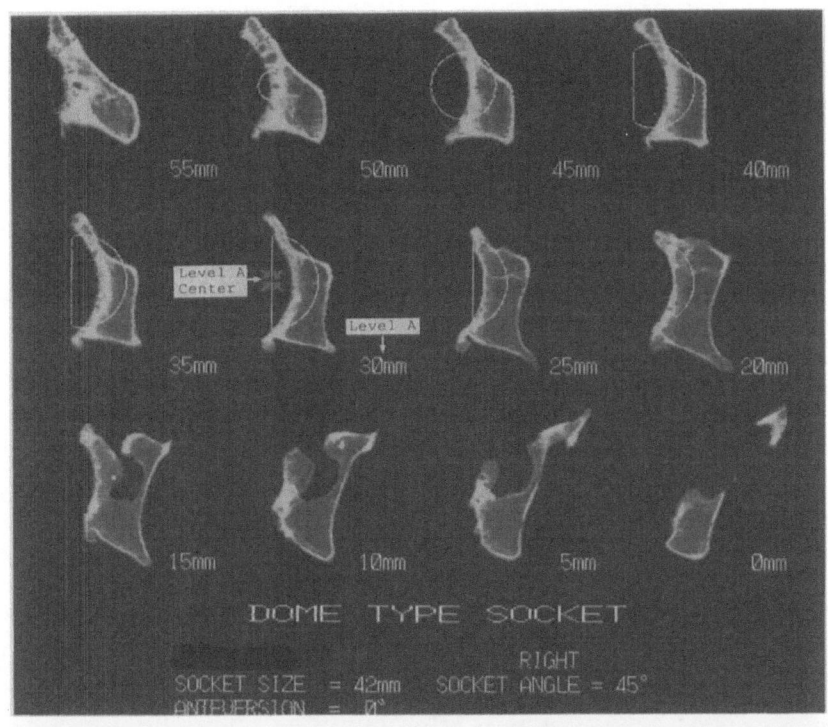

3

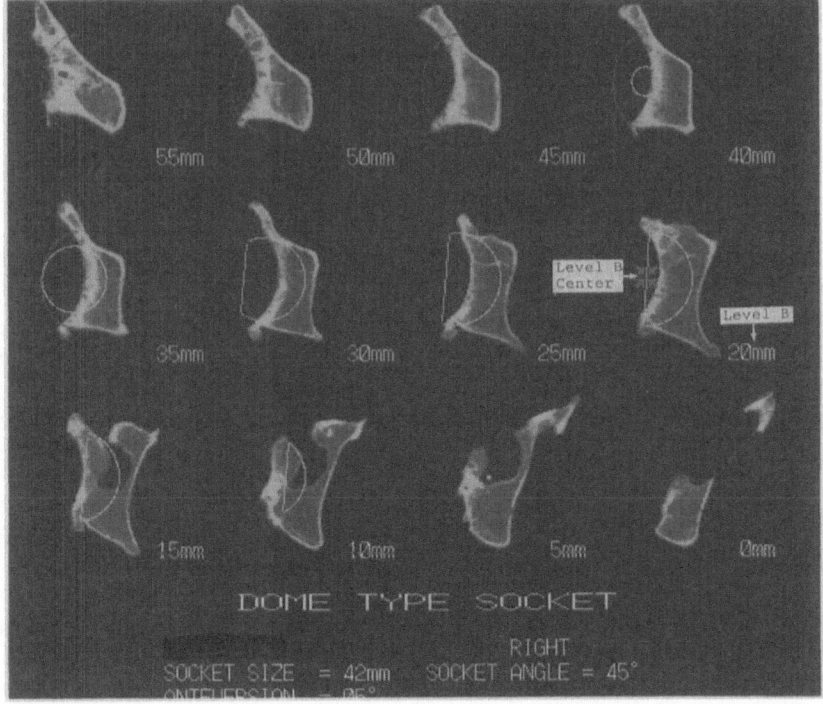

4

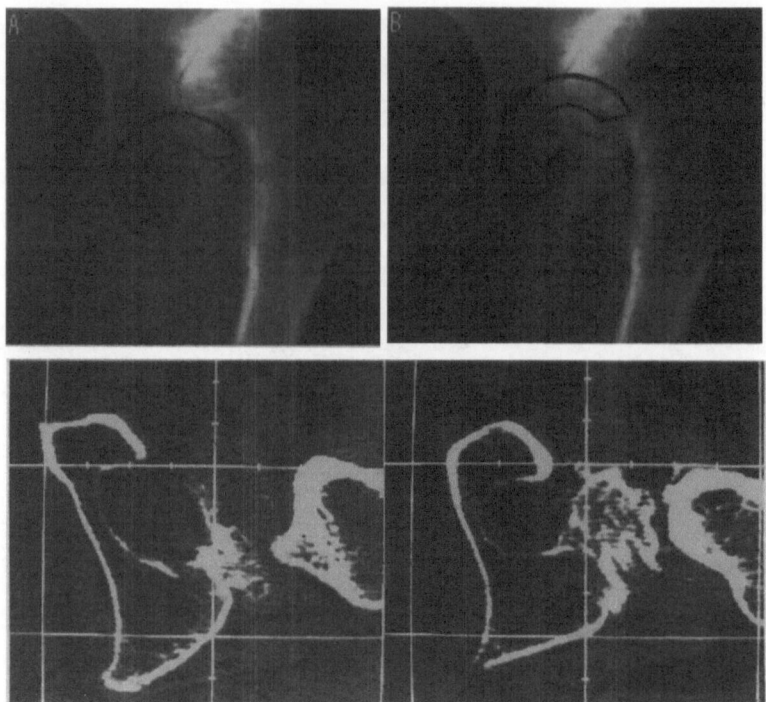

Fig. 5A,B. Case 2. *Upper panel*, preoperative planning AP roentgenogram of severely dislocated osteoarthritis of the right hip. **A** The socket is located at the original acetabulum. **B** The setting of the socket is 10 mm higher than the usual level. *Lower panel*, each CT image is a slice at the center level of the socket shown in the roentgenogram above. The original acetabulum (**A**) is hypoplastic, and the anterior wall is deficient

Severely Dislocated Hip

Total hip replacements for severely dislocated hips must overcome many problems, e.g., small deformed acetabulum, narrow femoral canal, muscle contracture, and so on. Careful preoperative planning is necessary for these patients. In principle, we set acetabular sockets in the original acetabulum, even if the hip is severely dislocated and considerable limb lengthening, of up to 3 cm, might result. However, the original acetabulum in these patients often

⟶

Fig. 6. If we set a 42-mm socket in the original acetabulum as shown in this simulation, the anterior part of the socket will not be contained in the acetabulum

Fig. 7. If the setting of the socket is 10 mm higher than the usual level, anterior covering of the socket can be accepted. The socket can be set in the acetabulum with a single superior bone graft, as shown in this simulation

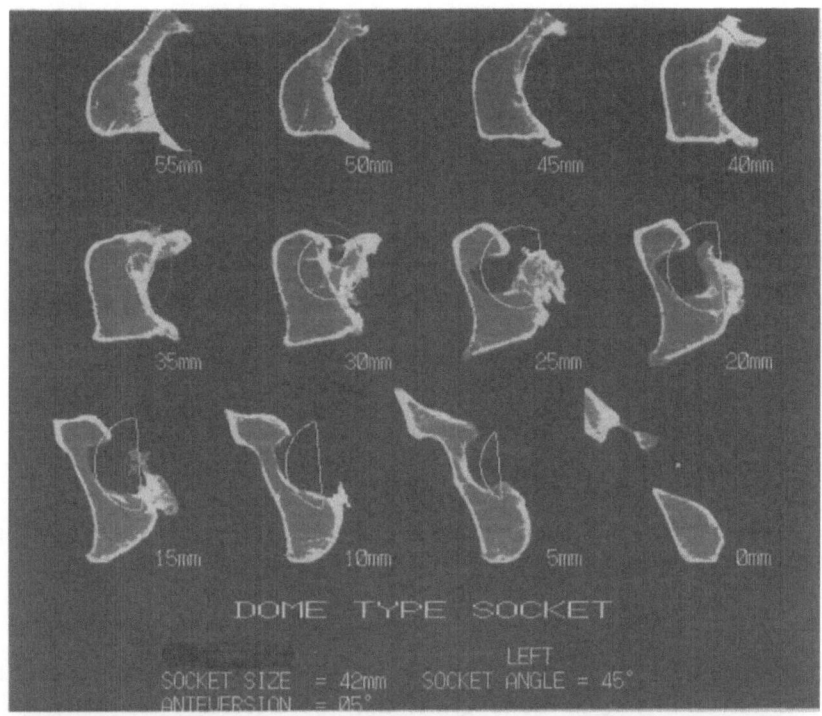

6

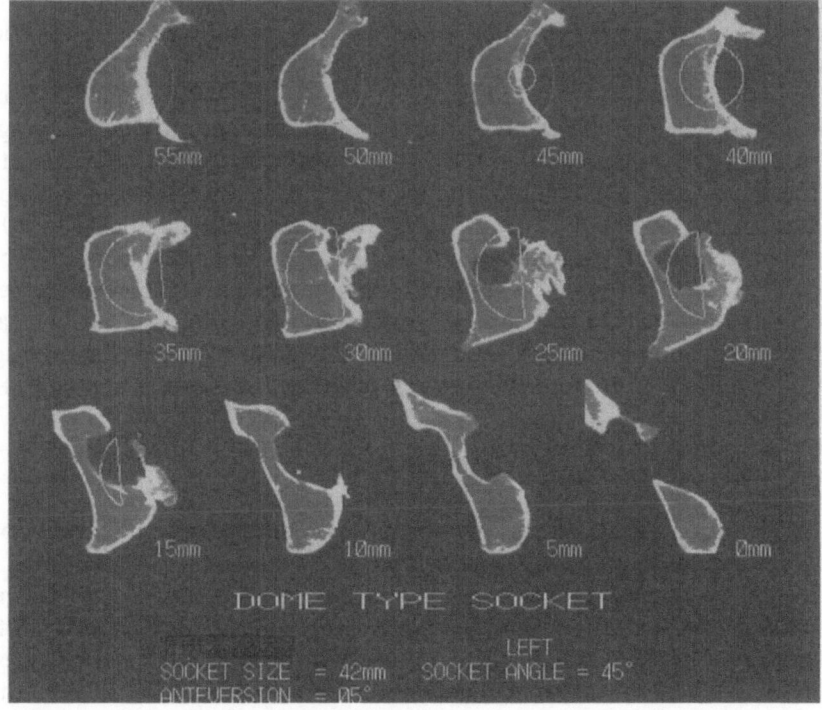

7

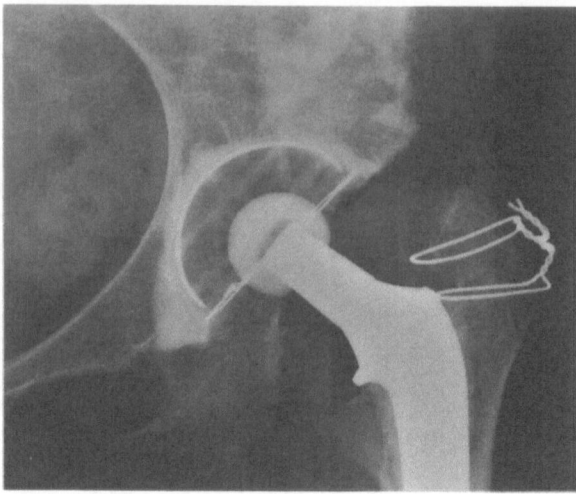

Fig. 8. Postoperative roentgenogram, case 2. The level of the socket setting was slightly higher than usual, and a bone graft from the femoral head was also performed

has no anterior wall and is so hypoplastic that secure fixation of the socket may be technically difficult, as bone grafts may be necessary not only on the superior but also on the anterior aspect of the socket. Therefore, we must find a more appropriate socket location to achieve anterior covering of the socket by the acetabulum.

A typical example is shown in Figs. 5, 6, and 7. This patient is a 56-year-old female with severely dislocated coxarthrosis of bilateral hip joints. The original acetabulum was hypoplastic, and the anterior wall was deficient. If we set the socket in the original acetabulum, the anterior part of the socket would not be contained in the acetabulum (Fig. 6). If the setting of the socket is 10 mm higher than the usual level, anterior covering of the socket can be accepted, as shown in the simulation, where a 42-mm diameter socket can be set in the acetabulum with a single superior bone graft (Fig. 7). The postoperative roentgenogram shows that the level of the socket setting was slightly higher than usual; a bone graft from the femoral head was also performed (Fig. 8).

Discussion

The aim of preoperative planning of total hip replacement is the selection of the most adequate shape, size, and position of the implants. Such planning may be affected by many factors, such as stature, basic disease, hip deformity, contracture, muscle strength, limb length discrepancy, and the condition of the opposite hip. The selection of the implants will be affected by the surgeon's basic concepts, and by anatomical and technical limitations such as the limit of limb lengthening. We must therefore have some basic principles in mind in order to make an adequate plan that takes these different factors into consideration.

We have many patients with hip osteoarthritis secondary to congenital dislocation of the hip joint in Japan. The size and shape of the acetabulum in these patients is so varied that preoperative morphologic evaluation of the hip joints is important for a successful operation. Although we have basic principles of socket setting, i.e., the socket is located in the original acetabulum, the size of socket should be as large as possible, and bone grafts should be used when adequate coverage of the socket cannot be achieved, it is usually impossible to identify the optimal size and position for the acetabular component preoperatively be roentgenographic evaluation alone. Preoperative evaluation of the hip using CT has been reported by several authors [2–5], but difficulties still remain regarding the extent of reaming and the size and location of the socket. We believe that our simulation system is quite useful for recognizing the three-dimensional relationship between the acetabulum and the socket and that use of this system could prove to be extremely helpful in the planning of THR.

References

1. Iida H, Yamamuro T, Okumura H, Ueo T, Kasai R, Tada K, Tsuji T (1988) Socket location in total hip replacement—preoperative computed tomography and computer simulation. Acta Orthop Scand 59(1):1–5
2. Barmier E, Dubowitz B, Roffman M (1982) Computed tomography in the assessment and planning of complicated total hip replacement. Acta Orthop Scand 53(4): 597–604
3. Lasda NA, Levinsohn EM, Yuan HA, Bunnel WP (1978) Computerized tomography in disorders of the hip. J Bone Joint Surg [Am] 60A(8):1099–1102
4. Mendes DG (1981) The role of computerized tomography scan in preoperative evaluation of the adult dislocated hip. Clin Orthop 161:198–202
5. Vannier MW, Totty WG, Stevens PM, Dye DM, Daum WJ, Gilula LA, Murphy WA, Knapp RH (1985) Musculoskeletal application of three-dimensional surface reconstructions. Orthop Clin North Am 16(3):543–555

15

—Overview—
Clinical Applications of Image Analyses in the Hip Joint

Hideo Okumura[1]

Summary. Various methods of image analysis are currently being utilized in orthopedic surgery. The generally accepted requirements for image analysis are: (1) value in clinical medicine, (2) convenience for clinical application, (short time required and low cost), (3) ease of performance and availability, and (4) precision. Solid models have been very useful in clinical medicine because they are substantial and convenient, and surgeons can use them for simulated operations. Solid models constructed automatically with a milling machine connected to computed tomographic data are very precise, but their high price is a disadvantage. Simulation of the three-dimensional fitting of a cementless stem in the femoral canal is a nice system for selecting a stem of adequate size to fit the dimensions and the shape of the medullary canal. Computer simulations in the preoperative planning of total hip arthroplasty are very valuable clinically and of great help in selecting the position of the socket in acetabular bone.

Key words: Computer simulation—Computed tomography—Finite element method—Hip joint—Image analysis—Preoperative plan—Solid model

Introduction

Various methods of image analysis are currently utilized in orthopedic surgery. In these analyses, slice pictures taken by computed tomography are utilized for the initial data and applied to image analysis because they can be reconstructed three-dimensionally, while plain X-ray pictures show only two-dimensional figures.

My comments on the image analysis methods presented in this session follow, together with my opinions on the requirements for image analyses.

---

[1] Department of Orthopaedic Surgery, Ehime University School of Medicine, Shigenobu-cho, Shitukawa Onsen-gun, Ehime, 791-02 Japan

149

Requirements for Image Analysis

The generally accepted requirements for image analysis are: (1) value in clinical medicine, (2) convenience for clinical application, (short time required and low cost), (3) ease of performance and availability, and (4) precision.

Comments on Presentations

Dr. Ohtsuka (see this volume) presented a solid model of the hip joint and performed a simulated operation. The solid model was substantial and was advantageous in that it could be used for the performance of simulated operations. The simulation used with the solid model fulfills the first three requirements, for image analysis listed above but it does not always fulfill the fourth requirement. The three-dimensional origin is not always decided accurately when the slices are piled up and the solid model is constructed three-dimensionally.

We experienced a revision case with a large acetabular bone defect (Fig. 1) and constructed a solid model using slice pictures taken by computed tomography (Fig. 2). When the slices were piled up, a small error was observed. However, this analytical method is thought to be very useful in clinical medicine because it is substantial and convenient.

Dr. Matsuno (see this volume) presented a three-dimensional milling system. The solid model was automatically constructed with a milling machine con-

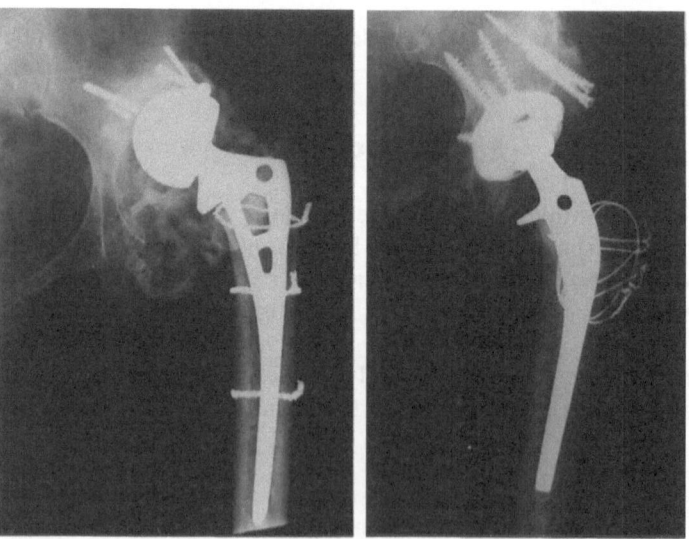

Fig. 1. Roentgenograms of a 51-year-old female with a large acetabular bone defect before (*left*) and after revision (*right*)

Fig. 2. Constructed solid mode of iliac bone using the slice pictures taken by computed tomography

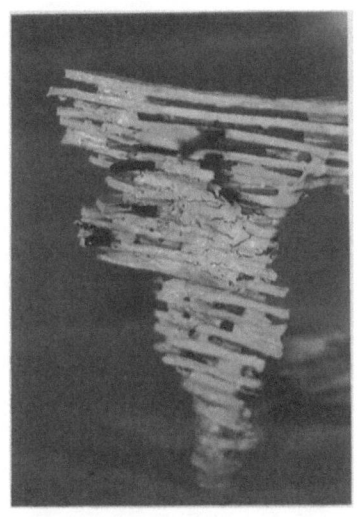

Preoperative X-P Planning

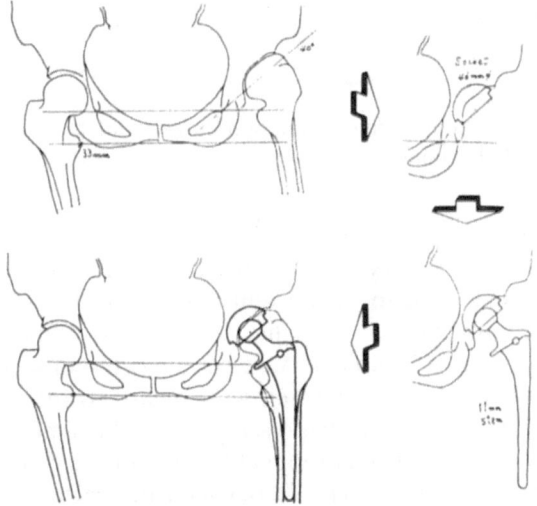

Fig.   3. Preoperative   planning drawn up on the basis of a plain X-ray picture

nected to computed tomographic data. It is thus very precise, but it may not be popular and thus readily available because of its high price. Further, it is possible that the medullary shape is not be incorporated, since the solid model expresses the surface shape only. In spite of its disadvantages, this system will probably have clinical applications in the future, due to its automatic control.

Dr. Yano (see this volume) presented a simulation of the three-dimensional fitting of a cementless stem in the femoral canal. The simulation is done on the

152   H. Okumura

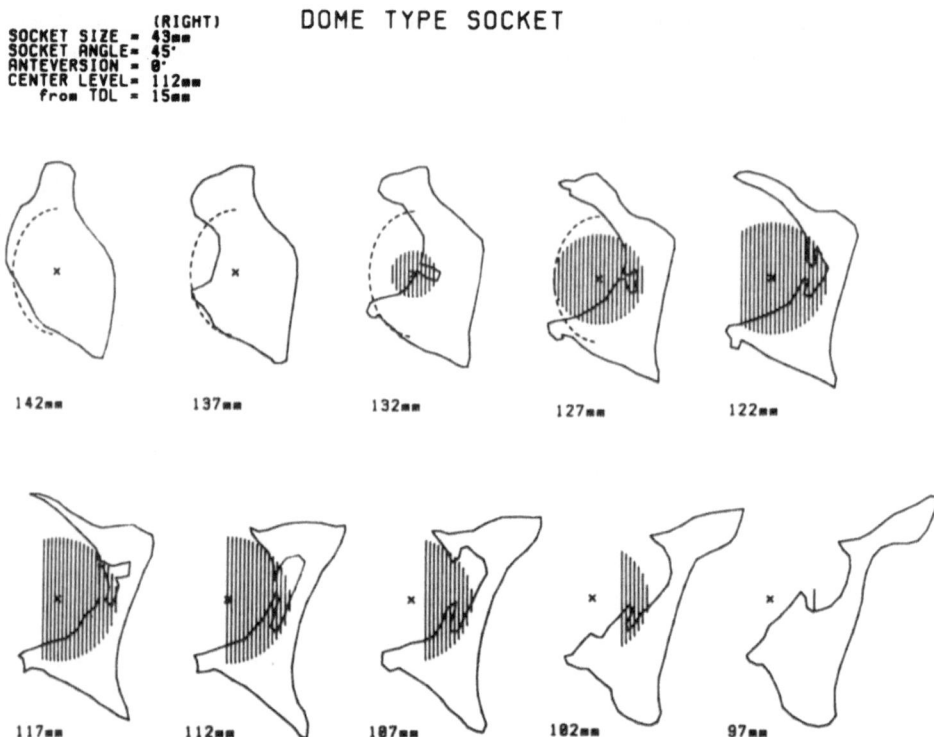

Fig. 4. Preoperative simulation carried out to select the position of the socket in the acetabular bone

screen of a personal computer. This is a nice system for selecting a stem of adequate size to fit the dimensions and the shape of the medullary canal [1]. The procedure is of value clinically, is convenient, and is easily performed. However, it is difficult to select the osteotomy level of the femur in this system, since this level is selected with regard to the position of the acetabular socket and with regard to leg discrepancy. The simulation must be carried out, at the same time that the preopeative planning is drawn up on the basis of plain X-ray pictures (Fig. 3). The fit between the calcar anteversion of the femur and the rotational degree of the stem is frequently a perplexing problem when the stem is set up in the operation. This simulation is very valuable, if the osteotomy level can be selected in the simulation. Further, the simulation is also beneficial in that the position of the stem in the medullary canal of the femur and the plane ratio of the canal filling of the stem can be observed, while these features cannot be obtained in a two-dimensional simulation.

Dr. Iida (see this volume) presented a computer simulation used in the preoperative planning of total hip arthroplasty. The simulation system is applied to select the position of the socket in the acetabular bone [2]. We have also

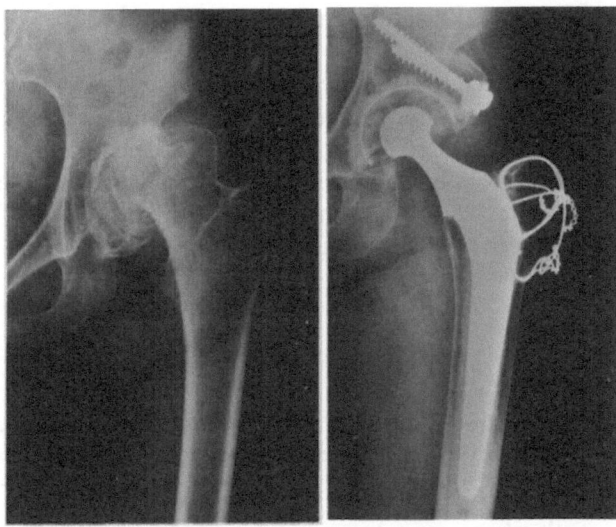

Fig. 5. Roentgenograms of a 58-year-old female before operation (*left*) and 5 years after Charnley total hip replacement (*THR*) with an acetabular bone graft (*right*)

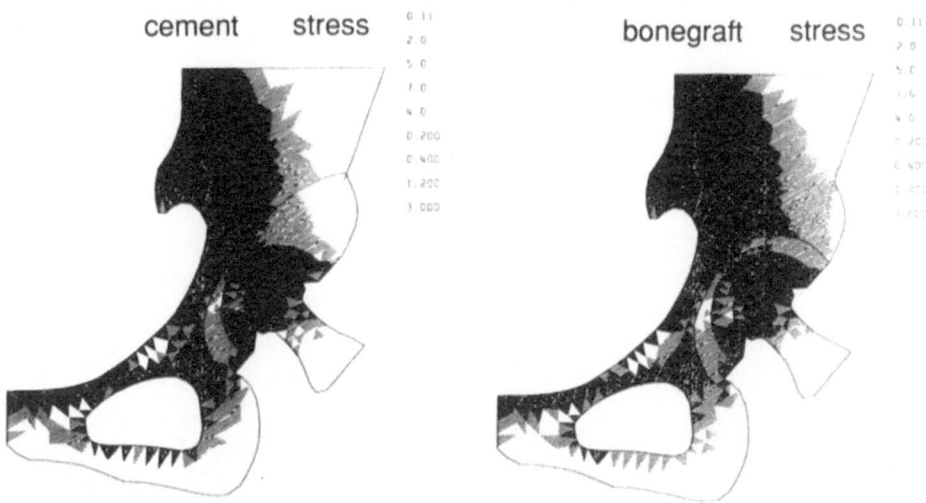

Fig. 6. Figure analyzed by finite element method in the case of a cemented implant (*left*) and a bone graft (*right*)

used this simulation system, particularly in patients with acetabular dysplasia (Figs. 4, 5). It is very valuable clinically and is very easy to perform. However, its use is limited in that it may be best employed as a supplemental device when the preoperative plan is drawn up.

Future Trends in Image Analysis

It is expected that, in the future, analyses of joint movement, contact areas of the joint, and the pressure of the contact area will be added to analyses already available with the solid models. It is also to be expected that the optimal congruity of the joint will be analyzed with these systems. We would also expect that biomechanical analysis with a finite element method (Fig. 6) would be connected to computer simulations of the socket and stem in total hip arthroplasty.

References

1. Toda M, Shiba M, Hashimoto T, Kimura S, Yano S (1992) Three-dimensional fitting of stem by computer analysis (in Japanese). Rinshou Seikeigeka 27:865–869
2. Iida H, Yamamuro T, Okumura H, Ueo T, Kasai R, Tada K, Tsuji T (1988) Socket location in total hip replacement. Acta Orthop Scand 59:1–5

B. Osteotomy

II Osteotomy

16

Biomechanical Analysis of the Dysplastic Hip: Pre- and Post-Pelvic and Femoral Osteotomy

Hirotsugu Ohashi[1], Kenji Hirohashi[2], Yoshinobu Hara, Itsuo Furuya, and Akira Shimazu[1]

Summary. Biomechanical analysis of the hip joint, using the rigid body spring model (RBSM) was performed to investigate the mechanism responsible for the progression of coxarthrosis and to obtain guidelines for treatment. As dysplastic change progressed, the abductor muscle strength required for one leg standing increased, as did pelvic displacement. After Chiari's operation, pelvic displacement decreased remarkably, while more abductor muscle strength was required, due to the verticalization of the abductor muscle and the shortening of its lever arm. Varus osteotomy appeared to preserve the efficiency of abductor muscle strength by lengthening its lever arm. Clinically, in all patients, pain was either relieved or disappeared as a result of the muscle strengthening exercises having been done. When the practical abductor muscle strength was reduced to 50%–70% of the theoretical value calculated with the RBSM, hip pain score was about 20 points. Thus, by assessing the muscle strength both practically and theoretically, in addition to making an X-ray diagnosis, we were able to select a suitable treatment plan for patients with dysplastic hips.

Key words: Dysplastic hip—Chiari's osteotomy—Varus osteotomy—Valgus osteotomy—RBSM—Abductor muscle strength—Muscle strengthening exercise

Introduction

Pelvic and femoral osteotomies are often used for treatment of the dysplastic hip; the intent usually being to improve bony structural deficits. Nevertheless, when the long-term results of Chiari's operation were reviewed, superolateral

[1] Department of Orthopaedic Surgery, Osaka City University Medical School, 1-5-7 Asahi-machi, Abeno-ku, Osaka, 545 Japan
[2] National Institute of Fitness and Sports, 1 Shiramizu-cho, Kanoya, Kagoshima, 891-23 Japan

migration of the femoral head was evident in 54.1% of patients after about 9 years [1] and this phenomenon was strongly correlated with the progression of osteoarthritic change. This phenomenon cannot be understood without an understanding of muscle strength around the hip joint. These osteotomies not only change the bony structure of the hip joint but also the direction of the muscles and the direction of the resultant force on the femoral head; moreover, muscle strength may change as time passes. In order to investigate the mechanism responsible for the superolateral migration of the femoral head, a biomechanical analysis, including both morphologic and muscle strength factors, is necessary.

The biomechanics of the hip joint have been investigated in terms of the lever theory, in cadaver experiments, photoelastic studies, numerical models, and so on. Pauwels [2] reported the basic biomechanical conditions of the hip in one leg standing in terms of the lever theory. Many studies have been undertaken to investigate the biomechanics of the normal hip. In the dysplastic hip, Bombelli [3] demonstrated the shearing force, which seemed to push out the femoral head laterally. He looked at the inclination of the weight-bearing surface, a curved area of dense bone in the ilium. When the weight-bearing surface had a craniolateral inclination, the shearing force appeared. This analysis seemed to explain the mechanism responsible for the migration of the femoral head; however, the resultant force against the hip joint was always the same and the muscle strength factor was not addressed.

Analyses of physical models are how easily available, using personal computers. Such models can deal with shape data, material data, and boundary conditions. The finite element model (FEM) is widely used for biomechanical analysis; however, using this model, it is difficult to analyze nonlinear contact problems which always exist in the natural joint. (The natural (biological) joint does not operate under linear conditions.)

We have applied the rigid body spring model (RBSM), proposed by Kawai and Toi [4] in 1977, to the biomechanical analysis of the hip joint. This element consists of rigid plates which cannot change their shape. The rigid plates are connected by two different types of springs, that is, a perpendicular one and a tangential one. With this model the computational time for nonlinear analysis of plane strain problems is reduced, thus making it suitable for analyzing biomechanical conditions in a joint where two bones contact each other with very little friction. Via the use of this model, the biomechanics of the hip can be analyzed in relation to shape and muscle strength.

The aim of this study was (1) to make a numerical model of the hip joint by using the RBSM, (2) to apply this model to normal and dysplastic hips to investigate the relationship between hip shape, muscle strength, and joint stability, (3) to investigate changes in biomechanical conditions after pelvic and femoral osteotomies, and (4) to compare the practical abductor muscle strength measured with a kinetic communicator (KIN/COM; RC2000; Chattanooga, Hixson, Tenn.) and the theoretical muscle strength calculated by the RBSM with the clinical symptoms.

Materials and Methods

Physical Model

The RBSM was employed for the biomechanical analysis of the hip joint [5]. On a bipedal standing anteroposterior (A-P) X-ray film of bilateral hips, points on the articular surface, pelvis, and femur were plotted with the digitizer. The hip joint was divided into small rigid plates, as shown in Fig. 1. The articular surface was divided into eight elements. The Young's modulus of bone was assumed to be $1500 \, kg/mm^2$ and that of articular cartilage was assumed to be $1.4 \, kg/mm^2$ [6]. The plates which represented bones were connected by very strong springs which allowed no motion between them. The springs in the articular surface were set up to resist compression only, not tension, friction, or rotation. The femur was fixed in all directions. The loading conditions simulated one leg standing, and the body weight and the abductor muscle strength were loaded to the pelvis. In the muscles around the hip, we took note of the abductor muscle, which plays an important role in the one leg standing position. The body weight was loaded at the midline on the uppermost level of the iliac wing. The direction of the abductor muscle was determined by the method reported by Ninomiya et al. [7]. The direction was assumed to be from the lateral edge of the greater trochanter to the lateral third on the line which connected the lateral edge of the ilium with the inner edge by the shortest distance. The body weight (to be exact, the body weight minus the weight of the supporting leg) was assumed to be $50 \, kg$ in all cases.

A personal computer (PC-9801 VM2; NEC) was employed and the software was modified for the hip joint in our department. By changing the abductor muscle strength, we calculated the displacement and the rotation of the pelvis. When the rotation of the pelvis was zero, we recorded the abductor muscle strength (theoretical value) and the extent of inferomedial pelvic displacement.

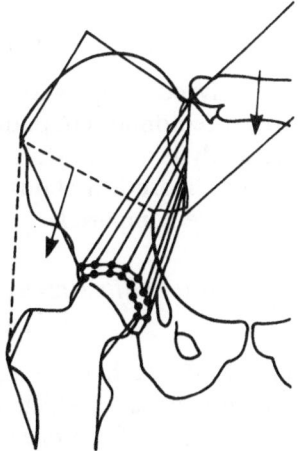

Fig. 1. Our analytical simulation model of the hip for the rigid body spring model (*RBSM*). The articular surface was divided into eight elements

Since 1984, we have given our secondary coxarthrosis patients muscle strengthening exercises, which we consider to be a conservative treatment for this condition [9]. These exercises strengthen the muscles around the hip joint and include squatting; sit-ups, holding the position; squatting with one leg; and stepping-up and -down on a 25-cm-high step. Eleven patients were treated with these exercises for 3 months. The maximum abductor muscle strength (practical abductor muscle strength) was measured by KIN/COM before and after this exercise. We also calculated the theoretical abductor muscle strength which would balance the pelvis in the RBSM.

The clinical symptoms were assessed in accordance with the Japanese Orthopaedic Association (JOA) [10] hip score. The pain score, in particular, was adopted for comparison with muscle strength, since pain may lower the level of the activities of daily life (ADL), thus bringing about muscle atrophy, which then may worsen the osteoarthritic changes. The pain score ranges from 40 (no pain) to zero (severe pain).

Statistical Analysis

Unpaired and paired Student's *t*-tests were used for statistical analysis.

Results

Biomechanical Analysis of Normal and Dysplastic Hips (Fig. 3a,b)

The theoretical abductor muscle strength was 2.20 ± 0.23 times body weight for normal subjects. This value increased as dysplastic change progressed. These values were significantly different in normal and dysplasia (type A), and in subluxation (type B) and dislocation (type C). The extent of pelvic displacement also increased as dysplastic change progressed, the displacement in type C being slightly less than that in type B2. The distribution of the data in each dysplastic hip was relatively more scattered than in the normal hip.

Biomechanical Analysis of Pelvic and Femoral Osteotomies

The theoretical abductor muscle strength required for one leg standing and the amount of displacement were calculated by using the RBSM (Fig. 4). Examples of Chiari's osteotomy and varus osteotomy are shown in Fig. 5.

After Chiari's operation, pelvic displacement decreased remarkably, while more abductor muscle strength was required for one leg standing. The muscle strength and extent of displacement were both significantly improved by the varus osteotomy, but were not changed by the valgus osteotomy.

Regarding preoperative condition, osteotomies tended to be selected; however, all these patients had been operated before we began this analysis. The relatively advanced cases, which required strong abductor muscles and showed a great extent of displacement, were often treated by Chiari's operation. The relatively mild cases were treated by varus osteotomy. Valgus osteotomy was performed in those patients who required strong abductor muscles and where the extent of displacement was relatively small.

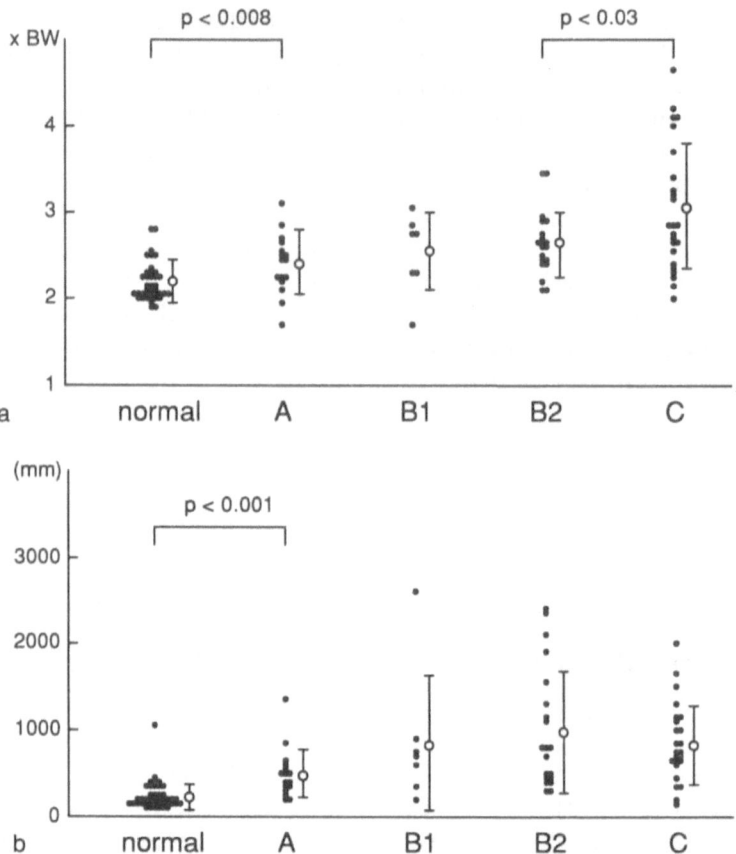

Fig. 3. a Theoretical abductor muscle strength in normal and dysplastic hips. Circles indicate means and the vertical levels are SD. **b** Extent of pelvic displacement in normal and dysplastic hips. Circles indicate means and the vertical levels are SD

Relationship Between Practical and Theoretical Abductor Muscle Strength and Clinical Symptoms

The individual data are listed in Table 1. Figure 6 shows the relationship between the practical and theoretical abductor muscle strength and the pain score component of the JOA hip score. The Y-axis is the pain score, and the X-axis indicates the percent of the practical muscle strength against the theoretical value. In all cases the pain was relieved or disappeared corresponding to the increase in practical muscle strength. The pain score was about 20 when the practical muscle strength was reduced to 50%–70% of the theoretical value, and even if this percentage is between 70% and 100%, the patient might have a pain score of about 30 points.

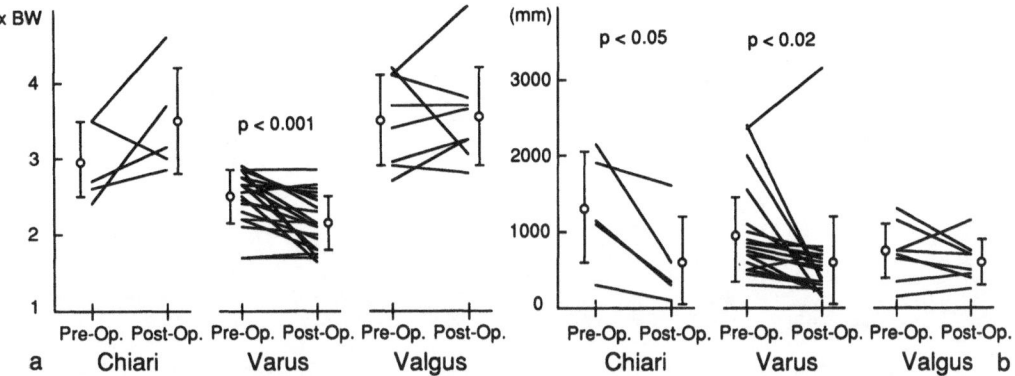

Fig. 4. **a** Theoretical abductor muscle strength required for one leg standing. This value decreased significantly after the varus osteotomy, while it increased after the Chiari's osteotomy. Circles indicate means and the vertical levels are SD. **b** Extent of pelvic displacement. This decreased remarkably after the Chiari's operation. Circles indicate means and the vertical levels are SD

Discussion

A joint consists of two or more bones that conduct only compression, not tension or friction. In the RBSM, displacement of the rigid plate is displayed by two parallel displacement components (u, v) and a rotational component (θ), and rigid plates are connected by perpendicular and tangential springs [4]. Thus, this model is suitable for the analysis of contact problems where two contact elements can separate freely. These conditions seem to resemble those in the natural joint.

The RBSM has been applied to many kinds of joints, for example, hip, knee, shoulder, wrist, and so on. With regard to the hip joint, the normal hip, dysplastic hip, osteotomized hip, and totally replaced hip have been analyzed. However, the loading conditions have not yet been standardized. Himeno et al. [11] reported the contact pressure distribution in the hip joint by loading the force on the femoral head against the fixed pelvis. They calculated the force acting on the femoral head via the equation of the moment of body weight and abductor muscle. In another model [12], the femur was fixed and springs were placed between the pelvis and the greater trochanter, acting like the abductor muscle. Body weight acted on the pelvis and the traction forces of the springs were calculated by using the RBSM. These forces were assumed to represent the abductor muscle strength, although they could be regarded as representing the eccentric muscle strength. In our model, in order to bring it close to the natural hip joint, we loaded the body weight and the abductor muscle strength on the pelvis and fixed the femur for all directions.

First of all, we applied our model of the normal hip to compare it with other biomechanical studies. The abductor muscle strength for one leg standing or

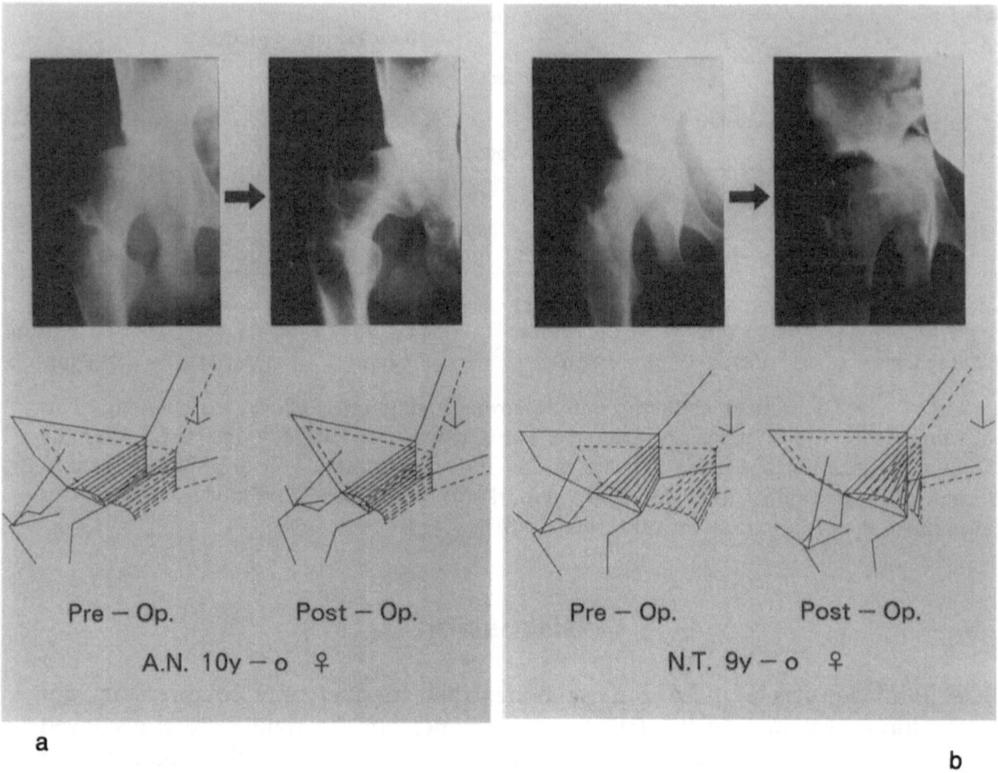

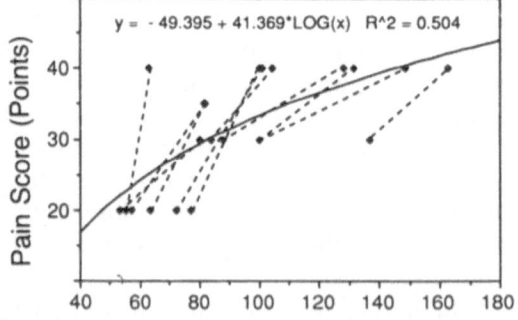

Fig. 5. a Analysis of the Chiari's operation. The length of the *left arrow* indicates the theoretical abductor muscle strength required for one leg standing and the *dotted line* represents the displaced pelvis. The pre-operative X-ray picture was classified as type B2. The theoretical abductor muscle strength was 2.60 × body weight (BW) and the displacement before operation was 1102 mm. These values were 2.86 and 328, respectively, after operation. **b** Analysis of the varus osteotomy. The pre-operative X-ray picture was classified as type B2. The theoretical abductor muscle strength was 2.88 × BW and the displacement before operation was 780 mm. These values were 2.35 and 603, respectively, after operation

y = - 49.395 + 41.369*LOG(x) R^2 = 0.504

Pain Score (Points)

Practical / Theoretical Abd. M. Strength x 100 (%)

Fig. 6. Relationship between the pain score and the percent practical muscle strength against the theoretical value.

Table 1. Theoretical and measured abductor muscle strength correlated with pain.

No.	Age	Sex	Theoretical value (× BW)	Measured value/Theoretical value			Pain		
				Pre. (%)		3 Months (%)	Pre.		Post.
1.	10	F	2.38	100	→	131.6	30	→	40
2.	15	F	2.21	88.1	→	128.3	30	→	40
3.	30	F	3.32	77.3	→	100	20	→	40
4.	40	F	2.48	56.9	→	82.2	20	→	35
5.	25	F	2.42	72.3	→	101.2	20	→	40
6.	16	F	2.11	137.1	→	163	30	→	40
7.	17	F	2.62	80.2	→	104.5	30	→	40
8.	32	F	2.51	63.4	→	81.6	20	→	35
9.	43	M	2.46	100	→	148.5	30	→	40
10.	39	F	2.56	53.3	→	84.3	20	→	30
11.	37	F	4.17	55.4	→	62.8	20	→	40

BW, Body weight

walking has been investigated. Inman [13] calculated the abductor muscle strength, which was found to be 1.65 times body weight for one leg standing. Osborne and Fahrni [14] estimated it to be 33 lb for a body weight of 20 lb. Pauwels [2] indicated that this strength was 2.75 times body weight, based on the lever theory. Merchant [15], who measured the abductor muscle strength in a cadaver pelvis, found the value to be 154 lb for a body weight of 150 lb in a neutral position.

Using another RBSM system which had springs between the greater trochanter and pelvis, it was reported that the abductor muscle strength was 1.97 times body weight [12]. In our model, the theoretical abductor muscle strength required for one leg standing was estimated to be 1.83 times body weight if the weight of the supporting leg was one-sixth of the total body weight.

Regarding the extent of pelvic displacement, a value of 236 mm was indicated even in the normal hip in our model. We believe that this impractical value was due to our incomplete model design, which did not have a joint capsule, ligaments, or muscles, except for the abductor muscle. Femoral head displacement was prevented only by the abductor muscle strength, thus the extent of displacement could not be assessed quantitatively. However, the extent of this displacement increased as the dysplasia progressed; we consider that this extent is related to the degree of stability of the hip joint.

Our hypothesis to explain the progression and prevention of osteoarthritic change is shown by two cycles in Fig. 7. In the case of the dysplastic hip, we found increased abductor muscle strength was necessary to balance the pelvis, and there was displacement to a great extent. The superolateral migration of the femoral head and subsequent joint degeneration is often seen clinically when osteoarthritic changes progress. Thus, if the hip is dysplastic, the shearing force, which pushes the femoral head superolaterally, is easily exerted when there is muscle weakness. The migration of the femoral head accelerates the

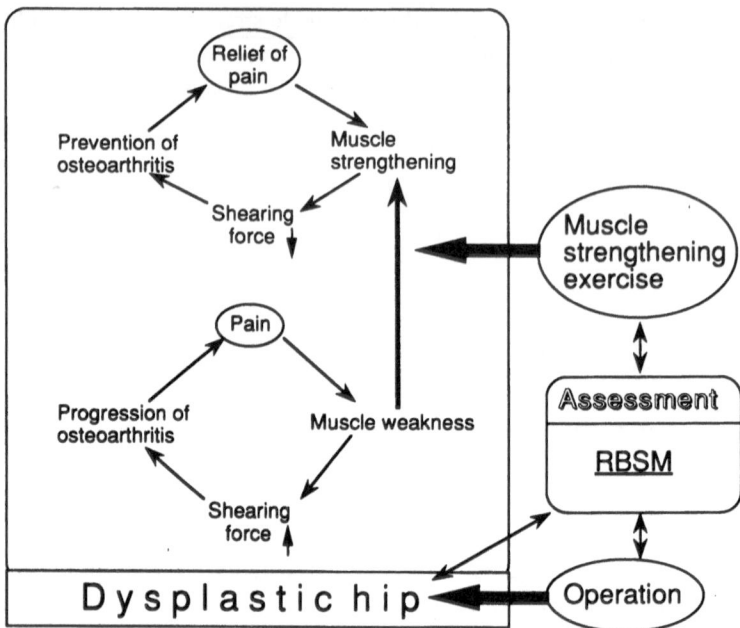

Fig. 7. Our hypothesis explaining the progression and prevention of osteoarthritic change. Two cycles are assumed. The prevention of osteoarthritis is attempted by operations or by muscle strengthening exercises. The RBSM occupies a central position in the assessment of these treatments

osteoarthritic changes and causes coxalgia, which, in turn, leads to muscle atrophy, resulting in muscle weakness. Muscle strengthening around the hip joint, on the other hand, may reduce the shearing force and prevent the progression of joint degeneration. A decrease in or loss of pain will follow.

From our study of the dysplastic hip, we obtained two parameters; the theoretical abductor muscle strength and the extent of pelvic displacement. We assumed that the required abductor muscle strength was an index which related the practical abductor muscle strength required to prevent the progression of osteoarthritic change and this to reduce coxalgia. The extent of displacement was assumed to indicate hip joint stability, which relates to the shearing force.

Chiari's pelvic osteotomy has been performed at our department since 1956. The basic concept of this procedure consists in reconstructing a congruent shelf above the dysplastic hip joint (without the necessity for bone grafting and its inherent risk) and in optimally correcting the pathologic position of the femoral head. The dome-shaped osteotomy provides congruent anterior coverage, and sufficient medialization provides lateral coverage and also shortens the body weight lever arm, thus improving the efficiency of the abductor muscles. However, the efficiency of the abductor muscle is disputable. Clinically, there are many different evaluations of changes in Trendelenburg's sign in the literature. Some authors have reported distinct improvement, or at least no deterioration, in muscle strength after the osteotomy [16,17], but contrasting views have been given by others [18,19].

Chiari's osteotomy has also been analyzed biomechanically. It was reported that this osteotomy reduced the abductor muscle strength required for one leg standing [12]. Our results, indicate that Chiari's osteotomy significantly reduced the extent of pelvic displacement, thus providing stability of the hip joint, while the theoretical abductor muscle strength was increased. This discrepancy is considered to be due to differences in the loading conditions. The accuracy of the analysis should be confirmed from the clinical results.

In reviews of the long-term results of Chiari's osteotomy, this procedure was reported to have failed to halt or reverse the degenerative process, and a significant increase in degenerative changes was found after an average follow-up of 15.5 years [20]. It has also been reported that the efficiency of Chiari's osteotomy was limited to about 20–30 years with regard to the progression of osteoarthritic changes [1].

While the results of Chiari's osteotomy are impressive in the light of improvements of bony structural deficits, changes in muscle direction must be borne in mind. After Chiari's osteotomy, the abductor muscle takes a more vertical direction and its lever arm is shortened. These muscular changes are thought to incline the force acting on the femoral head superolaterally and to reduce the efficiency of the abductor muscle. These are the adverse effects of Chiari's osteotomy. Because of this, it is considered that this procedure cannot prevent migration of the femoral head and osteoarthritic changes. However, muscle strengthening can prolong the duration during which this procedure is efficient.

Varus osteotomy is carried out to increase the efficiency of the abductor muscle by changing its direction medially and by lengthening its lever arm. It is also intended to increase the weight-bearing surface by improving joint congruency. Our analysis showed that the theoretical abductor muscle strength was significantly reduced after varus osteotomy; thus, this procedure is considered to preserve abductor muscle strength. It also reduced the extent of pelvic displacement, possibly due to the improvement of joint congruency.

The biomechanical effects of valgus osteotomy are rather difficult to understand. Two features were intended: (1) to lengthen the lever arm of the abductor muscle by medialization of the rotational center and (2) to stimulate roof osteophyte formation by stretching the capsule. Our analysis did not deal with joint capsule and ligament factors, and the valgus osteotomy did not appear to have an effect on the theoretical abductor muscle strength or on the extent of displacement.

In order to relate our analytical results to clinical problems, we applied our analysis to patients treated with the muscle strengthening exercises. Since these exercises change only muscle strength and not the bony structure of the hip joint, we compared the theoretical abductor muscle strength calculated with the RBSM with the practical abductor strength, measured by KIN/COM, and the clinical symptoms. Clinically, the muscle strengthening exercise either relieved or abolished the pain. We obtained the percentage of the practical abductor muscle strength against the theoretical value, and we attempted to relate the pain score to this percentage. One hundred percent seemed to be the lower limit for relief from coxalgia and this would be a theoretical goal of the muscle strengthening exercise.

To return to our hypothesis, dysplasia of the hip and muscle weakness are the factors which cause progression of osteoarthritic change. The dysplastic hip should be treated by operation or by muscle strengthening exercises. However, planning of the treatment is usually based mainly on the surgeon's experience, the type of operation often being selected on the basis of X-ray pictures. There are no criteria to indicate whether a patient should be treated by operation or by muscle strengthening exercises, and there is no theoretical goal for these exercises. Our results indicate that our RBSM for the hip joint could simulate the biomechanical conditions of the dysplastic hip and could display both the biomechanical changes that occur after pelvic and femoral osteotomies and the percent of the practical abductor muscle strength against the theoretical value, related to the pain score. Thus, we believe that by using our RBSM, the dysplastic hip could be assessed or classified, and then the type of operation and the goal for the muscle strengthening exercise could be selected. We actually teach all our patients with dysplastic hips these muscle strengthening exercises, even if they are candidates for operation. The patients are assessed by measuring the practical muscle strength and by the RBSM analysis, in addition to X-ray examination. On the basis of these practical and theoretical assessments, we plan our treatment for the dysplastic hip.

References

1. Hirohashi K, Kanbara T, Inose M, Sakamoto K, Shimazu A (1984) The Chiari's osteotomy—in cases under 15 years old (in Japanese). Seikeigeka MOOK 36: 99–116
2. Pauwels F (1980) Biomechanics of locomotor apparatus. Springer, Berlin Heidelberg New York, pp 76–105
3. Bombelli R (1983) Osteoarthritis of the hip. Springer, Berlin Heidelberg New York, pp 13–22
4. Kawai T, Toi Y (1977) A new element in discrete analysis of plane strain problems. Seisan Kenkyu 29:204–207
5. Ohashi H, Hirohashi K, Hara Y, Kanbara T, Machii Y, Sakamoto K, Shimazu A (1987) A biomechanical study of superolateral displacement of the femoral head in secondary coxarthrosis (in Japanese). Cent Jpn Orthop Traumat 30:1441–1450
6. Oonishi H, Hamaguchi T, Iwase R, Nishikubo K, Shikita T, Isha H, Hasegawa T(1978) Mechanical analysis of the human pelvis and its application to artificial hip joint: A three-dimensional analysis using the finite element method. Proceedings, 1978 Annual Meeting of Japanese Orthopaedic Biomechanics Research Society (JOBRS), pp 89–100

7. Ninomiya S, Tagawa H, Miyanaga Y, Seki N (1976) The relationship between the position of the artificial hip joint and the resultant force acting on the femoral head (in Japanese with English abstract). J Jpn Orthop Assoc 50:15–20

8. Hirohashi K, Kanbara T, Asada K, Okajima M, Hayashi M, Shimazu A, Kinoshita T, Kimata T, Oonishi H (1979) Clinical studies of pre-arthrosis and arthrosis of hip joint in second decade age group—especially referring to pre-operative condition of the hip joint (in Japanese). Hip Joint 5:53–63

9. Hirohashi K, Kanbara T, Hashimoto T, Okubo M, Hara Y, Machii Y, Tanaka N, Ohashi H, Ueda Y, Furuya I, Kanao K, Ichikawa N (1987) Muscle strengthening as a method of treatment for secondary coxarthrosis (in Japanese). Hip Joint 13:27–34

10. Japanese Orthopaedic Association (1971) Evaluation sheet for osteoarthritis of the hip (third plan) (in Japanese). J Jpn Orthop Assoc 45:813–833

11. Himeno S, Nishio A, Kawai T, Takeuchi N (1981) The instability of the hip joint and its contact pressure (First report: On the pediatric hip; in Japanese). Rinsho Seikei Geka 16:835–845

12. Shiba N (1991) Biomechanics of the Chiari pelvic osteotomy (in Japanese with English abstract). J Jpn Orthop Assoc 65:337–348

13. Inman VT (1947) Functional aspects of the abductor muscles of the hip. J Bone Joint Surg 29:607–619

14. Osborne GV, Fahrni WH (1950) Oblique displacement osteotomy for osteoarthritis of the hip joint. J Bone Joint Surg [Br] 32-B:148–160

15. Merchant AC (1965) Hip abductor muscle force: An experimental study of the influence of hip position with particular reference to rotation. J Bone Joint Surg [Am] 47-A:462–476

16. Graham S, Westin GW, Dawson E, Oppenheim WL (1986) The Chiari osteotomy: A review of 58 cases. Clin Orthop 208:249–258

17. Reynolds DA (1986) Chiari innominate osteotomy in adults: Technique, indications, and contra-indications. J Bone Joint Surg [Br] 68-B:45–54

18. Calvert PT, August AC, Albert JS, Kemp HB, Catterall A (1987) The Chiari pelvic osteotomy: A review of the long-term results. J Bone Joint Surg [Br] 69-B:551–555

19. Zlatic M, Radojevic B, Lazovic C, Lupulovic I(1988) Late results of Chiari's pelvic osteotomy: A follow-up of 171 adult hips. Int Orthop 12:149–154

20. Lack W, Windhager R, Kutschera HP, Engel A (1991) Chiari pelvic osteotomy for osteoarthritis secondary to hip dysplasia: Indications and long-term results. J Bone Joint Surg [Br] 73-B:229–234

17

Computer Simulation of Hip Osteotomies for Dysplastic Coxarthroses: Use of a Rigid Body Spring Model

Naoto Shiba, Hisashi Yamashita, Fujio Higuchi, and Akio Inoue[1]

Summary. We have used the rigid body spring model (RBSM) or discrete element analysis to investigate the biomechanical effect of hip osteotomies. A two-dimensional (2-D) two-element model was generated by antero-posterior (A-P) radiography. The pelvic and femoral elements were assumed to be rigid and interface springs were placed 1 mm apart along the articular contact area to simulate articular cartilage. Four subjects, who had a normal hip joint, early, advanced, and late stage of coxarthroses, respectively, were analyzed. The coxarthrotic subjects were analyzed preoperatively and at final follow-up. The resultant and modified resultant force of the femoral head, dislocation force, mean and maximum compressive stress on the joint, and X–Y and rotational deviations of the pelvic element were calculated. In the normal hip joint, the stress pattern on the joint surface was similar to the "sourcil" ("eyebrow"), as described by Pauwels [1]. In preoperative coxarthrosis, extreme stress was seen on the joint, especially where the joint space was narrowed. A dislocation force not observed in a normal hip joint was apparent, with the X–Y and rotational deviations being larger than in the normal hip joint. This tendency increased with progressing stages of coxarthrosis. In the final follow-up examination, all the simulated results were in contrast to those obtained preoperatively. A 3-D reconstruction computerized tomographic (CT) scan was used to visualize the hip joint and its operation, but this procedure is not clinically practical for every patient. Surgeons usually use only plane radiograms when planning hip operations, such as osteotomy or total hip arthroplasty. The RBSM is therefore clinically useful because of its simplicity, and 2-D RBSM can easily be utilized by using A-P radiograms for each patient. With this method the patient's past radiographic data can also be utilized. A comparative study of clinical symptoms and simulated results using 2-D RBSM suggested that the simulated results were well correlated to the clinical symptoms. RBSM is clini-

[1] Orthopedic Department, Kurume University School of Medicine, 67 Asahi-machi, Kurume, Fukuoka, 830 Japan

cally useful not only for evaluating each patient biomechanically but also for planning hip osteotomy.

Key words: Hip joint—Biomechanics—Computer simulation—Osteotomy —Coxarthrosis—Radiogram—Stress analysis

Introduction

The finite element method (FEM), which has been widely used [2–4] to analyze the biomechanics of the hip joint and to evaluate the biomechanical effect of hip operations, is recognized as the best method for biomechanical analysis. This method is well developed, and allows a precise three-dimensional (3-D) calculation involving the material properties or non-linearity of the interface conditions, using currently available software. However FEM requires a technician's time to generate a model, long calculation time, and a large-capacity computer. Moreover, FEM is too comprehensive to be utilized clinically for each patient, and it requires precise engineering knowledge to handle it.

We have used a computer simulation model based on the rigid body spring model (RBSM) or discrete element analysis technique [5–8]. The RBSM is simplified, with each element assumed to be rigid, and it is formulated by considering the rigid body to be in equilibrium with external loading. Spring units are placed on the interfaces, e.g., the articular contact surface, prosthesis and bone, prosthesis and cement, and bone and cement. There are three types of interface springs: (1) Compression resistance tension break spring, (2) tension resistance compression break spring, and (3) shear spring, (1) and (2) being normal springs. Normal springs are placed perpendicular to the interface, and shear springs are placed parallel to it. To simulate various interface conditions, these springs are combined. A tension resistance compression break spring is used to simulate musculo-ligamentous structures [5,7]. The reaction force between adjacent bodies is produced by spring systems distributed over the possible contact surface between these bodies. The load that is added to these rigid bodies is transmitted through these springs. Displacement of the simulated compressive and tensive springs can be described as a function of the displacement of the centroid of the associated rigid bodies, and it is expressed as X–Y and rotational deviation. The RBSM must rely upon an iterative process in order to achieve an equilibrium condition because of the difficulties encountered in the contact surface problem. A model can be generated by merely tracing a radiogram, and the calculation can be accomplished immediately using even a small 16-bit microcomputer. For these reasons, it is easy to utilize RBSM clinically.

The RBSM, which can produce the required information from plane radiograms, is clinically useful because of its simplicity. Osteotomy programs for the knee and hip joints have thus been developed and utilized clinically. Here we report our simulation program for hip osteotomy using the RBSM program, and discuss its efficiency.

Materials and Methods

Anteroposterior (A-P) radiograms of the hip joint were scanned by color image scanner (GT-4000; Epson Corp., Tokyo, Japan), and recorded with a 16-bit microcomputer (PC-9801 RX, NEC Corp., Tokyo, Japan). These radiographic data were compressed with a data compression device (Image Engine; Takaoka Seisakusyo Corp., Tokyo, Japan), to 1/30–1/50. The quality of the compressed radiographic data was adequate for clinical purposes. The data for 40–50 hip joints could thereby be saved on a floppy disk, and data for 5000 hip joints could be saved on an optical disk.

Using the RBSM program, a 2-D two-element simulation model was generated by tracing contours from A-P radiograms. Each bone segment (pelvis and femur) was assumed to be rigid.

Boundary Conditions

The boundary conditions were adopted from Ninomiya [9], although details were modified (Fig. 1a). On the assumption of the stance phase of the gait cycle, 5/6 body weitht [10] was loaded at the center of L5, and the distal end of the femoral element was fixed in all directions. The center of rotation (CR) [9,11] was defined by functional (abduction, neutral, and adduction) A-P radiograms as a rule. The femur was fixed in all directions, and the resultant force was loaded on the CR.

Interface Conditions

Compression resistance tension break springs were used to simulate the articular cartilage (Fig. 1b). These springs are characterized as supporting only compression in the direction normal to the spring. To simulate the joint contact problem, proper spring constants for the spring systems in the 2-D RBSM were employed, as decribed by An et al. [12], as follows:

$$Kd = (E(1 - v))/((1 + v)(1 - 2v)h), \quad Ks = G/h \ (N/mm)$$

where E, G, v, and h are the compression modulus of elasticity, shear modulus of elasticity, Poission's ratio of the cartilage, and the thickness of the cartilage, respectively.

The thickness of cartilage was defined as a joint space on the A-P radiogram. However, it was difficult to define the elastic modulus and Poisson's ratio for each patient. These values were estimated to be 15 MPa and 0.45, respectively, from the normal articular cartilage of the hip joint [9]. The spring constant of the shear spring (Ks) was assigned as zero, since the friction of normal human joints is negligible. These spring systems were place 1 mm apart along the

174

N. Shiba et al.

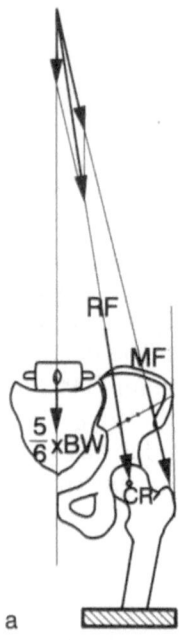

Fig. 1. a Boundary conditions of the hip simulation model. As a rule, the direction of abductor muscle force was estimated to be 1/3 of the width of the iliac bone in this equilibrium condition, *BW*, Body weight; *MF*, abductor muscle force; *RF*, resultant force; *CR*, center of rotation. **b** The rigid body spring model (*RBSM*) system. **c** Modified resultant force was calculated by determining the resultant force and dislocation resistance force

a

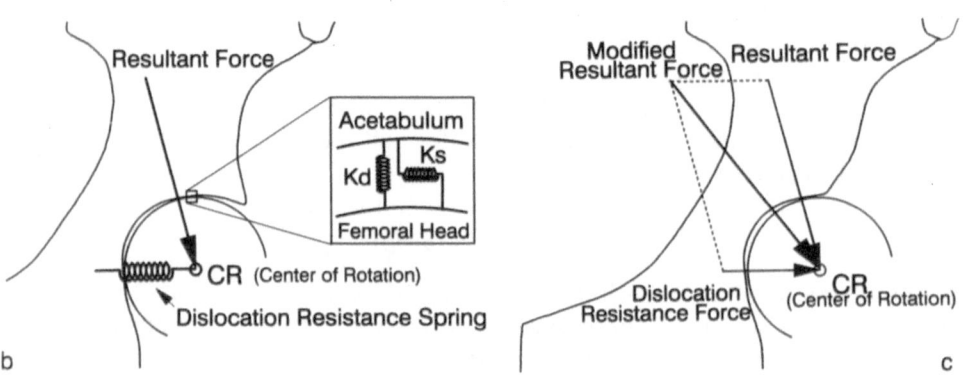

b

c

articular contact surface. A dislocation resistance spring (tension resistance compression break spring) was placed perpendicularly between the center of rotation of the femoral head and the pelvis (Fig. 2). The force produced by the dislocation resistance spring was considered to be a dislocation force, and the modified resultant force was calculated from the resultant force and the dislocation force (Fig. 1c). Using these characteristics, simulations were carried out.

Simulation Subjects

Simulation and analysis was performed for four subjects in this study using RBSM.

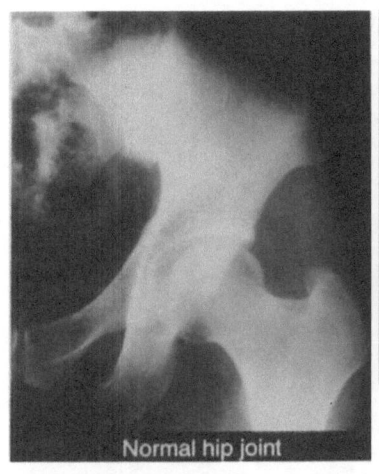

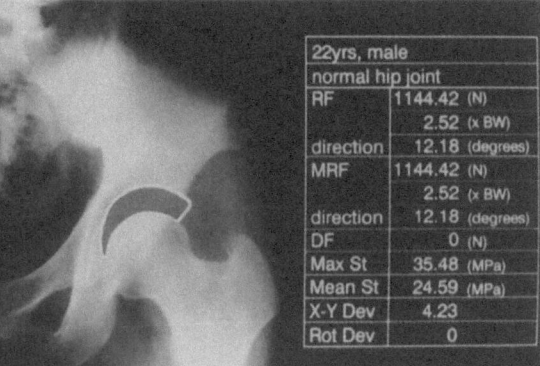

22yrs, male		
normal hip joint		
RF	1144.42	(N)
	2.52	(x BW)
direction	12.18	(degrees)
MRF	1144.42	(N)
	2.52	(x BW)
direction	12.18	(degrees)
DF	0	(N)
Max St	35.48	(MPa)
Mean St	24.59	(MPa)
X-Y Dev	4.23	
Rot Dev	0	

b

a

Normal hip joint

Fig. 2. a Radiogram of a normal hip joint. **b** Simulation results for a normal subject. *Painted area* indicates the compressive stress on the joint. *RF*, Resultant force; *MRF*, modified resultant force; *DF*, dislocation force; *Max St*, maximum stress on the joint; *Mean St*, mean stress on the joint; *X–Y Dev*, X–Y deviation of femoral head; *Rot Dev*, rotational deviation of the pelvis

Normal Hip Joint (Fig. 2a)

The subject was a 22-year-old male.

Early Stage of Coxarthrosis (Fig. 3a)

Chiari pelvic osteotomy [13,14] was carried out when this female patient was 45 years old. Preoperative and postoperative (4 years and 1 month) A-P radiograms were analyzed. The Japanese Orthopedic Association (JOA) [15] clinical evaluation scores for coxarthrosis for these two radiograms were 76 and 100 points, respectively (Table 1).

Advanced Stage of Coxarthrosis (Fig. 4a)

Chiari pelvic osteotomy was carried out when this female patient was 31 years old. Preoperative and postoperative (3 years and 2 months) A-P radiograms were analyzed. The respective JOA scores were 73 and 100 points.

Late Stage of Coxarthrosis (Fig. 5a)

Chiari pelvic osteotomy with intertrochanteric valgus osteotomy was carried out when this female patient was 30 years old. The preoperative and postoperative (4 years and 2 months) radiograms were analyzed. The respective JOA scores were 65 and 100 points. The resultant force and the modified

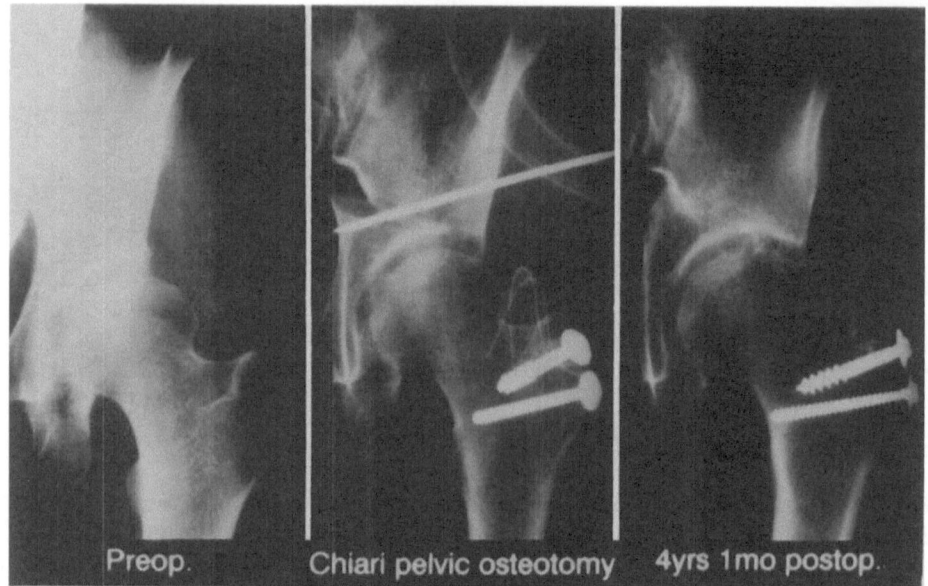

a   Preop.   Chiari pelvic osteotomy   4yrs 1mo postop.

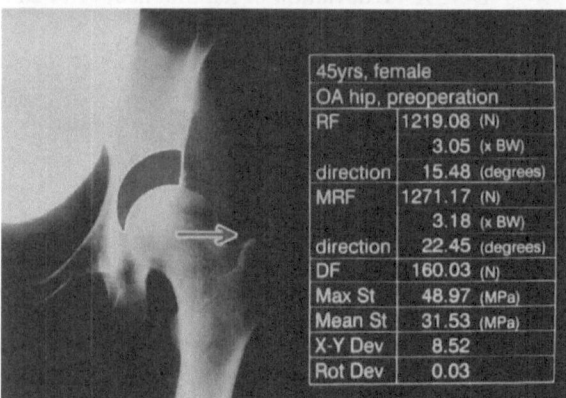

45yrs, female	
OA hip, preoperation	
RF	1219.08 (N)
	3.05 (x BW)
direction	15.48 (degrees)
MRF	1271.17 (N)
	3.18 (x BW)
direction	22.45 (degrees)
DF	160.03 (N)
Max St	48.97 (MPa)
Mean St	31.53 (MPa)
X-Y Dev	8.52
Rot Dev	0.03

b

49yrs, female	
4yrs1mo postoperation	
RF	1107.93 (N)
	2.77 (x BW)
direction	11.05 (degrees)
MRF	1107.93 (N)
	2.77 (x BW)
direction	11.05 (degrees)
DF	0 (N)
Max St	25.33 (MPa)
Mean St	18.83 (MPa)
X-Y Dev	3.02
Rot Dev	0

c

Fig. 3. a These radiograms show an early stage of cox-arthrosis. **b, c** Simulation results of early stage of cox-arthrosis; **b** preoperative, *arrow* on the femoral head indicates the magnitude of the dislocation force; **c** 4 years and 1 month postoperative

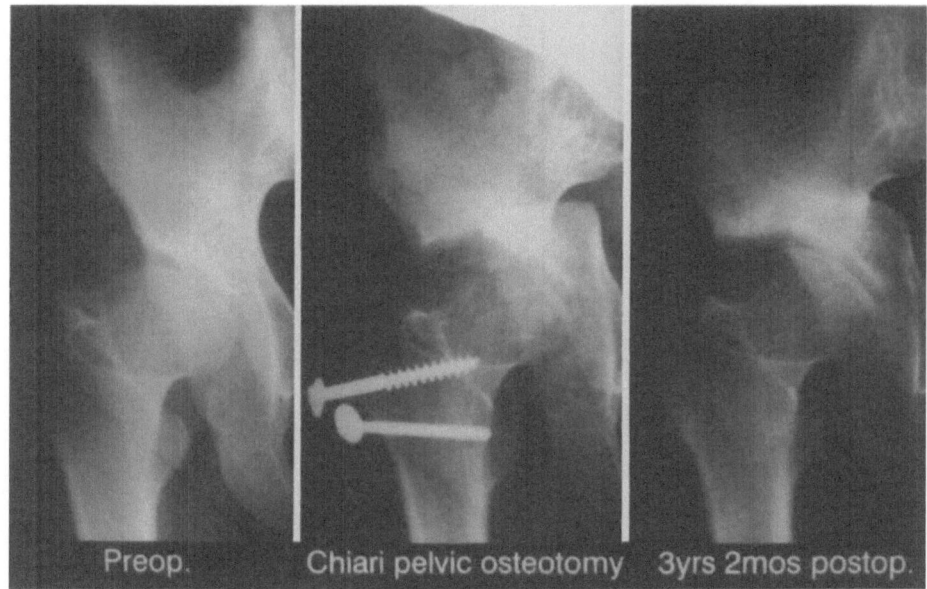

a

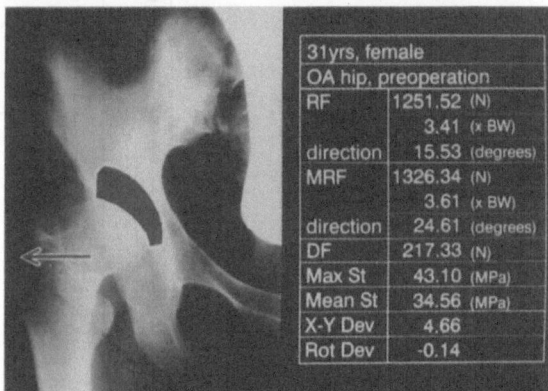

31yrs, female	
OA hip, preoperation	
RF	1251.52 (N)
	3.41 (x BW)
direction	15.53 (degrees)
MRF	1326.34 (N)
	3.61 (x BW)
direction	24.61 (degrees)
DF	217.33 (N)
Max St	43.10 (MPa)
Mean St	34.56 (MPa)
X-Y Dev	4.66
Rot Dev	-0.14

b

34yrs, female	
3yrs2mos postoperation	
RF	1141.62 (N)
	3.11 (x BW)
direction	11.08 (degrees)
MRF	1152.41 (N)
	3.14 (x BW)
direction	13.55 (degrees)
DF	50.58 (N)
Max St	25.08 (MPa)
Mean St	18.38 (MPa)
X-Y Dev	2.81
Rot Dev	-0.01

c

Fig. 4. a These radiograms show an advanced stage of coxarthrosis. **b, c** Simulation results of advanced stage of coxarthrosis; **b** preoperative; **c** 3 years and 2 months postoperative

Table 1. Clinical evaluation of coxarthrosis (*JOA* score) [15]

| Pain | Point | ROM | | | | Gait ability | Point | ADL | Easy | Difficult | Impossible |
		Flexion	Point	Abduction	Point						
None After a long walk, local discomfort and fatigue or dull discomfort, but no pain	40	Over 90°	12	Over 30°	8	Normal or almost normal Very slight limping after a long walk is classified as 'normal'	20	Sitting on chair	2	1	0
								Japanese sitting (Sitting down on the heels kneeling)	2	1	0
Mild occasional pain a) Severe pain occasionally, but once or twice a year. b) Local dull-discomfort on walking	30	60°–89°	9	20°–29°	6	Slight limping Possible to walk 30 min or 2 km a) Without crutch. b) No disturbance in daily outdoor activities	15	Bowing in the Japanese sitting position	2	1	0
								Standing up from the Japanese sitting position	2	1	0

Pain		Range of motion				Ability to walk		Activities of daily living	2	1	0
Moderate Pain on walking, but relieved by short rest	20	30°–59°	6	10°–19°	4	Severe limping Possible to walk 10–15 min or 500 m Easier walking with a crutch Taking frequent rests, possible to walk without crutch	10	Squatting Putting on and taking off socks	2 2	1 1	0 0
Severe Extreme pain on walking, relieved by rest. Spontaneous pain—occasional	10	Under 29°	3	Under 9°	2	Indoor activity is possible, but outdoor is difficult Two crutches are required for outdoor activities	5	Clipping toe nails Standing on affected leg	2 2	1 1	0 0
Extreme Spontaneous pain—continuous	0	Non-functional position, or functional position with ankylosis or poor ROM (less than 10°)	0			Difficult to walk	0	Ascending stairs Descending stairs	2 2	1 1	0 0

JOA, Japanese Orthopedic Association; ROM, range of motion

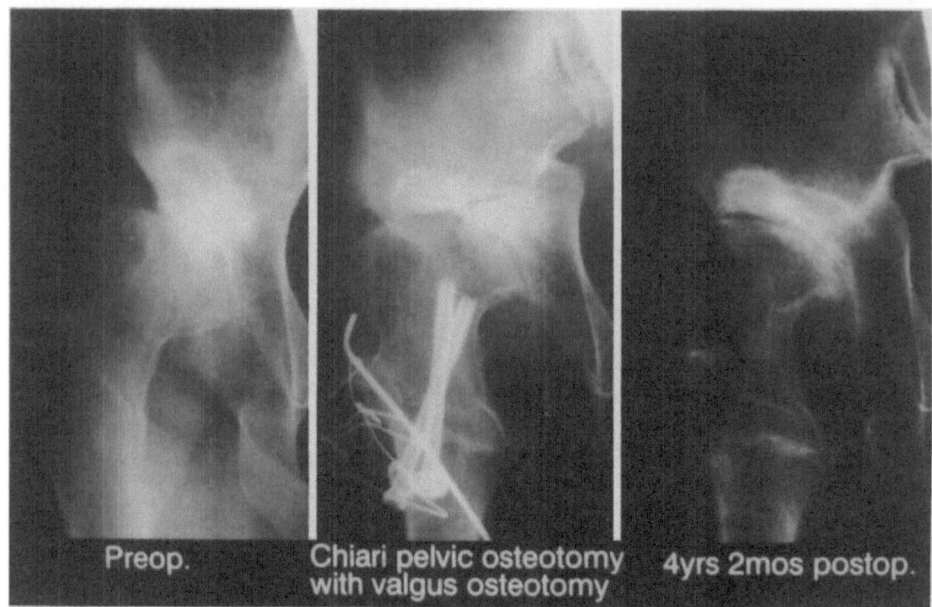

Preop.   Chiari pelvic osteotomy   4yrs 2mos postop.
                  with valgus osteotomy

a

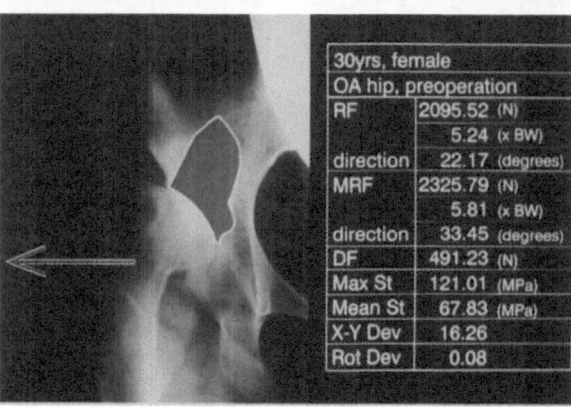

30yrs, female		
OA hip, preoperation		
RF	2095.52	(N)
	5.24	(x BW)
direction	22.17	(degrees)
MRF	2325.79	(N)
	5.81	(x BW)
direction	33.45	(degrees)
DF	491.23	(N)
Max St	121.01	(MPa)
Mean St	67.83	(MPa)
X-Y Dev	16.26	
Rot Dev	0.08	

b

34yrs, female		
4yrs2mos postoperation		
RF	2046.15	(N)
	5.11	(x BW)
direction	14.75	(degrees)
MRF	2128.11	(N)
	5.32	(x BW)
direction	21.59	(degrees)
DF	262.33	(N)
Max St	67.33	(MPa)
Mean St	51.66	(MPa)
X-Y Dev	11.76	
Rot Dev	0.07	

c

resultant force of the femoral head, mean and maximum compressive stress on the joint surface, dislocation force, and X–Y and rotational deviations were calculated. The distribution of the compression stress on the joint surface was then expressed.

Results

Normal Hip Joint (Fig. 2b)

The resultant force was 1144.42 N (2.52 × body weight), and its direction was 12.18° from the perpendicular. The modified resultant force was the same as the resultant force, because the dislocation force was 0 N. The maximum and mean compressive stress on the joint was 35.48 and 24.59 MPa, respectively. The X–Y and rotational deviation of pelvis was 4.23 and 0 respectively. The shape of compressive stress was similar to the "sourcil" ("eyebrow") as described by Pauwels [1], although the absolute value of the compressive stress might be larger than that obtained from a 3-D analysis.

Early Stage of Coxarthrosis

Preoperative (Fig. 3b)

The resultant force was 1219.08 N (3.05 × body weight), and its direction was 15.48°. The modified resultant force was 1271.17 N (3.18 × body weight), and its direction was 22.45°. The dislocation force was 160.03 N. The maximum and mean compressive stress on the joint was 48.97 and 31.58 MPa, respectively and the X–Y and rotational deviation of the pelvis was 8.52 and 0.03 respectively. All the results were worse than those for a normal hip joint.

Four Years and 1 Month Postoperative (Fig. 3c)

The resultant force was 1107.93 N (2.77 × body weight), and its direction was 11.05°. The modified resultant force was same as the resultant force, since the dislocation force was 0 N. The maximum and mean compressive stress on the joint was 25.33 and 18.83 MPa, respectively. The X–Y and rotational deviation of the pelvis was 3.02 and 0, respectively. All the results were improved compared to preoperative simulation results, although a small dislocation force

---

Fig. 5. a These radiograms show a late stage of coxarthrosis. **b, c** Simulation results of a late stage of coxarthrosis; **b** preoperative, a severe stress was seen on the joint, especially where the joint space was narrowed; **c** 4 years and 2 months postoperative. All the results were altered, although a dislocation force still remained. A relatively severe stress was observed on the medial side of the joint, and a sclerotic portion of the pelvis was seen in the same area (**a**). *OA*, Osteoarthritic

remained. These results showed the effect of medial displacement [13] and enlargement of the joint contact surface.

Advanced Stage of Coxarthrosis

Preoperative (Fig. 4b)

The resultant force was 1251.52 N (3.41 × body weight) and its direction was 15.53°. The modified resultant force was 1326.34 N (3.61 × body weight), and its direction was 24.61°. The dislocation force was 217.33 N. The maximum and mean compressive stress was 43.10 and 34.56 MPa, respectively. The X–Y and rotational deviation was 4.66 and, −0.14, respectively. These simulated results were worse than those for a normal hip or for early stage coxarthrosis.

Three Years and 2 Months Postoperative (Fig. 4c)

The resultant force was 1141.62 N (3.11 × body weight), and its direction was 11.08°. The modified resultant force was 1152.41 N, and its direction was 13.55°. The dislocation force was 50.58 N. The maximum and mean compressive stress was 25.08 and 18.38 MPa, respectively. The X–Y deviation and rotational deviation of the pelvis was 2.81 and −0.01, respectively. All the results were improved compared to the preoperative simulation results.

Late Stage of Coxarthrosis

Preoperative (Fig. 5b)

The resultant force was 2095.52 N (5.24 × body weight) and its direction was 22.17°. The modified resultant force was 2325.79 N (5.81 × body weight), and its direction was 33.45°. The dislocation force was 491.23 N. The maximum and mean compressive stress was 121.01 and 67.83 MPa, respectively, and the X–Y and rotational deviation was 16.26 and 0.08, respectively. The shape of the compressive stress was similar to the osteosclerotic and cystic region of the bone segment. All results were worse than for early or advanced stage.

Four Years and 2 Months Postoperative (Fig. 5c)

The resultant force was 2046.15 N (5.11 × body weight), and its direction was 14.75°. The modified resultant force was 2128.11 N, and its direction was 21.59°. The dislocation force was 262.33 N. The maximum and mean compressive stress on the joint was 67.33 and 51.66 MPa, respectively. The X–Y deviation and rotational deviation of the pelvis was 11.76 and 0.07, respectively. All the results were improved compared to the preoperative simulation results. The high stress area was ameliorated postoperatively, although all the results were still worse than those for a normal hip joint.

Discussion

Plane radiograms are still most important tool for clinicians, and surgeons usually use only these radiograms to plan hip operations such as osteotomy or total hip arthroplasty (THA). Three-dimensional CT has been developed and used to visualize the structures of the hip joint and its operation. Three-dimensional FEM analysis [3] and 3-D joint contact analysis using RBSM theory [16] have also been developed; although 3-D CT has been utilized for these programs, it is difficult to use this procedure clinically for all patients because of the expense and the exposure to radioactivity. We must also take care that these CT data are not taken under standing or loaded conditions, as the subjects are in the supine position during scanning.

In our 2-D RBSM for the hip joint, a dislocation resistance spring (tension resistance, compression break spring) was placed perpendicularly between the center of the femoral head and the pelvis and the dislocation force in each hip joint was calculated with this spring. The hip joint was assumed to be stable in each radiogram, and the spring constant of the dislocation resistance spring was estimated as 5000 N/mm; the error ratio in this spring constant was below 5%. The dislocation spring was used in place of a joint capsule or other musculo-ligamentous structure. To obtain an accurate anatomical simulation, these factors should be placed in their correct anatomical place. However, it is difficult to define the proper spring constant for each ligament for each subject, and the results become too complicated to be useful clinically if all factors are modeled. This complicated procedure might reduce the advantages of RBSM, which is simple. In our model, the dislocation resistance force was estimated to be exerted on the center of rotation of the femoral head as an effect of the capsule and musculo-ligamentous structures around the hip joint.

Experimental studies have been performed [17] to define the proper spring constant of the articular cartilage in the RBSM. Our purpose was to analyze each patient clinically, using only radiography, and we adopted the mathematical formula of An [12]. We believed that this method reflected the conditions of each subject. However, it was very difficult to define the material properties (elastic modulus and Poisson's ratio) for each subject, and so we estimated them from the data of normal subjects [9], only the thickness of the cartilage being reflected by a radiogram. This method could provide a more realistic stress pattern compared to that obtained using a uniform spring constant. However, to perform a more accurate simulation, the actual material properties should be determined in each case.

We also performed a comparative study of the clinical findings and the simulation results [18]. Fifty patients with unilateral coxarthroses (3 males and 47 females) were reviewed; the radiographic stages being advanced and late. The mean follow-up period was 4 years and 2 months (range, 2 years to 8 years and 4 months), and the mean operation age was 43 years (range, 16–63 years). They were divided into three groups by JOA score at the final follow-up: group

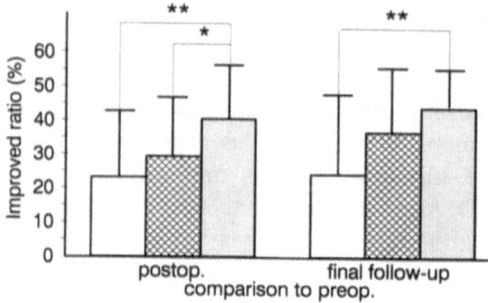

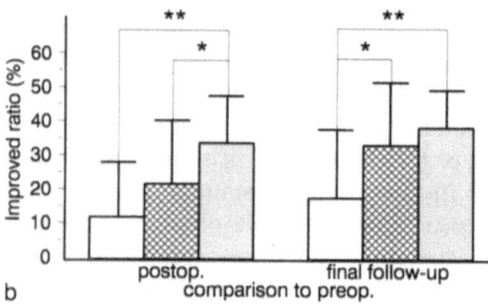

b

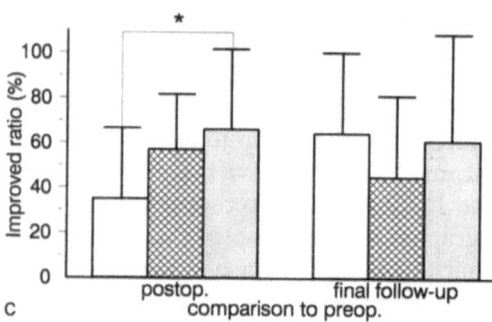

c

Fig. 6a–c. Each patient was analyzed preoperatively, immediately postoperatively, and at final follow-up. Postoperative simulation results and those of the final follow-up were divided by preoperative results for each subject and standardized. Statistical analyses were carried out using one-way analysis of variance (ANOVA). a Maximum compressive stress; b mean compressive stress; c dislocation force. Japanese Orthopedic Association (*JOA*) score [15], final follow-up—*white columns* indicate <79; *cross-hatched columns*, <80–89; *dotted columns*, >90. *$P <$ 0.05; **$P < 0.01$ (ANOVA)

A, 90–100 points; group B, 80–89 points; group C, less than 79 points. Analyses using RBSM were carried out using the preoperative, postoperative, and final follow-up A-P radiograms. The spring constant for the articular cartilage was estimated to be uniform, as Kd = 10, Ks = 0 (N/mm per unit), equivalent to the average spring constant for normal subjects. The results for each group were analyzed and compared statistically for each stage using one-way analysis of variance (ANOVA). The results showed that the simulated results and clinical findings were correlated (Fig. 6a–c). However, the radiogram findings, such as those of the Sharp angle, the center edge (CE) angle, and the acetabular head index (AHI), which are usually used, had no statistical correlation to the clinical findings. This study showed the simulation results and the clinical symptoms to be correlated, and demonstrated the efficacy of the RBSM for clinical evaluation.

Using the RBSM, useful biomechanical information can be obtained for evaluating patients clinically from their A-P radiograms. Another advantage of this method is that previously stored radiographic data can be analyzed biomechanically and utilized. RBSM is clinically useful not only for evaluating patients but also for planning hip osteotomies.

References

1. Pauwels F (1963) Principle et resultats d'Une Therapeutique Etiologique De la Coxarthrosie. In: Proceedings of IX'eme Congress de la Societe Internationale de Chirurgie Orthopedique et de Traumatrogie, Vienna, 1–7 Sept 1963, vol II
2. Brown TD, DiGioia AM (1984) A contact-coupled element analysis of the natural adult hip. J Biomech 17:437–448
3. Huiskes R, Weinans H, Riebergen B (1992) The relationship between stress shielding and bone resorption around total hip stems and effects of flexible materials. Clin Orthop 274:125–134
4. Rappaport DJ, Carter DR, Schurman DJ (1985) Contact finite element stress analysis of the joint. J Orthop Res 3:435–446
5. Horii E, Garcia-Elias M, An KN, Bishop AT, Cooney WP, Lincheid RL, Chao EYS (1990) Effect of force transmission across the carpus of procedures designed to treat Kienböck's disease. An analytic study. J Hand Surg 15-A:393–400 `
6. Kawai T (1977) A new element in discrete analysis of plane strain problems (in Japanese). Seisan Kenkyuu 29:204–207
7. Shiba N (1991) Biomechanics of Chiari pelvic osteotomy (in Japanese). J Jpn Orthop Assoc 65:337–348
8. Shiba N (1992) Simplified two-dimensional interface stress analysis (rigid body spring model) for a femoral stem: A comparative study to FEM. J Jpn Orthop Assoc 66:s1578
9. Ninomiya S (1975) The relationship between the position of the articular hip joint and the resultant force acting on the femoral head (in Japanese). J Jpn Orthop Assoc 50:15–20
10. Nordin M, Frankel VH (1989) Basic biomechanics of the musculo-skeletal system, 2nd edn. Lea and Feibiger, Philadelphia, pp 135–151
11. Bombelli R (1982) Osteoarthritis of the hip, 2nd edn. Springer Berlin Heidelberg New York, pp 13–38
12. An KN, Himeno S, Tsumura H, Kawai T, Chao EYS (1990) Pressure distribution on articular surfaces. Application to joint stability evaluation. J Biomech 23:1013–1020
13. Chiari K (1974) Medial displacement osteotomy of the pelvis. Clin Orthop., 98:55–71
14. Inoue A, Higuchi F, Shiba N (1990) Chiari pelvic osteotomy coxarthrosis in adults. J Orthop Surg Technique 5:105–111
15. Imura S, Matsunaga T (1990) Handbook of the hip joint—diagnosis and treatment (in Japanese). Nankodo, Tokyo, p 161

16. Blankevoort JH, Kuiper R, Huiskes R, Grootenboer JH (1991) Articular contact in a three-dimensional model of the knee. J Biomech 24:1019–1031
17. Ide T, Yamamoto Y, Tachiki S, Akamatsu N, Chao EYS (1990) Determination of spring constants for the rigid body spring model (in Japanese). Jpn J Orthop Biomech 12:125–131
18. Yamashita H, Shiba N, Higuchi F, Inoue A (1992) Clinical application of computer simulation using 2-D rigid body spring model for preoperative planning of hip osteotomies. In: Proceedings of the 19th Annual Meeting of the Japanese Hip Joint Assoc, Fukui, Nov 20 1992

18

Simulation of Rotational Acetabular Osteotomy: Comparison of Graphic and Organ Models

Katsuhiko Ebihara[1], Hiroyuki Sindo, Hirohiko Azuma[2], and Tokuhide Doi[3]

Summary. A three-dimensional graphic model and an organ model were produced for the simulation of rotational acetabular osteotomy. A solid model, which is easily made, was used as the three-dimensional graphic model, and a rapid prototype model represented the organ model. When these two model types were compared strictly in terms of practicality in simulating operations, it appeared that the rapid prototype model, which requires a shorter fabrication period, has virtually no limitations in regard to location of use, and enables the use of surgical equipment for simulations, was superior.

Key words: Surgical simulation—Rotational acetabular osteotomy—Soild model—Rapid prototype model—Lithography—Three-dimensional CT—Finite element analysis

Introduction

In most clinical cases we encounter in our daily practice, X-ray photographs may provide sufficient information for the planning of surgical operations, that would bring about successful results in these operations. However, although their number is limited, there are some cases for which preoperational planning is difficult, even for a very experienced surgeon. Our research group has been conducting a series of studies in which we have used three-dimensional computerized tomographic (CT) images [1,2] to produce a more complete coverage of the acetabulum. We have used this method for rotational acetabular osteotomy (RAO) in osteoarthritis due to acetabular dysplasia (Fig. 1). We describe

[1] Department of Orthopedic Surgery, Institute of Clinical Medicine, University of Tsukuba, 1-1-1 Tennoudai, Tsukuba, Ibaraki, 305 Japan
[2] Department of Orthopedic Surgery, Saitama Medical School, 38 Morohongo, Moroyama-machi, Irumagun, Saitama, 350-04 Japan
[3] Tokyo Metropolitan Prosthetic and Orthotic Research Institute, 3-17-3 Toyama-cho, Shinjuku-ku, Tokyo, 162 Japan

187

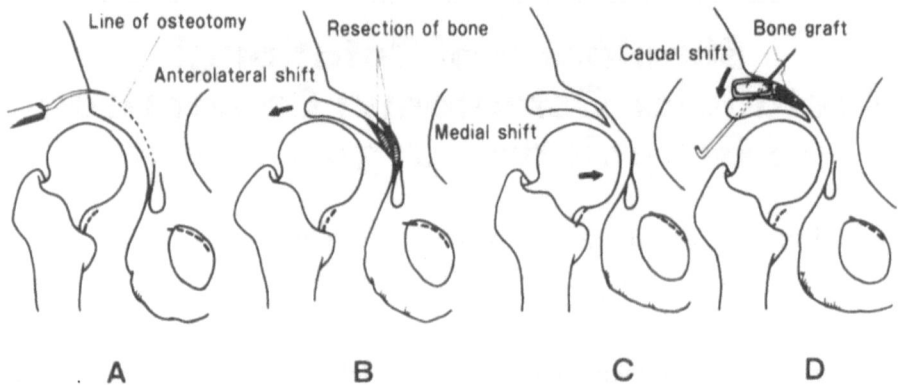

Fig. 1. Operative procedure for rotational acetabular osteotomy

here the fabrication of a three-dimensional graphic model (solid model) and an organ model (rapid prototype model) and their application to simulated RAO operation.

Methods

The two types of model hip joints produced by the methods outlined below, were used in simulated RAO operations. A Tagawa type chisel, with a bent angle of 55 mm, was used in the simulations.

Solid Model (Fig. 2)

CT images were taken at 4-mm intervals in the hip joint, and the contours of each slice were then input into the computer using a scanner. The data were transfered to an engineering work station, Apollo 590 Turbo (Hewlett Packard; Palo Alto, Calif.), and laid out on the Z-axis to create a wire frame model. The distance between each frame was covered with an automatic mesh system, and it was then smoothed with splines (surface model). Outside and inside the plane were further defined on the coordinate in order to produce three-dimensional expression, or, in other words, a solid model. To produce of this model, we used a solid modelar, Geomod I-DEAS, made by SDRC (Cincinnati, Ohio) [3].

Rapid Prototype Model (Fig. 3)

The three-dimensional data obtained from the CT images were converted to STL file format, and then input into Soliform, a rapid production system manufactured by Teijin Seiki (Tokyo, Japan). Different from conventional milling machines, which grind materials to fabricate a model, the Soliform

Fig. 2. Solid model of hip joint

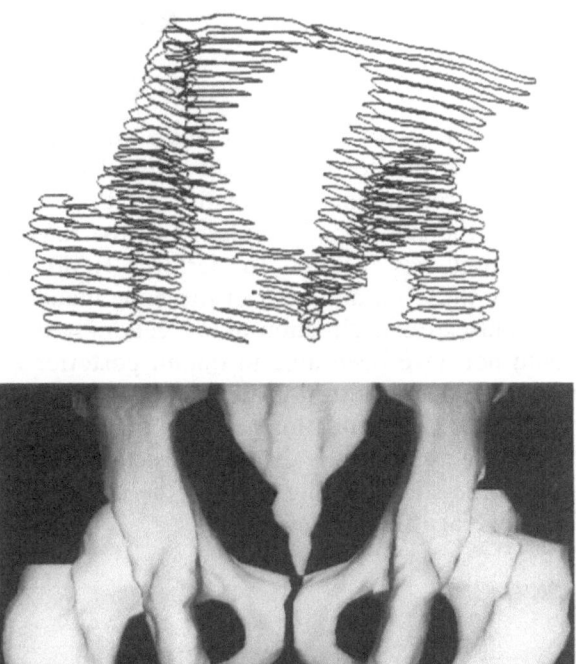

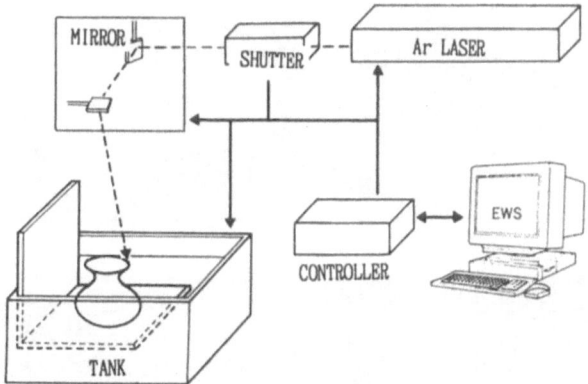

Fig. 3. Rapid prototype System. *EWS*, Engineering work station

system produces a model by forming plastic material into the section shapes and then piling them up. In this system, the input data are divided into so-called slice data by a given thickness, and plastic material in the tank is solidified separately into the shapes of each slice. When the first layer has solidified, the elevator brings down material equivalent to the thickness of the next layer. The second layer of the plastic is placed over the first layer while it solidifies. In our test operation, we selected a slice thickness of 0.3 mm [4].

Results

Solid Model

The patient was a 16-year-old male with subluxation of the right hip joint. Observation of the solid model showed us that the vertical diameter of the patient's acetabulum was small, and that there was significant incongruity between the shallow acetabulum and the femoral head (Fig. 4). The Tagawa type chisel was inserted to perform RAO simulation. The short distance between the front and back of the bone made it impractical to move it toward the front-external direction as we generally do for most patients, because we would not have been able to obtain posterior coverage and would have been concerned about posterior dislocation.

However, the simulation revealed that lowering the rotational bony fragment by as little as 30° to the external direction would enlarge the joint contact area to 1.4 times that existing prior to the operation, and this would allow reasonable coverage (Fig. 5).

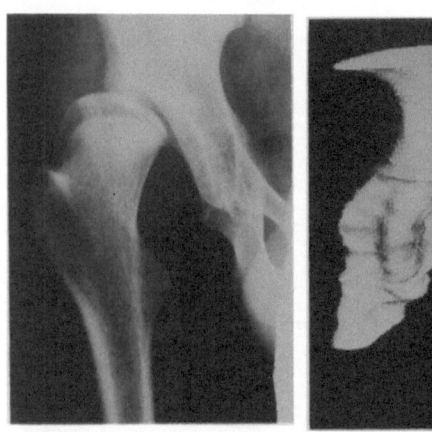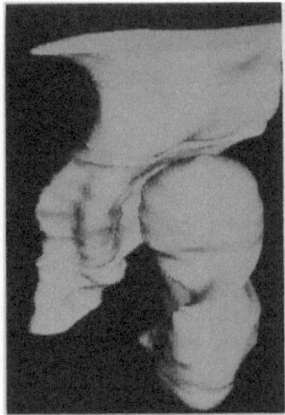

Fig. 4. Soild model of hip joint (posterior view)

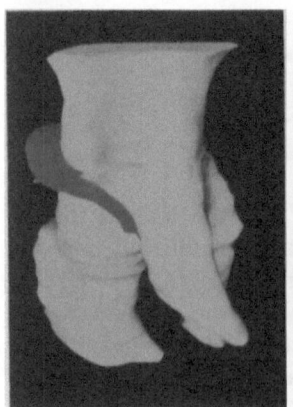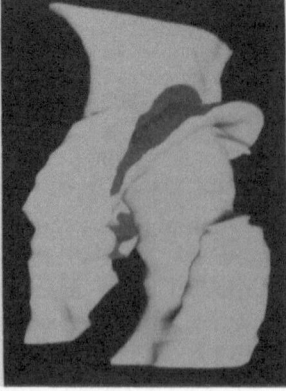

Fig. 5. Simulation of rotational acetabular osteotomy (*RAO*)

Rapid Prototype Model

The patient was a 35-year-old female with secondary osteoarthritis due to acetabular dysplasia. A rapid prototype model of the left hip joint was produced, and a RAO simulation was conducted (Fig. 6). Although the model was quite difficult to cut with the Tagawa type chisel, the use of a drill and bone saw enabled us to insert the chisel in the intended directions. the material was also strong enough to resist very strong pressure from the chisel. The conventional practice of shifting the bone toward the front external direction produced significant coverage with the patient's acetabulum (Fig. 7).

Discussion

Three-dimensional CT imaging [5,6], a recently-developed technique, has limitations, although it is a simple and effective way to demonstrate graphics. Because such models are expressed as a mass of cubes, called VOXEL, in order to reduce the amount of data which the machine has to handle, plane bone cutting of, for example, a hip joint, such as that employed in the Chiari method [7], which does not affect the cubes, can be simulated, whereas simulation of cutting bone over a curved, surface, such as occurs in RAO, is practically impossible (Fig. 8).

This limitation prompted our research group is interest in solid models, which offer more complete three-dimensional graphic expression and do not have restrictions in processing data as with VOXEL. With solid models originally used for the developement of industrial products, the designing and, processing of a substance of any shape can be precisely simulated, and, further, dynamic analysis by the three-dimensional finite element method is also possible. For these reasons we initially considered that the solid model would be an effective tool with which we could obtain accurate simulation of a planned surgical operation and dynamic estimates of the results. The rapid prototype model is another method of simulating operations. This method of fabricating a model by solidifying plastic materials with a laser, the rapid prototyping system, has been attracting attention among medical personnel since 3D Systems (Valencia, Calif.) produced a machine for this purpose, the SLA-1, in 1987. Different from conventional methods of grinding various kinds of materials with grinding tools, the rapid prototyping system produces a model by shaping plastic forms of slices (of the model to be produced) and piling them up. This methods offers a few advantages to us: Firstly, this system enables us to obtain models of very complex internal shapes, which models cannot be ground with any grinding tool. For instance, it is beyond the capacity of a milling machine to grind the shape of bone cysts and the medullary canals which exist inside the bone or to grind narrow joint spaces. The rapid prototyping system does not require any extra work on our part for creating models of any complex shape, since the only thing it has to do is to form plastic into the shapes of the slices demonstrated by CT graphics and pile them up. The second benefit of the

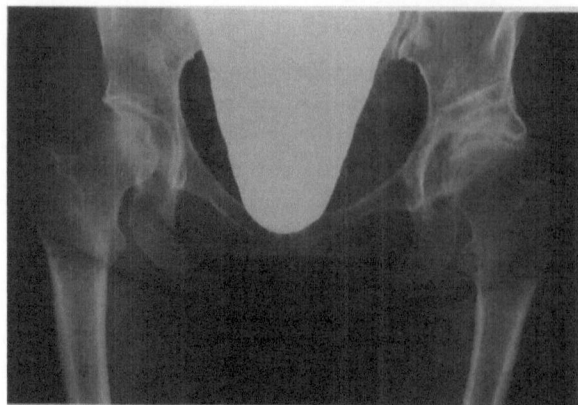

Fig. 6. RAO of right hip joint. (Life hip joint was already operated)

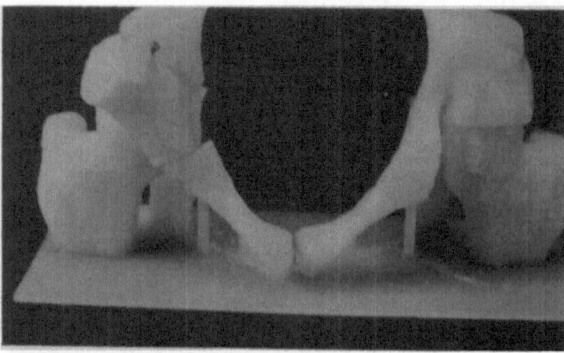

Fig. 7. Rapid prototype model (simulation of RAO)

rapid prototyping system is that it does not require replacement of jigs, nor does it produce waste due to its non-contact technology. Nor does it generate noise. These features make it possible to run the system unmanned, even at night.

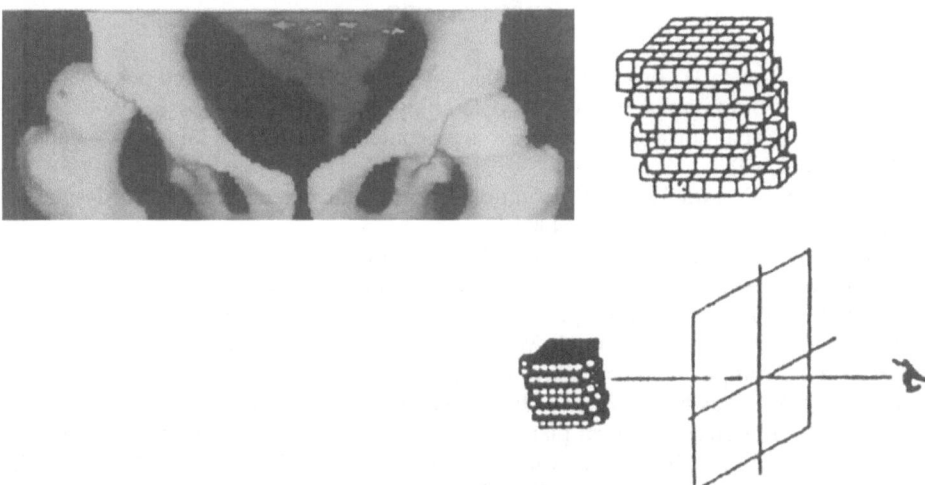

Fig. 8. Three-dimensional computerized tomographic (*CT*) imaging

With this system, an average model of a hip joint requires 7–8 h for production. Therefore, if the hip joint data are produced from the CT graphic information in the morning, the system can begin operating in the evening of the same day, and a completed model can be obtained the next morning. Lastly, we compared the advantage of the rapid prototype model over the solid model. To graphically simulate any given operation, we must create graphic models which provide perfect three-dimensional duplication of the actual targets of the operation. Use of a solid model, to our knowledge, is the best method of achieving this goal. However, since the operation of solid models requires the availability of a work station and highly sophisticated computer knowledge, it is not a method that is suitable for all surgeons, nor can it be used in all locations. The rapid prototype model, in contrast, can be fabricated in a relatively short time, it can be freely moved around, and, using such a model, it is even possible to conduct a simulated operation with an actual surgical knife. Thus, the surgeon can show the model to the patient to explain plans for treatment, and, provided the model is properly sterilized, it may even be used in the operation room for reference during the operation.

We therefore believe that, as far as simulated operations are concerned, the rapid prototype model is easier to use and more practical then the solid model.

Conclusion

Simulated RAO operations were conducted using solid and rapid prototype models. We concluded that the rapid prototype system provided a more practical model for simulation than the solid model.

References

1. Azuma H, Taneda H, Igarashi H, Fujioka M (1990) Preoperative and postoperative assessment of rotational acetabular osteotomy for dysplastic hips in children by three-dimensional surface reconstruction computed tomography imaging. J Pediatr Orthop 10:33–38
2. Ninomiya S, Tagawa H (1984) Rotational acetabular osteotomy for the dysplastic hip. J Bone Joint Surg [Am] 66:430–436
3. Ebihara K, Sindo H, Azuma H (1992) Simulation and finite element analysis of rotational acetabular osteotomy with solid modelar. Trans ORS 17:483
4. Ashely S (1991) Rapid prototyping system. Mech Eng 113:34–43
5. Bulk DL Jr, Mears DC, Kennedy WH, Cooperstein LA, Herbert DL (1985) Three-dimensional computed tomography of acetabular fractures. Radiology 155:183–186
6. Vannier MW, Marsh JL, Warren JO (1984) Three-dimensional CT reconstruction images for craniofacial surgical planning and evaluation. Radiology 150:179–184
7. Butter-Manuel PA, Guy RL, Reynolds DA (1991) Three-dimensional CT imaging in hip dysplasia. J Bone Joint Surg [Br] 83:686–687

19

Simulation for Hip Osteotomy Using Three-Dimensional Computerized Tomography

Ryoichi Izumida and Yoichi Ishinada[1]

Summary. Three-dimensional computerized tomography (3-DCT) has been developed to overcome the inherent difficulties involved in two-dimensional assessments of images in orthopedic surgery, and surgical planning and simulations based on 3-DCT have already begun. Since 1988, we have been attempting surgical simulations using the Nagoya University Craniofacial Surgical Planning System (NUCSS) [1]. However, this is a program developed for a mainframe computer. To popularize surgical simulation, it seemed necessary to develop a program for a simulation of actual osteotomy which could be run on the smaller computers commonly used. Under these circumstances, the joint development of SurgiPlan was undertaken with the Teijin company (Yokohama, Japan) [2,3]. SurgiPlan (Fig. 1), developed as an aid to orthopedic surgical planning, is run on a UNIX Graphics 3D work station (SPARC station-2; Japan Sun Microsystems, Tokyo). Its functions are roughly divided into two categories, 2- and 3-D data processing. Two-dimensional data processing includes that undertaken preliminary to 3-D simulation, such as analysis of original CT images and separation of individual bones. Osteotomy simulation, which is the main purpose of this program, is included in the 3-D data processing function. The evaluative function includes the measurement of distances and angles and the display of contact area of joints. To run this simulation program, consecutive CT data are fed into the computer. These CT data are then processed into two levels, bone and others, to extract bone data as well as to separate data for individual bones (e.g., pelvis and femur). Then a 3-D image is constructed, and surgical simulation of osteotomy is carried out. Simulation is repeated until a satisfactory result is obtained. It is finished with recording of the distance and angle over which the bone fragments were moved, and is utilized in actual operations. In orthopedic surgery, 3-DCT is indicated in disorders of the hip joint or spine, which disorders are frequently missed in 2-D evaluation as the structures overlap one another in lateral radiographs.

---

[1] Orthopedic Surgery Department, National Saitama Hospital, Suwa 2-1 Wako, Saitama, 351-01 Japan

To date, surgical planning and post-operative evaluation of these regions have been made on the basis of 2-D imaging. It is expected that 3-D surgical planning and simulation based on 3-DCT imaging will contribute much to clinical medicine.

Introduction

In orthopedic surgery, the classic diagnostic procedure involves two-dimensional (2-D) assessment of the image obtained by X-ray. Even after the introduction of computerized tomography (CT), 2-D imaging remained the mainstay of orthopedic diagnosis. However, the bones and joints to be treated are three-dimensional (3-D) structures, and even an experienced orthopedist has difficulty constructing a mental image of such a structure if it is severely deformed. Thus, 3-D CT (3-DCT) was developed, and it has already been utilized in surgical planning and simulations.

In 1988, we began research into the application of the Nagoya University Craniofacial Surgical Planning System (NUCSS), developed by the Computer Science Course in the Engineering Department of Nagoya University, to orthopedic surgery. To obtain flexibility, we also performed the research for and developed (jointly with the Teijin company) the SurgiPlan system, which is compatible with workstations.

SurgiPlan

SurgiPlan (Fig. 1), developed as an aid to orthopedic surgical planning, is run on a UNIX Graphics 3D work station (SPARC station-2; Sun Co.), and its function is roughly divided into two categories: 2- and 3-D data processing.

Fig. 1. SurgiPlan

Two-dimensional data processing includes the CT image analyzing function for observing X-ray CT images and the bone separating function as pre-processing for 3-D simulation. Bone region data are extracted and separated individually by the bone separating function.

For a hip joint operation, it is necessary to recognize the two bones at both sides of the facies articularis as two different regions. The fact that slice data values are classified by different colors shows that separation is smoothly performed. However, if bones are close to each other due to arthrosis and separation is not thereby automatically performed, it is necessary to draw a separation line with the computer mouse.

The 3-D data processing is divided into display control and simulation functions. The display control function includes such auxiliary functions for performing simulation as positioning of image coordinates and line of sight, grid display, and zoom designation for enlargement or contraction. The simulation function includes such functions as osteotomy, movement of bone pieces, measurement, and attribute designation. This simulation function is the main purpose of this system.

Osteotomy includes four methods of bone cutting, i.e., vertical, planar, closed line, and sphere. The vertical method, the first method, is used for cutting a bone on a plane vertical to a screen, and is not designated in depth, although a bone cutting line can easily be imaged. Therefore, this method is suitable for simulating osteotomy of the femur.

The planar method, the second method, is used for cutting a bone on a plane where three points are designated, and is effective for such methods as the Salter and Chiari operations, in which a bone is cut on a plane where the height of the outside of the pelvis is different from that of the inside. The closed line method, the third method, is used for cutting a bone by designating a dot sequence on the surface of a bone, and is actually invented by assuming Pemberton operation. The sphere method, the fourth method, is used for cutting a bone by utilizing a spherical surface with a designated radius and central point, and is a simulation of rotational acetabular ostectomy.

There are two methods of moving bone pieces, i.e., translate and rotate.

Translate is the function for moving a bone piece in any direction in parallel, in which the bone piece is moved by designating the start and end points of movement. Rotate is the function for rotating a bone piece around a rotation axis in any direction; this is executed by inputting a rotation axis and rotation angle.

This system has a distance and angle measuring function and a contact surface display function which serve as evaluation and measurement functions.

The contact surface display function designates a bone, measuring the distance between that bone and others in 3-D space, and color-mapping the surface of the designated bone with contour lines. The distance between the femoral head and acetabulum increases in the sequence of dark blue, green, sky blue, red, and yellow, indicated by the letters a, b, c, d, and e, respectively, in Fig. 2. In this case, dark blue represents the range 0–0.8 mm,

preop.

ostop.

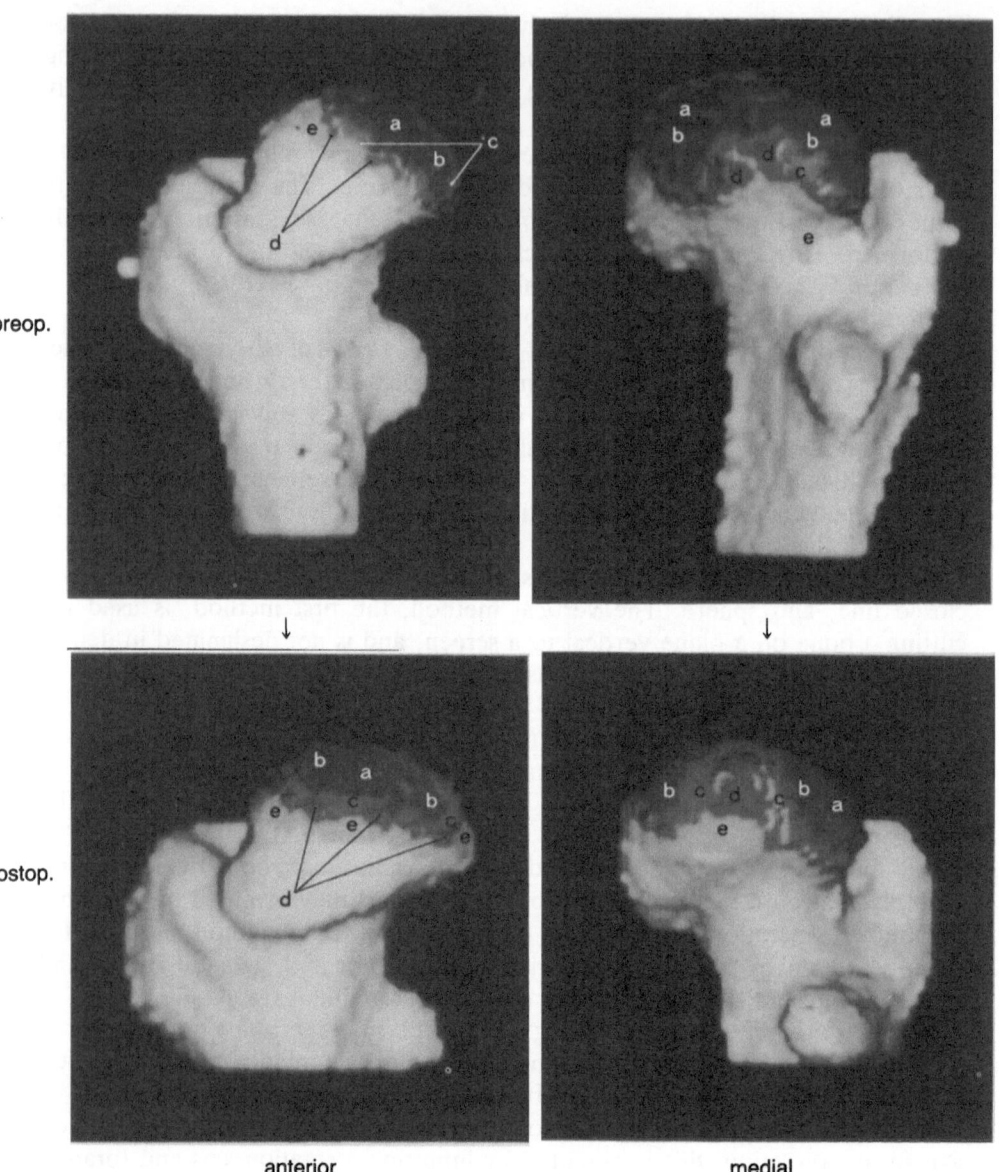

anterior   medial

Fig. 2. Adequate expansion of the load area is seen

green the range 0.8–2.3 mm, sky blue the range 2.3–3.9 mm, red the range
3.9–5.5 mm, and yellow the range 5.5 mm or more. Quantitative analysis of
contact surfaces is currently being studied to eliminate errors.

Finally, attribute is a function for designating the attribute for each bone
piece; this is an effective auxiliary function. It is possible to delete unnecessary

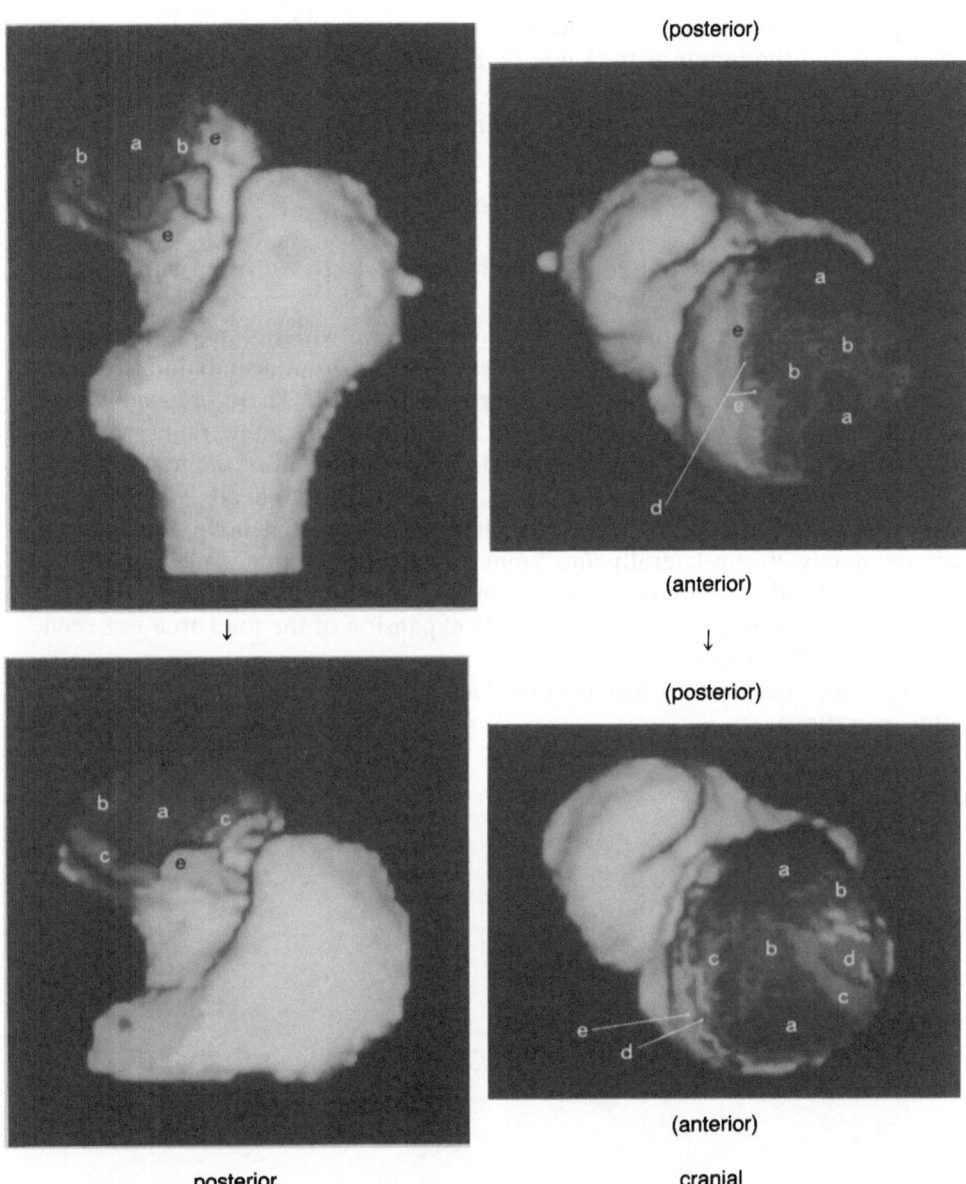

(posterior)

(anterior)

↓

(posterior)

↓

posterior

(anterior)

cranial

Fig. 2. *Continued*

bone pieces on an image or to display them again by using the function "display"/"erase" for bone pieces included in the above function. "Color" designation of bone pieces is a function for classifying bone pieces by color to highlight them, in which seven colors can be designated. For the "group" function, several bone pieces can be grouped and moved without changing their positional relation. For example, it is thus possible to move acetabular fragments and the femur at the same time. An actual bone cutting state can be reproduced with this function.

A case the authors have recently experienced is described below.

Case Report

This patient, a 26-year-old female, had unilateral coxarthrosis with femoral head deformation (Figs. 3, 4).

For simulation, the bone is cut by using a sphere with a diameter of 50 mm (Fig. 5). The outlined area in Fig. 6 shows the gouged-out acetabulum. For this case, the acetabular anteversion is approximately 20°. Therefore, simulation is performed by rotating bone pieces 25° on the axis in parallel with the acetabulum direction and passing through the central point of the femoral head (Fig. 7). In the picture shown in Fig. 8, the femoral head is deleted by using the erase function. In this case, it is estimated that bone pieces move approximately 20 mm laterally and 5 mm forward on the bone cutting plane.

As a result of performing evaluation before and after operation by using the contact surface display function, adequate expansion of the load area was seen, especially in the cranial view (Fig. 2).

Satisfactory results were also observed in the X-ray images (taken before and after operation).

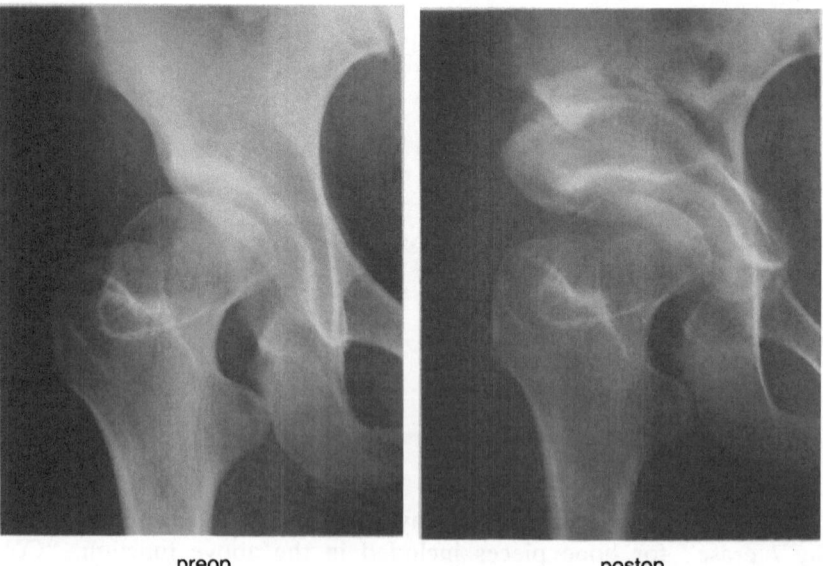

a   b

preop.   postop.

Fig. 3a,b. Radiographs of 26-year-old female with coxarthrosis and femoral head deformation

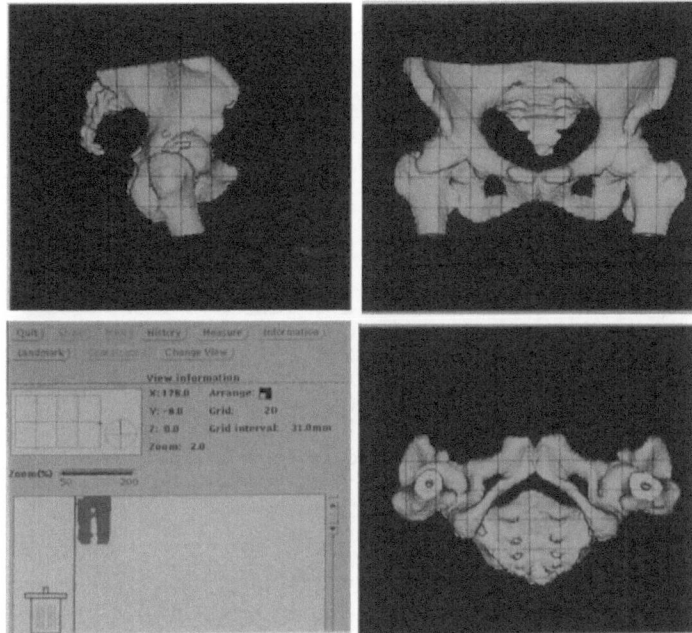

Fig. 4. Preoperative three-dimensional computerized tomographic (*3-D CT*) image

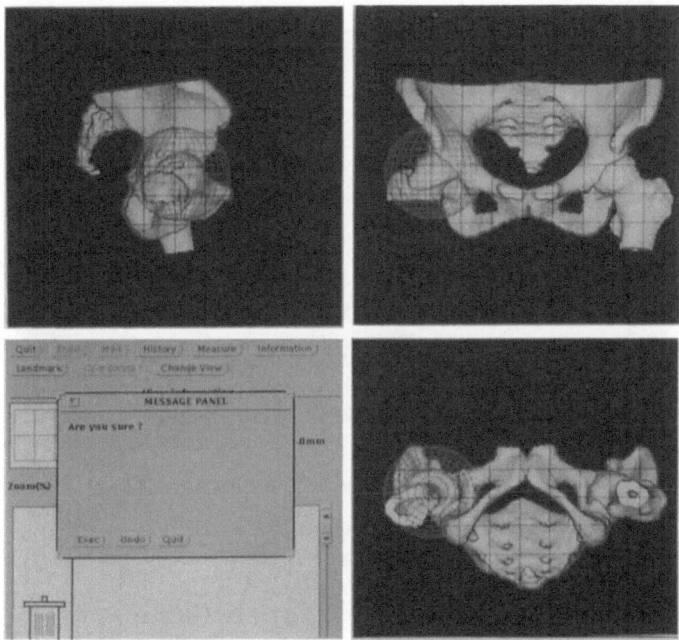

Fig. 5. Osteotomy simulation using "sphere" (diameter: 50 mm)

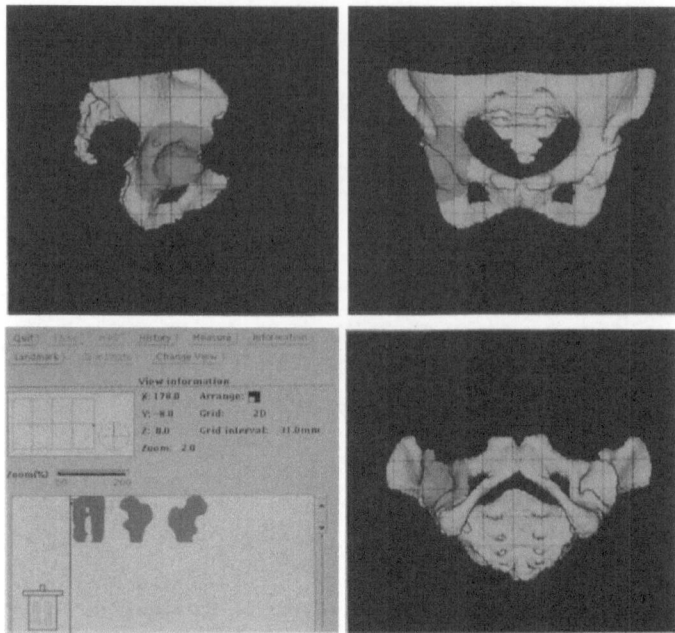

Fig. 6. 3-D CT image after cutting the pelvic bone (femoral head deleted)

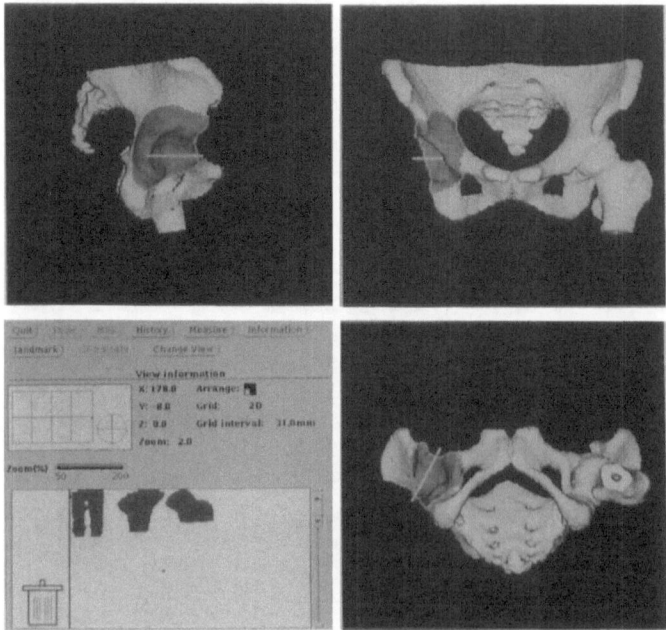

Fig. 7. This picture shows the axis around which the acetabular fragment rotates

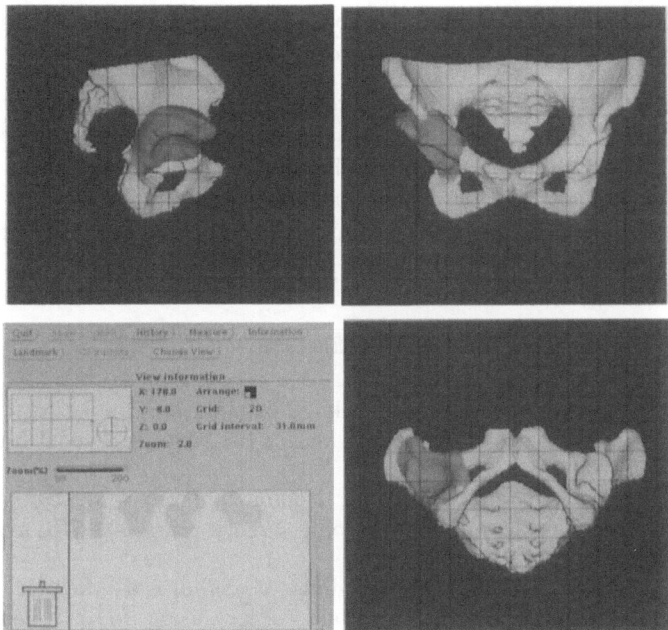

Fig. 8. Osteotomy simulation completed

Discussion

Computerized tomography provides much useful information in orthopedics because of its capacity to accurately delineate body structures at successive depths. However, it is not always easy to construct a mental 3-D image by putting such 2-D images from different planes together.

Computer graphics has advanced remarkably in recent years. Three-dimensional CT, a clinical application of computer graphics, which has been utilized in areas concerned with hard tissues (bones, joints, teeth, etc.), is attracting a great deal of attention, since the major purpose of surgery in this field is 3-D morphological modification of structures.

In orthopedic surgery, 3-DCT is indicated in disorders of the hip joint and spine, which disorders are frequently missed in 2-D evaluations, as the structures overlap one another in lateral radiographs. To date, as stated before, surgical planning and postoperative evaluation of these regions have been made on the basis of 2-D imaging.

Some problems with 3-DCT to be overcome in the future are:

1. Although simulations corresponding to various bone cutting methods can be performed, the difference between actual bone cutting and simulated bone cutting is great.

2. It is difficult, with this method, to obtain the actual feeling of movement of bone pieces in the Salter and Chiari operation.

To overcome these problems, it is necessary to introduce virtual reality (VR) and a solid model. For problem (2), it is necessary to accumulate basic data for the movement of bone pieces by collecting 3-DCT data before and after operation. We are accumulating such data at present.

We believe that 3-D surgical planning and simulation based on 3-DCT imaging will contribute much to clinical medicine.

References

1. Soyama Y, Yasuda T, Yokoi S, Toriwaki J, Izumida R, Fujioka M (1989) A hip joint surgical planning system using 3-D images (in Japanese). Jpn J Med Electron Biol Eng 27:70–78
2. Izumida R, Ishinada Y, Matsumoto M, Henmi O, Fujioka M, Ozaki S, Moriya Y (1992) A computer system for hip joint surgery using 3-D CT images (1) (in Japanese). J Joint Surg 11:146–150
3. Izumida R, Ishinada Y, Henmi O, Kawakubo M, Fujioka M, Mano M, Yamamoto S (1992) A computer system for hip joint surgery using 3-D CT images (2) (in Japanese). J Joint Surg 11:960–968

20

Biomechanical Analysis After Chiari's Pelvic Osteotomy by Means of Superimposed Pelvic Computed Tomography Scan

Genji Fujii, Minoru Sakurai[1], Hidemitsu Kumeta[2],
and Kanichi Funayama[3]

Summary. Changes in the bony and muscular structure of the pelvis after Chiari's pelvic osteotomy were investigated in 18 hips (18 females patients) by means of pre- and postoperative computed tomography. After operation, the distance between the centers of the femoral heads decreased by an average of 0.7 cm and that between the anterior-superior iliac spines increased by an average of 1.1 cm. The iliac wing of the operated side was displaced laterally, but that of the non-operated side remained unchanged. Femoral head coverage improved from 53.3% to 88.1% postoperatively. On simulation, the direction of the vector of the gluteus medius and the resultant force changed vertically, from a preoperative value of 21.6° to 16.7° and from 14.9° to 11.1°, respectively, after operation. The resultant force on the femoral head decreased from a preoperative value of 3.42 W to 2.92 W after operation. In our method of Chiari's osteotomy the inferior fragment was medialized, while the superior fragment was simultaneously displaced laterally; consequently biomechanical improvement was achieved.

Introduction

In Japan, osteoarthritis of the hip joint is generally secondary to acetabular dysplasia or congenital dislocation of the hip. The anatomical characteristics of the dysplastic pelvis are inward wing ilium [1,2], steep acetabulum, and anterolateral defects in the coverage of the femoral head.

Reconstructive surgery of the pelvis is performed by various methods, one of which is Chiari's pelvic osteotomy. In this operation, medial displacement of

---

[1] Department of Orthopaedic Surgery, Tohoku University School of Medicine, 1-1 Seiryo-machi, Aoba-ku, Sendai, 980 Japan
[2] Department of Orthopaedic Surgery, Nagano Central Hospital, 1570 Nishi-tsuruga, Nagano, 380 Japan
[3] Department of Orthopaedic Surgery, Sendai Red Cross Hospital, 2-43-3 Yagiyama-honcho, Taihaku-ku, Sendai, 982 Japan

G. Fujii et al.

the inferior fragment results in a reduction of the resultant force on the hip joint and better coverage of the femoral head [3–6]. To investigate the anatomical and biomechanical effects of this operation, the authors examined changes in the relevant bony and muscular structures with the aid of pre- and postoperative anteroposterior (A-P) radiographs and superimposed computed tomography (CT) scans of the pelvis.

Patients

Forty-five patients with acetabular dysplasia were treated by Chiari's pelvis osteotomy at Tohoku University Hospital between 1984 and 1988. Eighteen hips were studied by pre- and postoperative A-P radiographs and CT scans of the pelvis (Table 1). All patients were female, with the average age at operation being 36.6 years (range, 14–53 years). The average follow-up period was 2.7 years (range, 4 months to 3.5 years). In ten hips, congenital dislocation of the hip was involved; in the remaining eight, there was acetabular dysplasia. None of the patients had had any previous surgery.

Table 1. Preoperative measurement of 18 patients.

Case	Age (years)	CE angle (°)	Sharp angle (°)	AHI (%)	ISD (cm)	ICD (cm)	IA op (°)	IA non-op (°)	CT cover (%)	Gl. med vector (°)	RF vector (°)	RF (W)
1	17	−7	53	44	20.1	18.8	64	65	47	22	15	3.5
2	40	6	52	58	21.3	19.1	62	66	67	24	17	3.4
3	14	−13	59	37	17.6	17.1	72	65	52	23	17	3.6
4	32	−2	50	51	22	17.6	62	59	55	18	14	3.5
5	32	1	53	52	23.8	17.7	55	56	60	15	11	3.3
6	50	−7	45	44	22.8	17.1	59	58	43	16	11	3.3
7	40	6	51	64	21.5	19.7	65	64	59	24	17	3.2
8	44	5	50	58	19	18.7	71	70	59	22	16	3.3
9	45	10	48	64	22.4	17	60	58	62	21	14	3.1
10	23	1	49	57	20.3	18.6	68	63	57	26	18	3.5
11	38	−23	56	31	19.7	18.7	68	67	45	25	18	3.7
12	37	−4	51	46	22.6	19.6	60	60	54	23	16	3.7
13	53	11	45	63	—	16.8	—	—	55	24	14	3.2
14	35	8	45	63	20.6	17.3	63	62	58	20	14	3.3
15	48	−5	51	47	20	16.8	63	63	48	20	13	3.3
16	21	7	49	60	17.9	16.7	65	65	40	21	14	3.7
17	43	−1	50	52	20.8	18	64	63	54	22	14	3.4
18	43	−5	55	48	19.5	18.2	66	65	45	23	16	3.6
Mean	36.4	−0.7	50.7	52.2	20.7	18.0	63.9	62.9	53.3	21.6	14.9	3.42
SD	11.3	8.7	3.8	9.6	1.7	1.0	4.3	3.7	7.3	3.0	2.1	0.19

CE, Center-edge angle; AHI, acetabular head index; ISD, inter-spinous distance; ICD, intercapital distance; IA, iliac angle; CT cover, computed tomography coverage of femoral head; Gl. med., gluteus medius; RF, resultant force; op, operated side; non-op, non-operated side

Methods

Our method of operation was a slight modification of the original Chiari's pelvic osteotomy. We performed a dome-shaped pelvic osteotomy through the anterolateral approach, with the patient on a traction table.

The center-edge (CE) angle of Wiberg, Sharp's acetabular angle, and Heyman's acetabular head index (AHI) were measured by A-P radiographs. Computed tomography scans of the pelvis were taken at 10-mm intervals from the level of the center of the L-5 vertebral body to the symphysis pubis, with the patient in the supine position and the extremities in neutral rotation.

The following parameters were measured on CT scans:

Inter-Spinous Distance (ISD)

This is the distance between the inner walls of both anterior superior iliac spines (Fig. 1). The average distance in normal Japanese females is 23.8 cm ± SD 1.9 cm [2].

Inter-Capital Distance (ICD)

This is the distance between the centers of both femoral heads (Fig. 1). The average distance in normal Japanese females is 17.1 cm ± SD 1.1 cm [2].

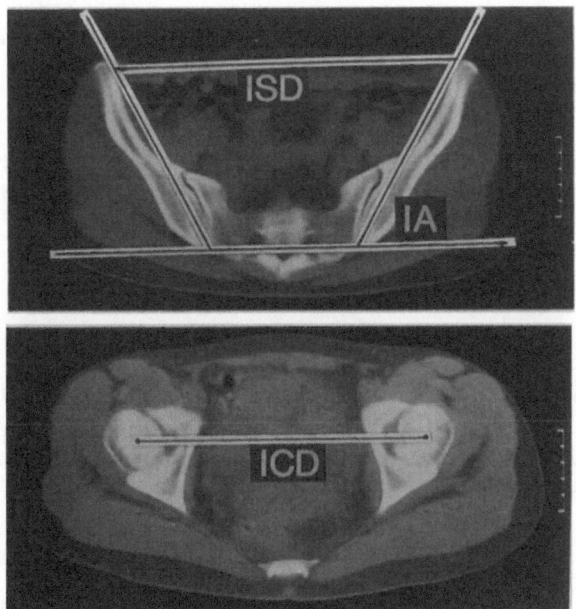

Fig. 1. Measurement of computerized tomography (*CT*) parameters; inter-spinous distance (*ISD*), inter-capital distance (*ICD*), and iliac angle (*IA*)

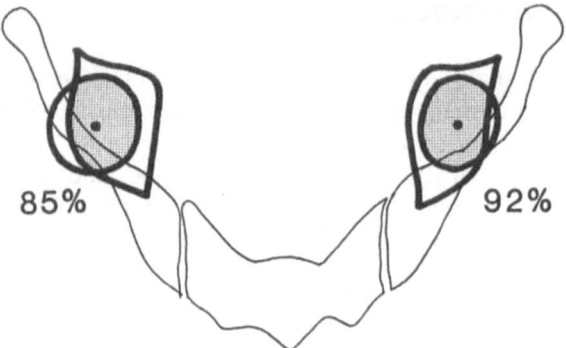

Fig. 2. Measurement of femoral head coverage. *Shaded* area of the femoral head is covered by the acetabulum

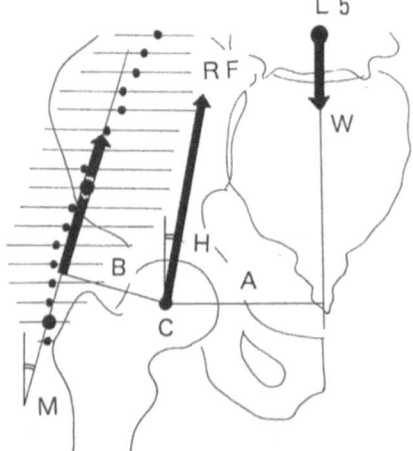

Fig. 3. Parameters for computerized tomography (*CT*) measurement. *W*, Vector of body weight (lies on the center of L5 vertebra); *C*, center of the femoral head; *A*, lever arm of body weight; *M*, direction of the gluteus medius vector; *B*, lever arm of gluteus medius; *RF*, resultant force; *H*, direction of RF vector

Iliac Angle (IA)

The iliac angle is the angle formed by the line joining the inner walls of the anterior superior iliac spine (ASIS) and the posterior point of the sacroiliac joint and the line joining the posterior point of both sacroiliac joints in the horizontal plane (Fig. 1). The average angle in normal Japanese females is 60.8° ± SD 3.6° [2].

Coverage of the Femoral Head

The coverage of the femoral head is measured by superimposing two planes of CT scans at the level of the acetabulum and the center of the femoral head (Fig. 2). The average coverage of the femoral head in normal Japanese females is 84.7% ± SD 8.0% [4].

Direction of the Gluteus Medius Vector (M)

The contours of the gluteus medius muscle were defined at two levels, ASIS and the greater trochanter. The center of the contour of the muscle was identified. The direction of the vector is the line joining these two centers in the frontal plane (Fig. 3). The average vector in normal Japanese females is 15.9° ± 4.3° [4].

Resultant Force on Femoral Head (RF)

As shown in Fig. 3, the gravity of the body lies vertically at the center of the L-5 vertebral body, and the center of rotation of the hip is the center (C) of the femoral head on the CT scan. In the frontal plane, the distance between the gravity line and the point C is the lever arm of hody weight (A) and the distance between the gluteus medius vector and point C is the lever arm of the gluteus medius (B). The resultant force acting on the femoral head (RF) and its direction (H) are calculated by the following equation (Fig. 3):

$$RF = \sqrt{(A/B)^2 + 2(A/B) \times \cos(M) + 1}$$

$$H = \arctan \left\{ \frac{A/B \times \sin(M)}{1 + A/B \times \cos(M)} \right\}$$

The average RF and M in normal Japanese females is 2.94 × W + SD 0.20 × W and 10.4° ± SD 2.9°, where W is the weight.

Results

Sharp's Acetabular Angle

The mean Sharp angle for the 18 patients before operation was 50.7° (range, 45°–59°). Acetabula were abnormally steep compared to those of the normal Japanese female. The postoperative mean value was 37.4°. There was improvement of the Sharp angle after Chiari's pelvic osteotomy in all cases (Tables 1, 2).

Center-Edge (CE) Angle

The preoperative CE angle was abnormal in all cases, but improved from a mean of −0.7° to 37.6° postoperatively (Tables 1, 2).

Acetabular Head Index (AHI)

The AHI improved, from a mean preoperative value of 52.2% to a postoperative value of 89.2%, that is, better than the normal coverage of the femoral head (80.6% ± SD 6.7%) (Tables 1, 2)

Inter-Spinous Distance (ISD)

The mean ISD was 20.7 cm preoperatively, 3.1 cm less than the normal value. Postoperatively, the mean ISD increased to 21.8 cm. (Tables 1, 2).

G. Fujii et al.

Table 2. Postoperative measurement of 18 patients.

Case	CE angle (°)	Sharp angle (°)	AHI (%)	ISD (cm)	ICD (cm)	IA op (°)	IA non-op (°)	CT cover (%)	Gl. med vector (°)	RF vector (°)	RF (W)
1	37	39	96	21	17.6	60	64	90	19	12	2.8
2	36	43	86	22.9	18.8	57	65	87	20	13	3
3	42	39	93	18.7	16.2	63	66	97	18	13	2.8
4	31	37	86	23.1	17.1	57	59	95	11	8	2.9
5	39	38	84	24.5	17.2	50	56	73	14	8	3.1
6	26	36	76	23.5	16.7	56	59	83	8	6	3
7	50	34	100	22.9	18.9	60	62	96	16	11	2.9
8	38	38	91	19.7	17	63	70	97	18	12	2.5
9	45	34	93	23.3	16.4	58	59	83	15	11	2.7
10	21	43	83	21.6	18.2	65	62	99	21	13	3
11	30	45	80	20.8	18	59	68	94	21	14	3
12	23	44	83	23.4	19.1	56	59	78	18	12	3.3
13	44	33	100	22.9	16.5	57	58	82	20	12	2.9
14	61	26	100	22.5	16.6	58	62	97	13	8	2.7
15	36	34	84	21.4	16.7	54	63	76	19	11	3
16	45	32	92	18.9	16.2	61	61	83	16	12	3.2
17	38	38	90	21.8	17.4	58	62	89	17	12	2.9
18	34	40	88	20.3	17.3	59	66	86	16	12	2.9
Mean	37.6	37.4	89.2	21.8	17.3	58.4	62.3	88.1	16.7	11.1	2.92
SD	9.8	4.8	7.1	1.7	0.9	3.5	3.7	8.1	3.5	2.2	0.19

Inter-Capital Distance (ICD)

The mean preoperative ICD was 18.0 cm, 0.9 cm more than the normal value. Postoperatively, the ICD was reduced in all cases, to a mean of 17.3 cm. Medialization of the femoral head, i.e. medial reduction of the ICD, was accomplished with a mean value of 0.7 cm (Tables 1, 2).

Iliac Angle

The iliac angle of the operated side increased from a preoperative mean value of 63.9° to 58.4° postoperatively. But there was no significant change in the iliac angle on the non-operated side (Tables 1, 2).

Coverage of Femoral Head

Coverage improved from a mean preoperative value of 53.3% to 88.1% postoperatively (Tables 1, 2).

Direction of Vector of Gluteus Medius (M)

The mean preoperative vector (M) was 21.6°, 6° horizontal compared to normal. After operation, the vector changed vertically to 16.7° (Tables 1, 2).

Resultant Force of the Femoral Head (RF)

The resultant force decreased from a mean preoperative value of 3.42 × W to 2.92 × W after operation and the vector of resultant force decreased from a mean of 14.9° to 11.1° (Tables 1, 2).

Discussion

In acetabular dysplasia, not only is the acetabulum dysplastic but the total shape of the pelvis is also distorted [1,2]. The femoral head is displaced laterally and the iliac wings are directed more inwardly compared to normal [2]. Inward wing ilium is characteristic of acetabular dysplasia. It can be clearly visualized on CT scans of the pelvis by measuring the ISD, IA, and ICD [1,2].

By measuring CT scans and conventional radiographic parameters, we examined the precise effects of Chiari's medial displacement osteotomy on the shape of the pelvis. Chiari's osteotomy reduces the resultant force on the femoral head by medializing the inferior fragment; this procedure also increases

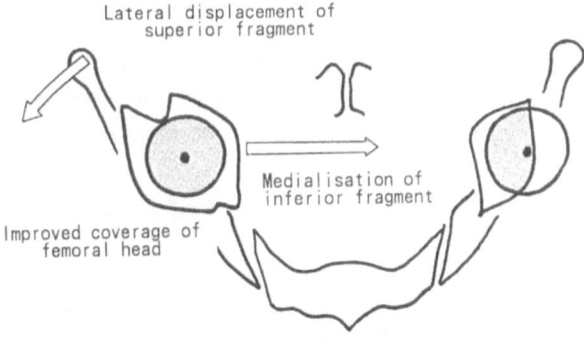

Fig. 4. Changes in the structure of the pelvis after Chiari's pelvic osteotomy (from [4] with permission)

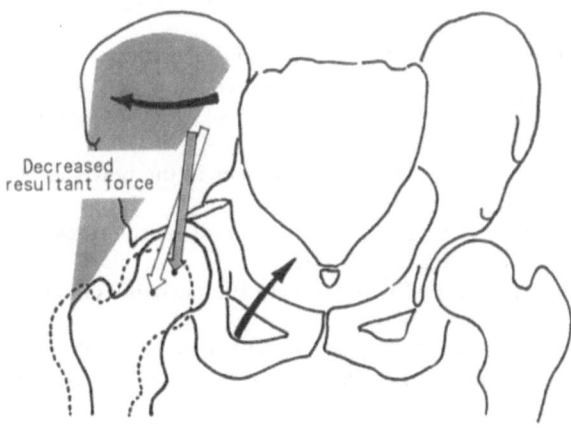

Fig. 5. Biomechanical effect of Chiari's pelvic osteotomy (from [4] with permission)

the area of the weight-bearing surface [3,5]. In our patients, the ICD was certainly reduced medially, by a mean 0.7 cm, and ISD was enlarged laterally, by a mean 1.1 cm. The inferior fragment was medialized, while the superior fragment was lateralized simultaneously. Our results coincide with the those found by Miura on conventional A-P radiographs [6]. The iliac angle of the operated side decreased, while that of the non-operated side remained unchanged. The lateral displacement of the superior fragment resulted from hinge-type movement at the sacroiliac joint of the operated side.

To analyze the biomechanical effects of Chiari's osteotomy, we measured the resultant force on the hip joint. Pauwels [7] measured this force by means of the balance, and found it to be $3.0 \times W$ around the femoral head. Ninomiya et al. [8] calculated the resultant force on A-P radiographs, with the findings being $2.87 \times W + SD\ 0.19 \times W$. Miyagishima et al. [9] calculated the resultant force as $2.8 \times W + SD\ 0.2 \times W$ by computerized simulation, using a rigid body spring model. In their simulation, the vector of the abductor muscle was not fixed. Rydell [10] measured the resultant force in two patients as $2.3 \times W$ and $2.8 \times W$, using a direct method. In this study, we measured the gluteus medius only as the main abductor muscle in the supine position without weight-bearing. The resultant force we obtained is consistentwith previous findings.

The anatomical changes that occurred in the pelvis after Chiari's osteotomy resulted in biomechanical improvement of the hip joints (Figs. 4, 5). However, since in this osteotomy the femoral head is covered by the interposed capsule, and not by the hyaline cartilage, histological and chemical changes in the interposed capsular tissue must be examined.

References

1. Funayama K, Koike M, Miyagishima J, et al (1984) Hip-shelf operation (modified Spitzy) for young adult women. In: Ueno R (ed) The hip—clinical studies and basic research. Elsevier, Amsterdam, pp 105–109
2. Kumeta H, Funayama K, Miyagishima J, et al (1986) Characteristics of the pelvis in acetabular dysplasia (inward wing ilium) (in Japanese). Rinsho-Seikeigeka, 21:67–75
3. Chiari K (1974) Medial displacement osteotomy of the pelvis. Clin Orthop 98:55–71
4. Fujii G, Kumeta H, Funayama K, et al (1988) Biomechanical analysis of the acetabulo-plasities by means of superimposed computed tomography (in Japanese). Hip Joint 14:117–124
5. Kusswetter W (1985) Changes in the pelvis after Chiari and Salter osteotomies. Int Orthop 9:139–146
6. Miura, T (1963) Study on the Chiari's transverse pelvic osteotomy (in Japanese). Niisei-kaisi 36:959–983
7. Pauwels F (1973) Atlas zur Biomechanik der gesunden und kranken Hüfte. Springer, Berlin Heidelberg
8. Ninomiya S, Tagawa H, Miyanaga Y, et al (1976) Simulation of the force acting on the hip joint (in Japanese). Nissei-kaisi 50:15–20

9. Miyagishima J, Funayama K (1985) Simulation of the force acting on the hip joint by means of rigid body spring model (RBSM) (in Japanese). Seikei-geka Kiso-kagaku 12:406–408
10. Rydell NW (1966) Forces acting on the femoral head prosthesis. Acta Orthop Scand [Suppl 88] 37:82–85

21

—Overview—
Alteration in Neck Axis Direction
After Transtrochanteric
Rotational Osteotomy

Mitsuo Tomihara and Seisuke Tanaka[1]

Summary. The rationale of transtrochanteric rotational osteotomy of the femoral head for idiopathic avascular necrosis (described by Sugioka [1–4]) is to reposition the necrotic antero-superior part of the femoral head to a non-weight-bearing locale. To do this, the femoral head and neck segment is rotated anteriorly around its longitudinal axis. Sugioka recommended that in a patient with an extensive lesion, intentional varus positioning, in addition to the rotation, can be achieved by inclination of the osteotomy plane. However, the anteversion angle may also increase or decrease according to the direction of the inclination and the degree of rotation. We examined alterations in the neck axis direction after rotation around a general axis and devised a formula. Using this formula, we performed rotational osteotomy with intentional varus positioning of 20°, without alteration of the anteversion angle.

Key words: Osteonecrosis—Computer graphics—Computed tomography—Microcomputer—Rotation matrix—Osteotomy—Femoral neck axis

Introduction

In recent years, computer graphics have provided three-dimensional observation of figures derived mainly from computerized tomographic (CT) scans, biomechanical analysis, implant designing, surgical simulation, and substantial models. Several osteotomies of the hip can be simulated using an algorithm to rotate figures around a general axis in a three-dimensional space [5]. Using this algorithm, we have also simulated transtrochanteric rotational osteotomy, Southwick's osteotomy, and intertrochanteric extension osteotomy [6–8].

Sugioka [1] devised transtrochanteric rotational osteotomy of the femoral head as femoral head-preserving surgery to treat idiopathic osteonecrosis. In

[1]Department of Orthopaedic Surgery, Kinki University School of Medicine. 377-2, Ohno-Higashi, Osaka-Sayama, Osaka, 589 Japan

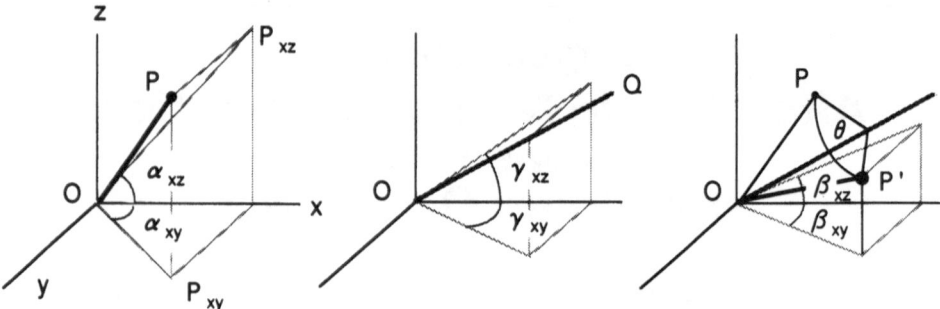

Fig. 1. Rotation of the neck axis (*OP*) around a general rotation axis (*OQ*). OP' is the neck axis after rotation

this procedure, the necrotic superior and anterior femoral head is rotated anteriorly around the longitudinal neck axis so that the weight-bearing force is transferred to the intact posterior surface. The direction of the longitudinal neck axis does not change when it is rotated around its axis. Sugioka [2,3,9] recommended that, in a patient with an extensive lesion, intentional varus positioning, in addition to rotation, could be achieved by inclination of the osteotomy plane. However, the anteversion angle may also increase or decrease according to the direction of the inclination and the degree of rotation. A method to obtain the varus position after rotation was also described by Sugioka [4]. In this study, we examined alterations in the neck axis after rotation around a general axis.

Methods

The point where the femoral diaphysis axis intersects with the neck axis will be made the origin $O(0,0,0)$, with the head center as P, and the length of OP as 1 (Fig. 1). The point where the head center P is projected on plane xy will be made P_{xy}, and the point projected on plane xz, P_{xz}. If angles which OP_{xy} and OP_{xz} make with axis x are set as α_{xy} and α_{xz} respectively, the coordinates of $P(x_0,y_0,z_0)$, the center of femoral head, can be calculated as follows:

$$a_1 = \tan \alpha_{xy}$$

$$a_2 = \tan \alpha_{xz}$$

$$x_0 = 1/\sqrt{1 + a_1^2 + a_2^2}$$

$$y_0 = x_0 a_1$$

$$z_0 = x_0 a_2$$

If lines which rotation axis OQ projects on planes xy and xz meet with axis X to form angles γ_{xy} and γ_{xz}, then the direction cosine of OQ, or (n_1,n_2,n_3), can be calculated as follows:

$$t_1 = \tan \gamma_{xy}$$

$$t_2 = \tan \gamma_{xz}$$

$$n_1 = 1/\sqrt{1 + t_1^2 + t_2^2}$$

$$n_2 = n_1 t_1$$

$$n_3 = n_1 t_2$$

Point P' (x_1, y_1, z_1), which is a point made by rotating the head center P with the angle of θ around the OQ axis, can be calculated, using the followings:

$$r_{11} = n_1^2 + (1 - n_1^2)\cos \theta$$

$$r_{12} = n_1 n_2 (1 - \cos \theta) + n_3 \sin \theta$$

$$r_{13} = n_1 n_3 (1 - \cos \theta) - n_2 \sin \theta$$

$$r_{21} = n_1 n_2 (1 - \cos \theta) - n_3 \sin \theta$$

$$r_{22} = n_2^2 + (1 - n_2^2)\cos \theta$$

$$r_{23} = n_2 n_3 (1 - \cos \theta) + n_1 \sin \theta$$

$$r_{31} = n_1 n_3 (1 - \cos \theta) + n_2 \sin \theta$$

$$r_{32} = n_2 n_3 (1 - \cos \theta) - n_1 \sin \theta$$

$$r_{33} = n_3^2 + (1 - n_3^2)\cos \theta$$

$$[x_1 \ y_1 \ z_1] = [x_0 \ y_0 \ z_0] \begin{bmatrix} r_{11} & r_{12} & r_{13} \\ r_{21} & r_{22} & r_{23} \\ r_{31} & r_{32} & r_{33} \end{bmatrix}$$

Therefore, angles β_{xy} and β_{xz}, which form between axis x and lines projected on planes xy and xz by the neck axis OP' after rotation, can be calculated as follows:

$$\beta_{xy} = \tan^{-1}(y_1/x_1)$$

$$\beta_{xz} = \tan^{-1}(z_1/x_1)$$

The above can be calculated instantly by incorporating into the worksheet of an aggregate chart-style database. Conversely, it is also possible to calculate the rotation axis to obtain a certain varus positioning, without changing the anteversion angle.

At surgery, a Kirschner wire (wire 1; Fig. 2) is first pierced through the neck axis. After this is confirmed by X-ray, a second wire (wire 2; Fig. 2) is pierced through the rotation axis obtained through calculation, then another two wires (wire 3; Fig. 2) are pierced through a plane vertical to the second wire; this plane then becomes the osteotomy plane.

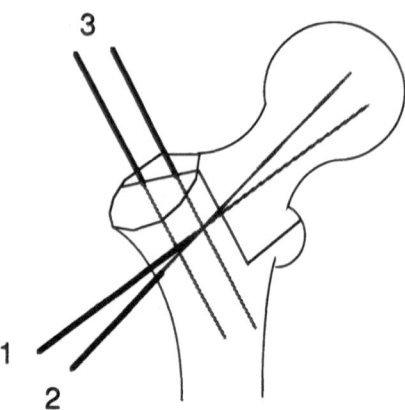

Fig. 2. Direction of the Kirschner wires and the osteotomy plane

Case Study

The patient was 31-year-old male who complained of pain in the right hip; this seemed to be related to the short-term use of steroids (Fig. 3). Magnetic resonance imaging (MRI) revealed avascular necrosis of bilateral hips, and the stereogram showed that the range of necrosis extended from the summit to the antero-inferior portion (Fig. 4). The direction of the neck axis was $-40°$ on plane xy, and $60°$ on plane xz, indicating an anteversion angle of $40°$ and an apparent collo-diaphyseal angle of $150°$ (Fig. 5). When a rotational osteotomy of the femoral head was simulated on this stereographic processing, the lesion was released from the load portion at a $60°$ anterior rotation (Fig. 6b), and by adding a further $20°$ varus positioning, the lesion was further released from the load portion (Fig. 6c).

Meanwhile, giving a $60°$ anterior rotation at $-15°$ axis on plane xy and $40°$ axis on plane xz would, according to our formula (Table 1), make the direction of the neck axis $-40.4°$ and $38.6°$, respectively. That is, the anteversion angle remained more or less unchanged with an intentional varus positioning of $21.4°$ (Fig. 6d). The degree of leg shortening could be calculated from changes in the value of the z coordinates of the femoral head center. In this case, the leg was predicted to be shortened by 20 mm.

To reduce leg shortening, the diaphysis side was fixed by sliding it down and inwards. Post-operative X-ray revealed an inversion of $17°$, with leg shortening of 13 mm (Fig. 7). It was assumed from the lateral image that the anteversion angle remained unchanged. The prognosis of the patient 6 months after operation is satisfactory.

Discussion

If a plane of osteotomy is established with a posterior opening angle of $10°$, using Sugioka's original method [2] (Fig. 8a), the rotation axis in our patient

Fig. 3. Preoperative
radiograph

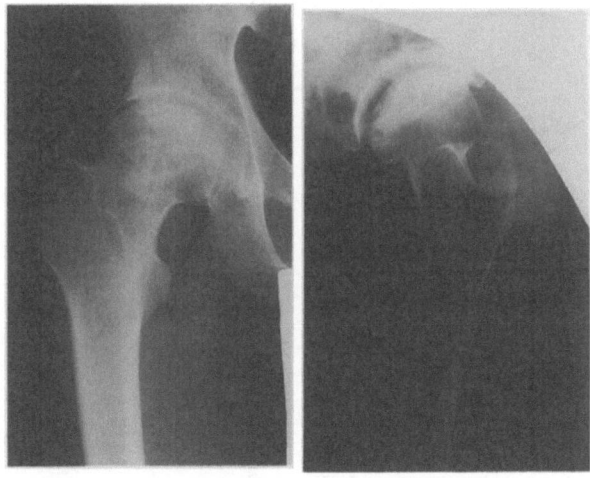

Fig. 4a,b. Stereographic processing
of computerized tomographic (*CT*)
figures. **a** View from anterior and
20° below, **b** view from anterior and
20° above

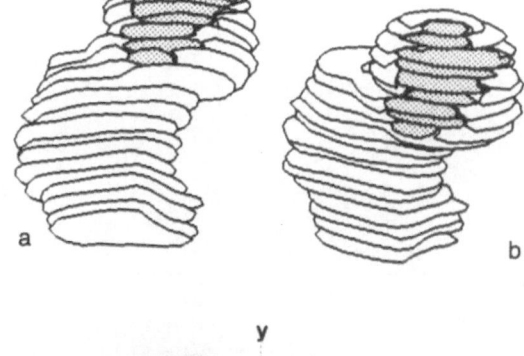

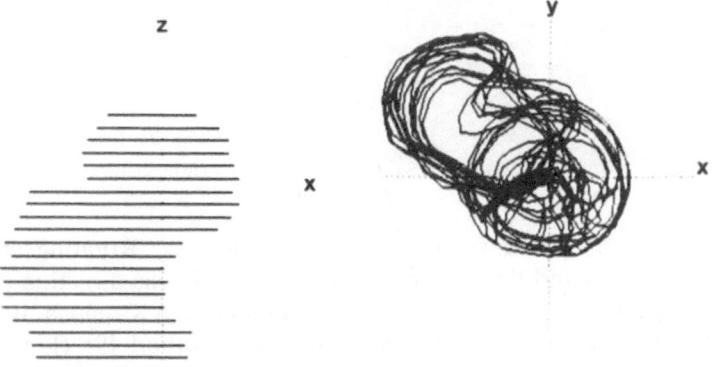

Fig. 5. Direction of the neck axis on planes xy and xz

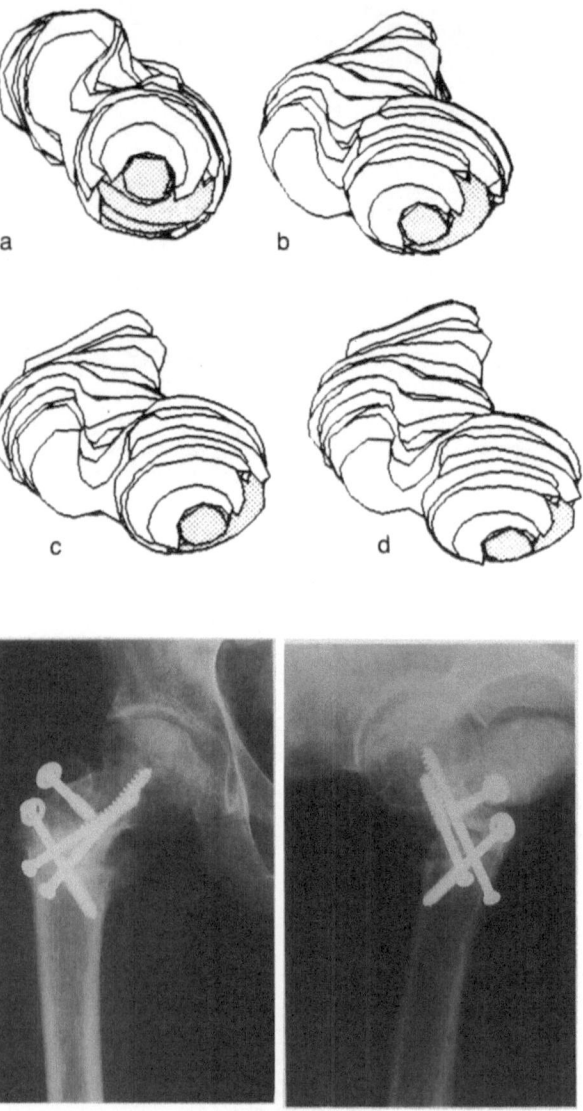

Fig. 6a–d. Simulation of the rotaional osteotomy. **a** View from the cephalad direction before rotation, **b** 60° rotation around the neck axis, **c** further addition of a 20° varus positioning, **d** 60° rotation around an axis of −15° on plane xy and 40° on plane xz

Fig. 7. Postoperative radiograph

would be −23.7° on plane xy and 54.3° on plance xz. Rotating 60° in the anterior direction around this axis will generate a neck axis direction of −33.2° and 49.2°, respectively, and a reduced anteversion of 6.8° and a 10.8° varus positioning (Fig. 9a). Using Sugioka's improved method [9] (Fig. 8b), the rotation axis would be −16.5° and 31.9° on each plane, generating a neck axis direction of −48.1° and 36.8°, respectively, an increased anteversion of 8.2°, and a 23.1° varus positioning (Fig. 9b). With this method, there seems to be a 40° varus positioning, with no change of the anteversion angle, this angle changing only in the 90° anterior rotation (Table 2). With our formula, we

Table 1. Direction of the neck axis after rotation.

Neck axis		Rotation axis		Rotation angle		Neck axis after rotation	
α_{xy}	−40	γ_{xy}	−15	θ	60	β_{xy}	−40.4
α_{xz}	60	γ_{xz}	40			β_{xz}	38.6
				r_{11}	0.782		
a_1	−0.839	t_1	−0.267	r_{12}	−0.620	b_1	−0.852
a_2	1.732	t_2	0.839	r_{13}	0.062	b_2	0.800
				r_{21}	0.470		
x_0	0.461	n_1	0.750	r_{22}	0.520	x_1	0.650
y_0	−0.386	n_2	−0.201	r_{23}	−0.713	y_1	−0.554
z_0	0.799	n_3	0.630	r_{31}	0.410	z_1	0.520
				r_{32}	0.597		
				r_{33}	0.698		

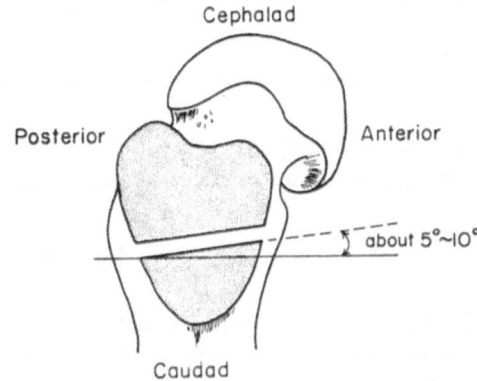

Fig. 8a,b. Sugioka's methods for obtaining intentional varus positioning. **a** Original method (from [2] with permission), **b** improved method (from [9] with permission)

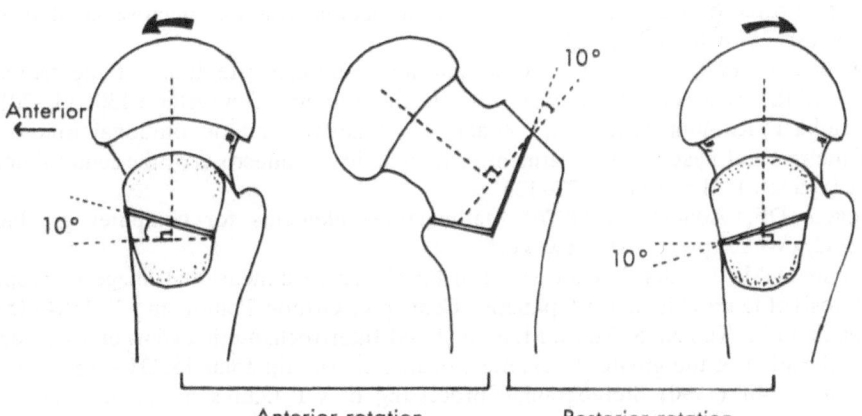

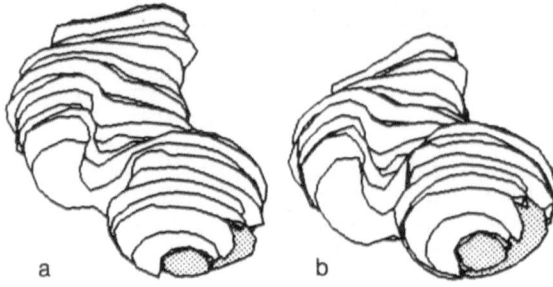

Fig. 9a,b. Simulation of Sugioka's method. **a** Original method, **b** improved method

a   b

Table 2. Direction of the neck axis after rotation according to Sugioka's methods.

	Neck axis		Rotation axis		Rotation angle	Neck axis after rotation	
	α_{xy}	α_{xz}	γ_{xy}	γ_{xz}	θ	β_{xy}	β_{xz}
Original method [2]	−40	60	−27.7	54.3	60	−33.2	49.2
Improved method [9]	−40	60	−16.5	31.9	60	−48.1	36.8
					90	−41.1	18.6

obtained a rotation axis to produce a certain intentional varus positioning, with no change in the anteversion angle at various rotation angles (Table 1).

References

1. Sugioka Y (1973) Transtrochanteric anterior rotational osteotomy of the femoral head for avascular necrosis in adults (in Japanese). Cent Jpn J Orthop Traum Surg 16:574–576
2. Sugioka Y, Eguchi M, Kaibara N, Kitajima T (1976) Transtrochanteric anterior rotation osteotomy of the femoral head for idiopathic avascular necrosis in adults (in Japanese). Hip Joint 2:23–32
3. Sugioka Y (1978) Transtrochanteric anterior rotational osteotomy of the femoral head in the treatment of osteonecrosis affecting the hip. Clin Orthop 130:191–201
4. Sugioka Y, Kitajima T, Katsuki I, et al (1981) Transtrochanteric rotational osteotomy of the femoral head in the treatment of idiopathic osteonecrosis of the femoral head (in Japanese). Hip Joint 7:127–133
5. Rogers DF, Adams JA (1976) Mathematical elements for computer graphics. MacGraw-Hill, pp 60–64, New York
6. Tomihara M, Tanaka S, Oka M, et al (1985) Three-dimensional image of aseptic necrosis of femoral head (in Japanese). Cent Jpn J Orthop Traum Surg 28:1584–1585
7. Tomihara M, Tanaka S, Miura M, et al (1989) Intertrochanteric extension osteotomy for elevation of the greater trochanter (in Japanese). Hip Joint 15:231–235
8. Tomihara M (1990) Stereographic processing of CT figures in the hip joint (in Japanese). Med J Kinki Univ 15:533–556
9. Sugioka Y (1984) Transtrochanteric rotational osteotomy in the treatment of idiopathic and steroid-induced femoral head necrosis, Perthes' disease, slipped capital femoral epiphysis, and osteoarthritis of the hip. Clin Orthop 184:12–23

IV

Stress Analysis in Total Hip Arthroplasty

22

Biomechanical Analysis of a New Type of Surface Total Hip Replacement

H. Miura, J. Arima, K. Hayashi, T. Mashima, T. Inadome, and Y. Sugioka[1]

Summary. In this study we introduced a new type of surface replacement. The basic concept of the femoral component was to prevent stress shielding and avascular necrosis of the femoral head by minimizing the resurfaced area in comparison to the currently available surface replacement. A small peg was added to the center of the implant in order to provide the initial rigid fixation. In developing the new type of surface replacement, the fixation angle, covering angle, and effect of the femoral component peg on stress distribution, as well as the open angle of the acetabular component, were analyzed, using the rigid-body spring model and the finite element method. Two-dimensional finite element models and rigid-body spring models of the femoral head with the component were developed. The new and currently-used components were compared, and the effects of the fixation angle and the peg on stress distribution were analyzed. The optimal resurfaced area was also analyzed by changing the coverage angle from 120° to 180°, since an extremely small implant would lead to severely limited motion and failure of the fixation. For the acetabular component, the open angle should be decreased in order to prevent both socket-neck impingement and increase in the amount of resected bone in the acetabulum. However, the safety zone of the open angle must be also determined to avoid dislocation of the component. Rigid body spring models consisting of the acetabular and femoral components were developed with an anti-dislocation spring at the edge of the acetabular component. The direction of the resultant force that generates the force in the anti-dislocation spring was analyzed. Severe stress shielding was seen in the central portion of the femoral head with the current type of surface replacement, while the new type showed a similar pattern of stress distribution to that of the normal femoral head up to a coverage of 180°. Stress concentration was seen along the peg, while stress shielding occurred immediately below the implant with a long peg. Full coverage of the femoral head by the component and insertion of the implant

---

[1] Department of Orthopaedic Surgery, Kyushu University, 3-1-1 Maidashi, Higashi-ku, Fukuoka, 812 Japan

parallel to the axis of the femoral neck could result in abnormal stress distribution, including stress shielding of the central portion of the femoral head and stress concentration at the medial side of the neck. A dislocation force was generated when the resultant force was applied within 30° of the edge of the acetabular component, which meant that there was a possibility of dislocation with an open angle of less than 120° under physiological loading conditions.

Key words: Biomechanics—Surface total hip replacement—Femoral component—Acetabular component—Finite element method—Rigid-body spring model—Stress-shielding

Introduction

Although there has been great enthusiasm for cementless total hip arthroplasty, due to problems encountered with cement fixation, the cementless procedure is not without problems, including thigh pain and bone resorption caused by stress shielding. The controversy over stem size and stiffness remains unresolved. Some surgeons choose cement fixation only for the femoral component. Considering this history of total hip arthroplasty, it may be difficult to expect the development of long-term rigid fixation.

Indeed, stemmed total hip arthroplasties have shown excellent long-term results, but insertion of the femoral stem into the medullar canal of the femur has caused serious problems, including bone resorption due to abnormal stress distribution. Especially in young patients, high revision rates after primary total hip replacement have been reported [1]. Osteotomy, which is not indicated for severely progressed osteoarthrosis, or arthrodesis, is unwillingly performed on young patients. An ideal surface replacement of the total hip replacement could substitute for these procedures in such high-risk groups. Although some orthopedic surgeons have attempted surface replacement and early results have been encouraging, the revision rate is higher than that for conventional total hip replacement, due to aseptic loosening of both the femoral and acetabular components [2]. Femoral side failures include severe bone resorption and necrosis of the femoral head, which may be caused by the shape of the femoral component.

Development of stable surface total hip replacement is considered to be possible with the newest technology in design, materials, and fixation methods. Re-challenge for surface total hip replacement should be continued. On the basis of our earlier experience with surface replacements [3, 4], a new type of surface replacement indicated for young patients with severe osteoarthritis of bilateral hips has been developed since 1990. The basic concept of the femoral component of the new model is to prevent stress shielding and avascular necrosis of the femoral head by minimizing the resurfaced area in comparison with the currently available surface replacement, which covers most of the

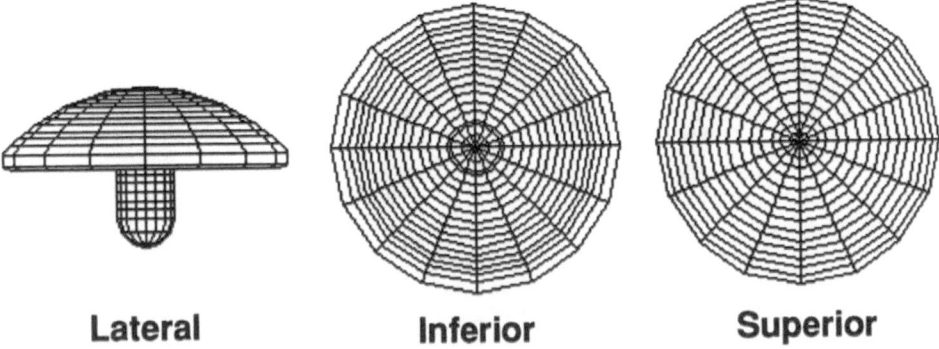

Lateral   **Inferior**   **Superior**

Fig. 1. Basic design of femoral component of new surface total hip replacement (THR)

surface of the head. The range of motion of the hip is allowed to the minimal level required for daily activity. A small peg is added to the center of the implant in order to provide the initial rigid fixation (Fig. 1). In developing the new type of surface replacement, the fixation angle [5], covering angle, and effect of the femoral component peg on stress distribution, as well as the open angle of the acetabular component, were analyzed, using the rigid-body spring model and the finite element method.

Methods

Two-dimensional finite element models (FEM) and rigid-body spring models of the femoral head with a component were developed. A load of 2000 N was applied to the rigid-body spring model and calculated stress on the femoral head was distributed on the surface of the FEM model as nodal point loading. Both perfect bonding and no bonding at the interface between implant and bone were modelled. For the latter model, gap elements with no friction at the interface were used. The distal end of the model was assumed to be a condition of zero deformation. The material properties of the implant and bones used in this study are shown in Table 1. Analyses were performed using COSMOS/M (Structural Research and Analysis Corporation, Santa Monica, Calif.) and a custom-made rigid-body spring model program. The new and currently used components were compared, and the effects of the fixation angle and the peg

Table 1. Material properties.

	Elastic modulus (N/mm^2)	Poisson's ratio
Cancellous bone	1 600	0.26
Cortical bone	17 000	0.29
Implant	200 000	0.29

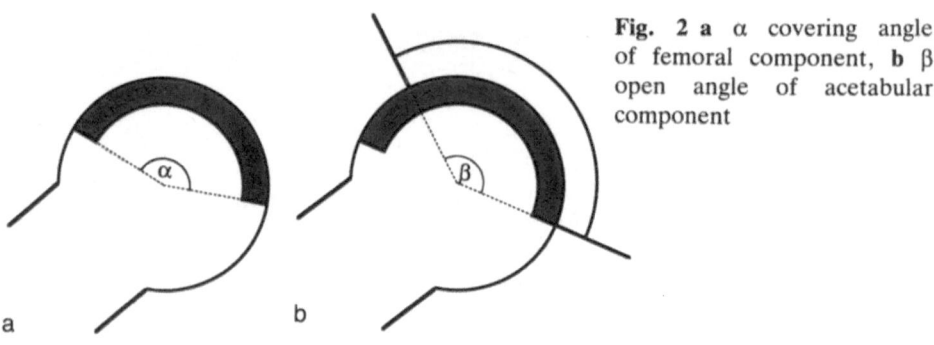

Fig. 2 a α covering angle of femoral component, **b** β open angle of acetabular component

on stress distribution were analyzed. The optimal resurfaced area was also analyzed by changing the coverage angle from 120° to 180° at 20° increments, since an extremely small implant would lead to severely limited motion and failure of the fixation. For the acetabular component, the open angle should be decreased in order to prevent both so-called socket-neck impingement and the increase of the amount of resected bone in the acetabulum, which are the disadvantages of larger head diameter. However, because a decreased open angle of the acetabular component would lead to dislocation of the component, the safety zone of the open angle must be also determined.

Fixation Angle of the Femoral Component

The currently used component is designed to be inserted along the axis of the femoral neck; however, the direction of the resultant force on the femoral head is not generally equivalent to that of the axis of the femoral neck. The fixation angle of the femoral component was analyzed using the FEM and rigid-body spring model in order to obtain a more physiological stress distribution.

Covering Angle of the Femoral Component (Fig. 2a)

The optimal resurfaced area should be determined, since an extremely small femoral component would lead to increased pressure at the interface and subsidence of the component. Four FEM models with coverage angles of 120°–180° were constructed and compared with the currently used component and normal femoral head. The new models were analyzed both with and without a peg.

Effect of the Peg

The peg is considered to be indispensable in obtaining fixation rigidity; however, it can influence the stress distribution of the femoral head. FEM models with a small, middle, and large peg, and without a peg, were developed to analyze the effect of the peg on stress distribution.

Open Angle of the Acetabular Component (Fig. 2b)

A rigid-body spring model consisting of the acetabular and femoral components was developed with an anti-dislocation spring at the edge of the acetabular component. Compressive springs, normal and tangential to the articular surface, were distributed over the contact surface between the two components, but the effect of shear stress on the contact surface was neglected. The anti-dislocation spring is a tensile spring when is able to resist only tensile force. The resultant force was gradually shifted from the center to the edge of the acetabular component and the direction of the resultant force that generates the force in the anti-dislocation spring was calculated. Acetabular components with open angles of 100°–180° at 20° increments were analyzed to observe whether the dislocation force was generated under physiological loading conditions.

Results

Fixation Angle of the Femoral Component

Perfect Bonding at the Interface

Severe stress shielding, which we believe is the primary cause of femoral side failures, was seen in the central portion of the femoral head with the current type of surface replacement, while the new type showed a similar pattern of stress distribution to that of the normal femoral head. Stress concentration was seen along the peg, while stress shielding occurred immediately below the implant with the peg (Fig. 3).

No Bonding at the Interface

The rigid body spring model showed that the inferomedial portion of the current type did not contribute to fixation rigidity while peak pressure was observed at the proximal corner. In contrast, even stress distribution was seen in the new model which it was inserted parallel to the direction of the resultant force. The FEM revealed stress concentration at the inferomedial and superolateral portions of the femoral head in the current type. In the case of the new model, stress distribution close to the normal pattern was observed when it was inserted parallel to the direction of the resultant force, but stress concentration was seen at the superolateral portion when it was inserted parallel to the axis of the femoral neck (Fig. 4).

Covering Angle of the Femoral Component

Severe stress shielding was seen in the central portion of the femoral head with the current type of surface replacement, while the new type without the peg showed a similar pattern of stress distribution to that of the normal femoral

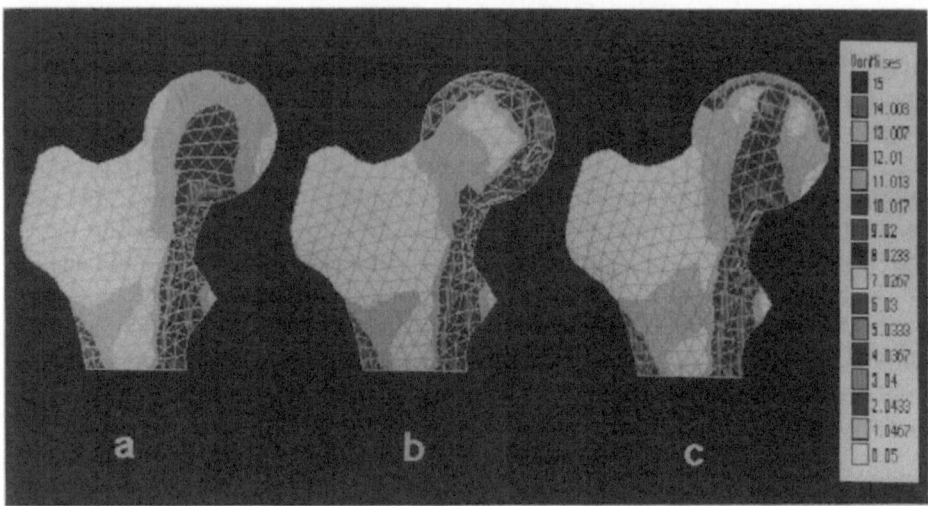

Fig. 3. von Mises stress (perfect bonding at the interface) **a** normal femoral head; **b** current type; **c** new type

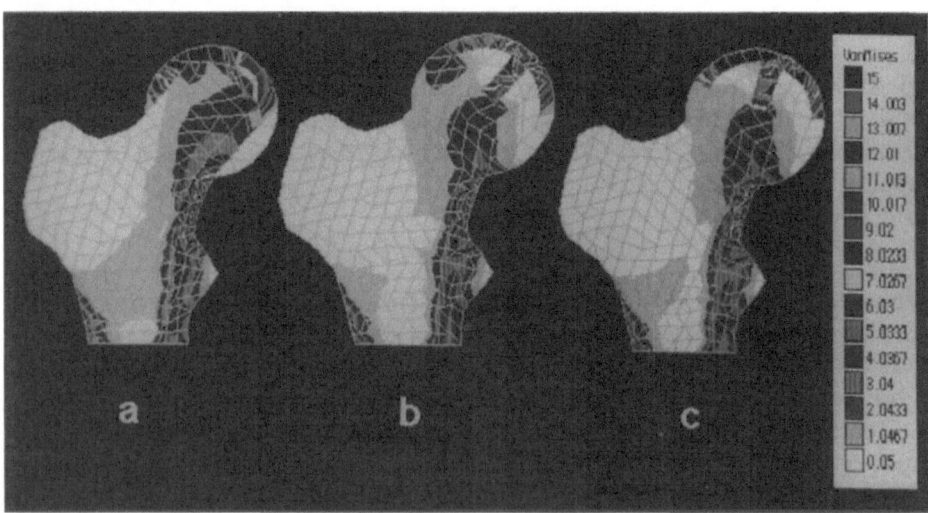

Fig. 4. von Mises stress (no bonding at the interface) **a** current type; **b** new type inserted along axis of femoral neck; **c** new type inserted along direction of resultant force

head, up to a coverage of 180°. The new type with the peg showed stress concentration along the peg and subtle stress shielding immediately below the implant (Fig. 5). Compared to von Mises stress at the central portion of the normal femoral head, the currently used component showed a reduction of approximately 60%, while the new model without the peg showed slight reduction up to a coverage of 180°. In contrast, the new type with the peg demonstrated an increase of percent stress due to the stress concentration along the peg.

Effect of the Peg

Basically, stress concentration was seen along the peg. In the case of a long peg, severe stress concentration along the peg and stress shielding immediately below the implant were observed (Fig. 6).

Open Angle of the Acetabular Component

When the direction of the resultant force was applied within 30° from the edge of the acetabular component, the dislocation force in the anti-dislocation spring was generated without regard to the open angle. If each acetabular component was inserted with an inclination of 45°, an acetabular component with an open angle of less than 120° created the possibility of dislocation under physiological loading conditions (Fig. 7). Acetabular components with an open angle of more. than 140° were considered to be stable.

Discussion

Ten- to 15-year follow-up studies of cemented total hip replacements have generally revealed a less than 10% revision rate for aseptic loosening [2]. For patients under 40 years of age, however, the revision rate increased to more than 50% [6]. Furthermore, a recent study [1] has shown a rate of 20%–50% for radiographic loosening; this may lead to higher failure rates in the furture.

On the other hand, Wagner surface replacement showed a 26% failure rate in a 3- to 6-year follow-up study [7]. Femoral side failure of the surface replacement can be caused by bone resorption and necrosis of the femoral head and by lack of fixation rigidity. Acetabular problems include aseptic loosening due to increased frictional torque [7] and the relatively thin polyethylene layer.

Surface total hip replacements using a socket and cup [3,4] have been performed in 138 hips in 127 patients (109 females and 18 males) in our institute since 1972. The mean age these patients was 44.5 years. Eighty-three hips in 87 patients were followed-up, and, at an average of 9 years after initial surgery, 57% of these patients have already had revision surgery. However, the average follow-up period for the remaining patients is 15.7 years, so this procedure achieved our initial aim of time-saving. Most of the revision patients showed a sinking of the femoral component and adduction contracture of the hip. Operative findings demonstrated severe bone resorption of the femoral head and no fixation rigidity, which findings were equivalent to the findings of the FEM analysis.

On the basis of our earlier experiences, we have developed a new type of surface total hip replacement, the basic concept of which is to prevent avascular

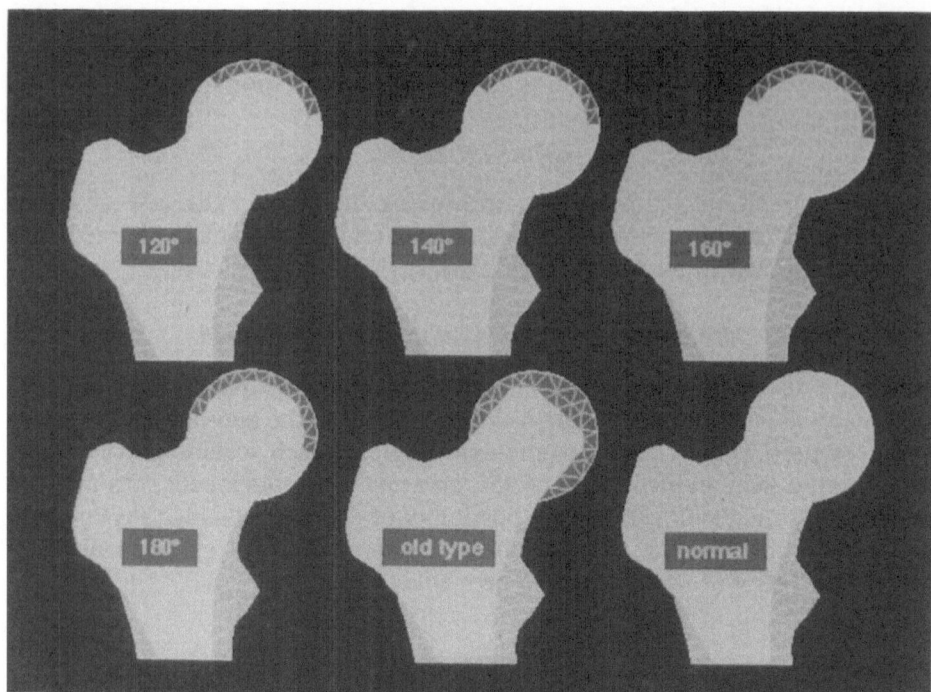

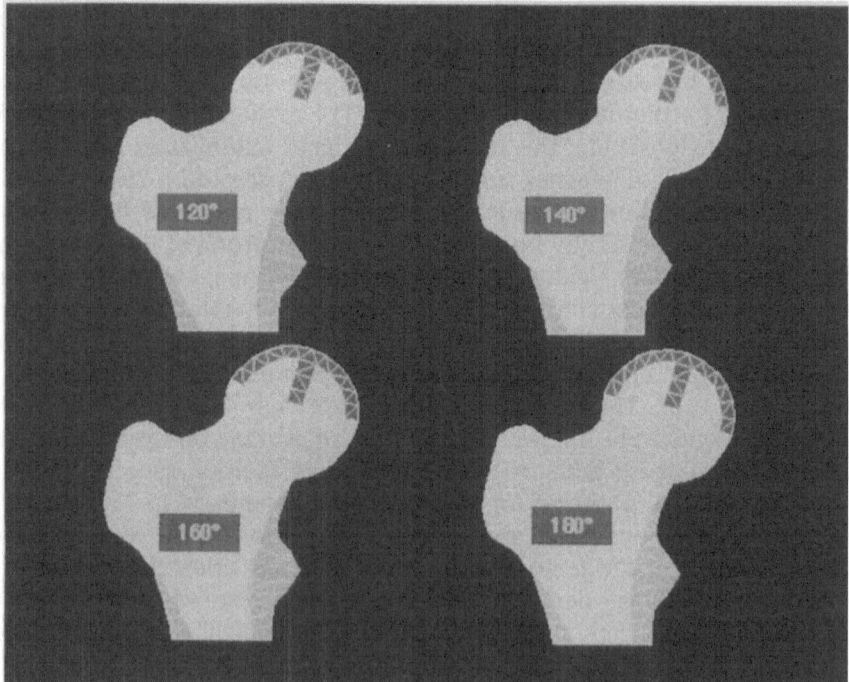

Fig. 5a,b. Finite element method (*FEM*) models and von Mises stress **a** without peg, **b** with peg

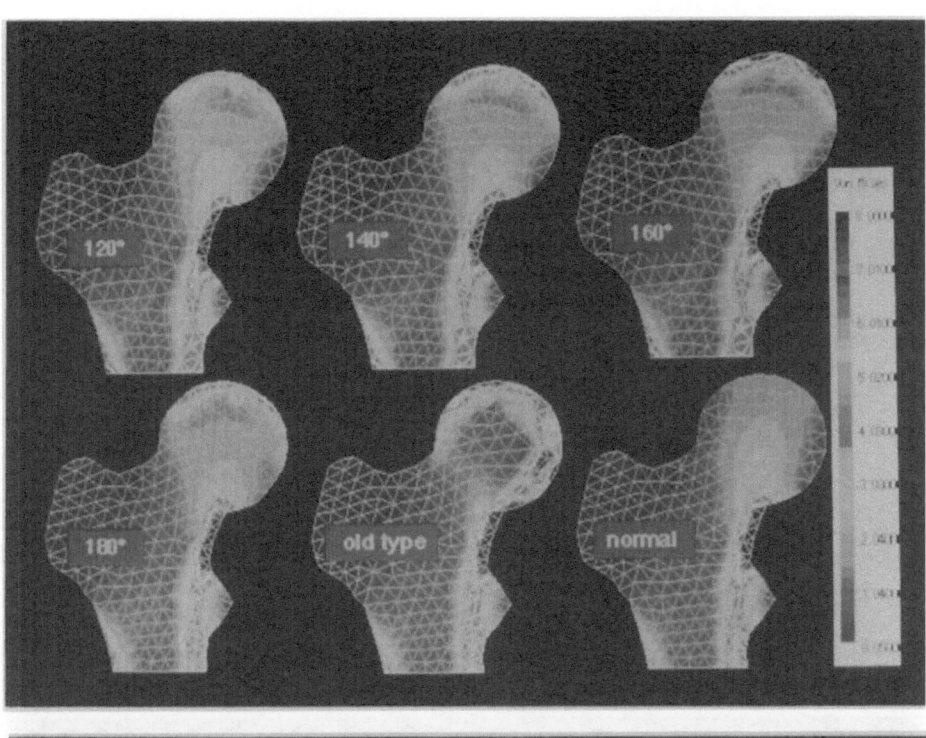

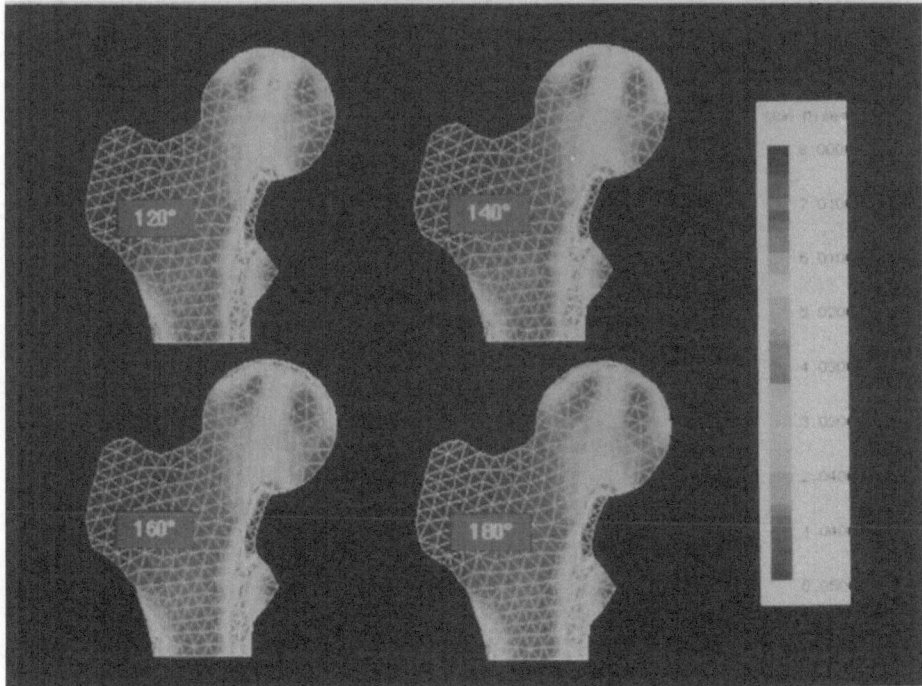

Fig. 5. *Continued*

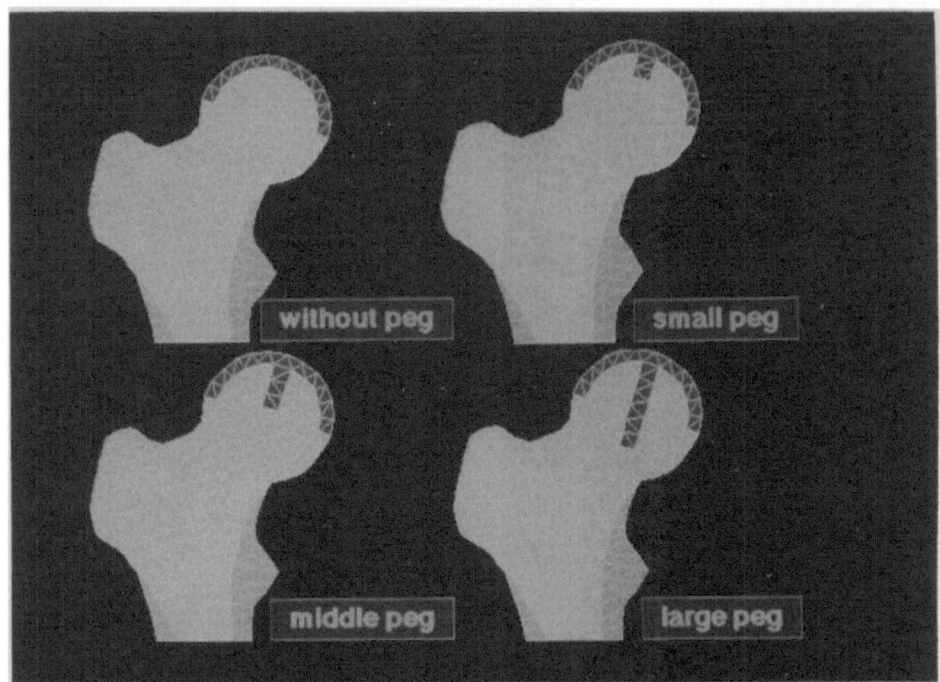

a

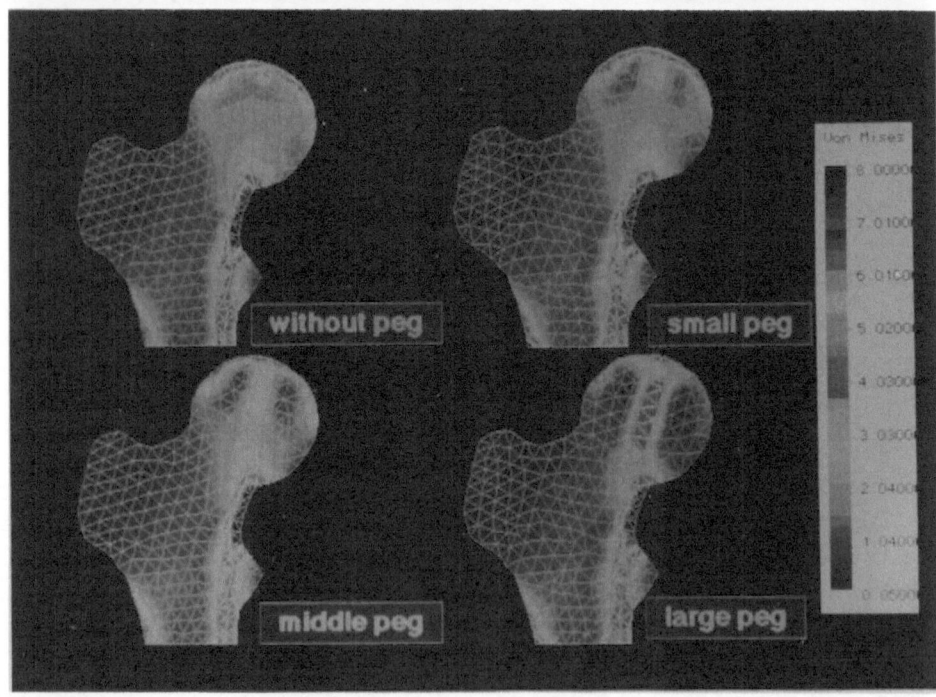

b

Fig. 6a,b. Effect of peg length **a** FEM model, **b** stress distribution on each model

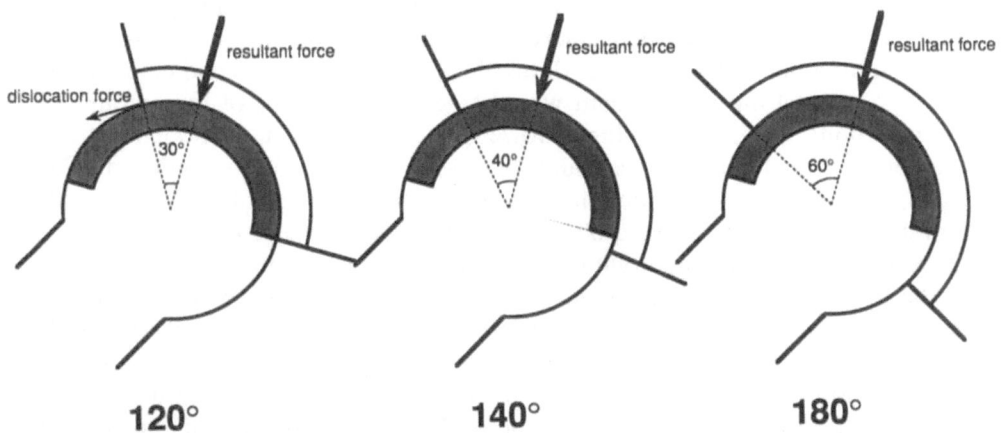

Fig. 7. Open angle of acetabular component. An open angle of less than 120° created the possibility of dislocation under physiological loading conditions

necrosis and bone resorption of the femoral head by minimizing the replaced area of the femoral head. Our results suggest that full coverage of the femoral head by the component and insertion of the implant parallel to the axis of the femoral neck could result in abnormal stress distribution, including stress shielding of the central portion of the femoral head and stress concentration at the medial side of the neck. The current type of surface THR is designed to be inserted along the axis of the femoral neck. However, the implant should be placed along the direction of the resultant force on the femoral head. Otherwise, an eccentric load can cause tilting of the component and variation in the stress distribution pattern. Because of its minimal resurfaced area, the femoral component of the new model can protect the nutrient vessels to the femoral head and can be inserted parallel to the axis of the resultant force. Therefore, a more physiological stress distribution and the possibility of long-term rigid fixation may be expected.

An extremely small femoral component, however, would lead to increased pressure at the interface and subsidence of the component. Coverage of the femoral head can be maximized up to the hemisphere without stress shielding, which feature is seen in the current type.

Although use of a peg is necessary to increase fixation rigidity, the use of a long peg should be avoided because it causes surrounding stress concentration and stress shielding under the component. Some minor changes in the shape of the peg might be necessary to prevent stress shielding.

Rigid-body spring model analysis for the open angle of the acetabular component showed that dislocation was not likely to occur in the component when the open angle was more than 140°. Compared to the stem type of femoral component, the surface type with a larger head diameter is not more likely to dislocate, since the head must translate to as much as a half of the head diameter for this to occur. Considering the possibility of socket-neck impingement [6], an open angle of 140° is desirable. Generally, larger head diameter of the surface total hip replacement would lead to an increase of the amount of resected bone in the acetabulum. However, with a smaller open angle of the acetabular component, the amount of resected bone can be reduced. For example, when a 10-mm-thick acetabular component is combined with a 40-mm-diameter femoral head, a 60-mm acetalubar component is necessary. Use of an acetabular component with a 140° open angle can minimize the diameter of the component to 56 mm. In addition, if the center of the rotation is moved laterally, an adequate high density polyethylene (HDP) thickness can be used. Furthermore, eccentric design of an acetabular component with a thick part in the weight-bearing area and a thin part in the non-weight-bearing area could minimize the diameter of the component. Thus, the problems of the surface total hip replacement, namely, a large acetabular component and a relatively thin polyethylene layer, could be prevented by using an acetabular component with a small open angle.

In conclusion, this study indicated that a minimal replacement area of the femoral head and placement of the implant along the direction of the resultant force on the femoral head were essential for obtaining physiological stress distribution and long-term rigid fixation. The optimal covering angle of the femoral head was 180°. Although a peg was essential for fixation rigidity, a long peg should be avoided. An acetabular component with an open angle of 140° is desirable for preventing both socket-neck impingement and dislocation.

New technologies in materials, design, and fixation methods can solve problems related to the surface total hip replacement, and efforts to develop an ideal surface replacement should be continued.

Acknowledgments. This work was supported, in part, by a Grant-in-Aid for General Scientific Research (Grant No. 02404061) from the Ministry of Education, Science, and Culture, Japan.

References

1. Dorr LD, Takei GK, Conaty JP (1983) Total hip arthroplasties in patients less than 45 years old. J Bone Joint Surg [Am] 65-A:474–479
2. Bradley GW, Freeman MAR, Revell PA (1987) Resurfacing arthroplasty. Clin Orthop 220:137–141
3. Nishio A, Eguchi M, Ogata K (1982) Socket and cup surface replacement. Orthop Clin North Am 13:843–856
4. Nishio A, Eguchi M, Kaibara N (1987) Socket and cup surface replacement of the hip. Clin Orthop 134:53–58

5. Miura H, Arima J, Hayashi K, Mashima T, Inadome T, Sugioka Y (1991) Biomechanical analysis of surface total hip replacement (in Japanese). In: Proceedings of
   the 1991 annual meeting of the Japanese Society for Orthopaedic Biomechanics.
   13:349–352
6. Charnley J (1979) Low friction arthroplasty of the hip. Springer, Berlin Heidelberg
   New York
7. Ma SM, Kabo JM, Amstutz HC (1983) Frictional torque in surface and conventional hip replacement. J Bone Joint Surg [Am] 65-A:366–370

23

Biomechanical Study of Femoral Stem Load Transfer in Cementless Total Hip Arthroplasty, Using a Non-Linear Two-Dimensional Finite Element Method

H. Oomori, S. Imura[1], H. Gesso[2], Y. Tanaka, K. Ichihashi, H. Takedani, Y. Okumura, and A. Boh[1]

Summary. To investigate the effects of the bonding characteristics and the stiffness of the stem on stem fixation and stress shielding of the femur, stress analyses of femoral stems in cementless total hip arthroplasty were performed by a two-dimensional finite element method (FEM), using non-linear elements in the stem-bone interface. The value and limitations of this method were examined and compared to a clinical radiographic follow-up study. The results of these analyses suggested that a coated surface to stimulate bone ingrowth was effective for the stable fixation of stems. However, it appeared to be better to limit the extent of the coated surface to the proximal region of the stem to avoid stress shielding of the proximal femur. With more flexible stems, stress shielding was avoided, but these stems tended to produce higher proximal interface stresses than the stiffer stems. The FEM analyses in this study were very useful for the prediction of stem fixation and bone changes around the stem in the comparative study of our postoperative radiographic evaluation. However, these analyses were not of value in quantitatively predicting serial bone changes in the femur due to stem migration.

Key words: Cementless total hip arthroplasty—Femoral stem—Load transfer—Stress analysis—Two-dimensional finite element method—Non-linear interface elements—Radiographic study

Introduction

The fixation of implants in cementless total hip arthroplasty depends on the bone ingrowth surrounding the implant. Thus, micromotion in implants cannot be avoided in the immediate and early stages after surgery before bone in-

---

[1] Department of Orthopaedic Surgery, Fukui Medical School, Shimoaizuki 23, Matsuoka-cho, Yoshida-gun, Fukui, 910-11 Japan
[2] Tsuruga Women's College, Kizaki 78-2-1, Tsuruga, Fukui, 914 Japan

growth occurs [1,2]. Micromotion in the bone-implant interface produces implant loosening [3]. Further, bone atrophy of the proximal femur with inserted stems, due to stress shielding, will create serious problems in the long-term fixation of implants. To overcome these problems, the biomechanical study of load transfer at the bone-implant interface and to the femur is very important.

Various investigators [4–7] have already shown that stress shielding of the femur is influenced by stem stiffness and by the bonding characteristics of the stem. However, very few of these studies were related to stem fixation.

A finite element method (FEM) has recently been playing an important role in the stress analysis of total hip arthroplasty. However, almost all FEM studies have used a linear analysis that assumes a fixed boundary around implants, and the results were not found to compare with clinical findings.

Against this background, we carried out this study to investigate the effects of bonding characteristics and stem stiffness on stem fixation and the stress shielding of the femur. We used a two-dimensional FEM employing non-linear elements, and we assumed a non-bonded condition at the stem-bone interface. We also attempted to determine the value and limitations of this FEM analysis by carrying out a comparative study of clinical radiographic evaluation.

Method

Our two-dimensional finite element model is shown in Fig. 1. It was created from anteroposterior radiographs after surgery, and consists of triangular elements. The front-plate represents the mid-frontal plane of the stem inserted into the proximal femur. The side-plate was used to connect the medial and

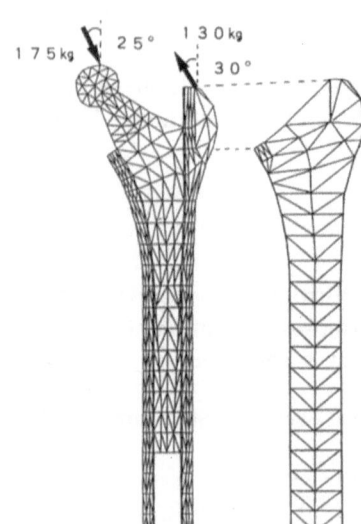

Fig. 1. Two-dimensional FEM model, this consists of a front-plate (*left*) and a side-plate (*right*). The front-plate represents the cross section, including the bone and stem structure, and the side-plate represents three-dimensional cortical bone integrity. The loading condition simulates a single-legged stance, based the work of Scholten et al. [11]

Fig. 2. Using non-linear interface elements (*D*), the non-bonded condition of the stem-bone interface is assumed to transfer only normal compressive stresses; $\sigma_\theta = 0$, frictionless; σ_r, compressive stress; *A*, cortical bone; *B*, cancellons bone; *C*, stem.

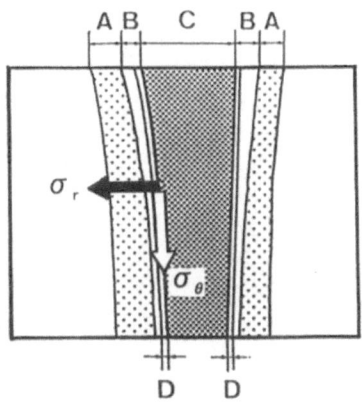

Table 1. Material properties of FEM model.

Material	Elastic modulus (Pa)	Poisson's ratio
Cortical bone	17.3×10^9	0.29
Cancellous bone	324.6×10^6	0.29
Stem (CoCrMo)	196.1×10^9	0.30
Stem (Ti6Al4V)	110.0×10^9	0.30
Cement (PMMA)	2.1×10^9	0.19

PMMA, Polymethyl methacrylate; *FEM*, finite element method

lateral cortices of the front-plate to simulate the tubular structure of the femur [8–10]. The model had 332 nodal points and 617 elements. The thickness of the elements was uniform. The thickness of the front-plate was 10 mm, and that of side-plate 5 mm. Loading conditions simulated a static single legged stance, based on the work of Scholten et al. [11].

The stem-bone interface in this model had elements that were a uniform 1 mm in width (Fig. 2). Utilizing these elements, we assumed that the non-bonding interface transferred only normal compressive stresses. When the non-linear interface elements were applied, slip and tensile separation could occur on the stem-bone interface. Calculations using these elements were iterated until the interface stresses had converged. The bonding interface was assumed to transfer compressive, tensile, and shear stresses in the fixed boundary condition of the usual FEM. These analyses were carried out using the FEM program developed by Gesso [12].

Table 1 shows the material properties of this model. The side-plate had the same material properties as the cortex.

From these FEM analyses, we obtained, the distribution of the stem-bone interface stresses and the von Mises stresses at the cortical bones. By comparing various models, we used the former stresses to predict stem fixation,

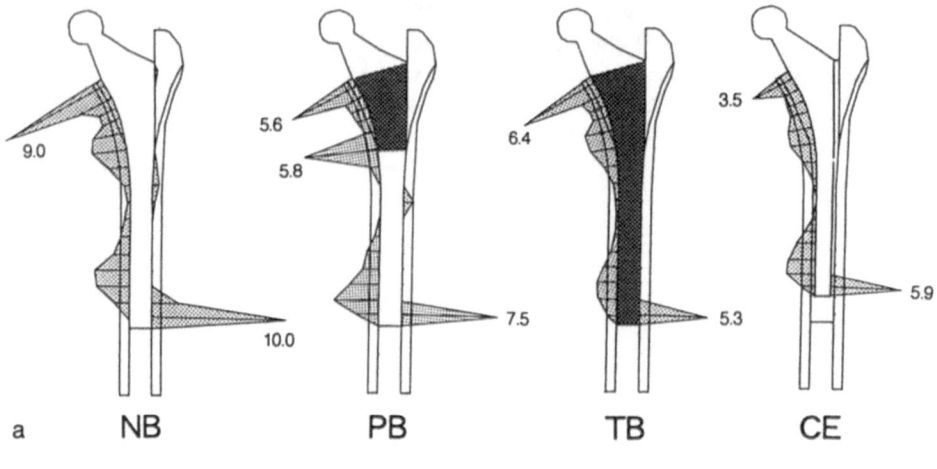

a   NB   PB   TB   CE

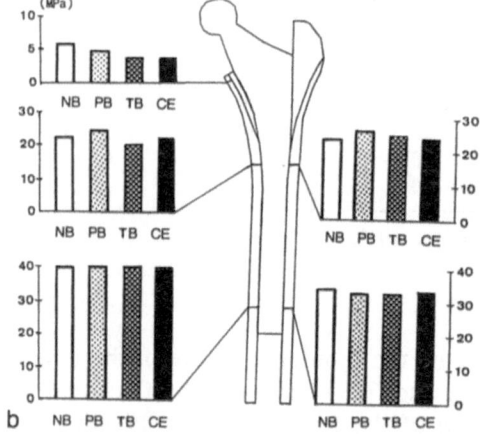

b   NB PB TB CE   NB PB TB CE

Fig. 3a,b. Effects of bonding characteristics around stems on load transfer. **a** Compressive stem-bone interface
stresses. The peak stresses are indicated
in the Fig. **b** von Mises stresses at the
cortical bones. *NB*, Non-bonded; *PB*,
proximally bonded; *TB*, totally bonded;
CE, cemented

and the latter to predict the degree of stress shielding of the femur with the
inserted stem.

To investigate the effects of the bonding characteristics of stems on load
transfer, we analyzed four models (Fig. 3a), namely, cementless non-bonded,
proximally bonded, and totally bonded stems, and a cemented stem model.
A bonded interface was assumed, representing the cemented stems and the
complete bone ingrown cementless stems at the porous-coated areas. With the
smooth-surfaced cementless stems, the non-bonded interface was assumed to
be frictionless; we used the non-linear elements as stated above. All stems used
were Omnifit (Osteonics Corp., Allendale, N.J.), made of CoCrMo alloy. The
cementless stems were assumed to be inserted in a press-fit manner, and the
cemented stem to be inserted with a cement mantle of about 2 mm.

To investigate the effects of stem stiffness on load transfer, we analyzed two
stem shapes with different diameters in the distal portion (Fig. 4a). One model

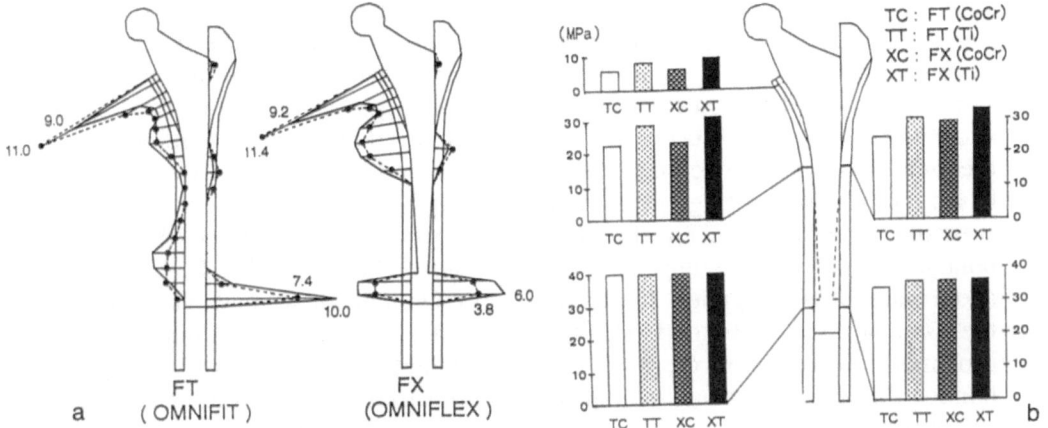

Fig. 4a,b. Effects of stem stiffness on load transfer. **a** Compressive stem-bone interface stresses in the two stem shapes (Omnifit [*FT*] and Omniflex [*FX*]). *Solid lines* represent CoCrMo alloy; 196 Gpa, and *dashed lines* Ti alloy; 110 Gpa. **b** von Mises stresses at the cortical bones for four stems with differing stiffness, *TC*, Omnifit, CoCrMo alloy; *TT*, Omnifit, Ti alloy; *XC*, Omniflex, CoCrMo alloy; and *XT*, Omniflex, Ti alloy

Table 2. Characteristic features of three stem types that were used in our clinic.

Stem type	Collar	Surface	Material
HS1	Collared	Smooth	CoCrMo
Omnifit	Collarless	Proximally coated	CoCrMo
Omniflex	Collarless	Proximally coated	Ti6Al4V

had a stiffer shape, representing an Omnifit stem, and the other a more flexible shape, representing an Omniflex (Osteonics) stem, the latter having a more tapered stem than the former. For the two stem models, two different materials, namely, CoCrMo alloy and Ti alloy were used. In these analyses the stem-bone interfaces were assumed to be non-bonded.

To compare the results of FEM analyses with clinical cases, we carried out a radiographic study of the femoral side in primary cementless total hip arthroplasties performed in our hospital. Table 2 summarizes the characteristic features of the three different stem types that were used in our clinic.

In the radiographic study, a total of 110 patients who had been followed for a minimum of 1 year were evaluated. The average follow-up period was 4 years (range, 13–99 months). This series included 22 patients with an HS1 stem (Osteonics), 66 with an Omnifit stem, and 22 with an Omniflex stem. Bone changes around the stems, i.e., endosteal bone formation and calcar resorption and atrophy were observed in the last follow-up radiographs, compared to the findings in the immediate postoperative radiographs. We divided the patients into two groups according to the stability of stem fixation (stable and unstable), and we graded four degrees of severity of stress shielding for the three stem types. With regard to stem fixation, we defined a radiographically unstable stem as one with definite evidence of either progressive migration or subsidence and with extensive radiolucent line formation more than 1 mm in width around the stem. With regard to stress shielding, the four degrees of severity were based on the work of Engh et al. literature [13].

In the FEM analyses for the above three stem types, we used the same bone geometry and loading conditions as those shown in Fig. 1. CoCrMo alloy material was applied to the HS1 and Omnifit stems and Ti alloy to the Omniflex stem. Bonded stem-bone interfaces were assumed only at the proximal porous coated areas of the Omnifit and Omniflex stems, and non-bonded interfaces were assumed at the smooth surfaced areas in the same way as outlined in Fig. 3a.

Results

The compressive interface stresses for the four models with different boundary conditions around stems of the same shape are shown in Fig. 3a. In each model, these stresses were generated predominantly on the medial side, and less on the lateral side, except at the distal end of the stem. The stress patterns were considered to be due to the bending effect on the loaded stem. Of all these models, the non-bonded stem generated the highest interface stresses, these being 9.0 MPa on the proximal medial side and 10.0 MPa on the distal lateral side. The proximally bonded, totally bonded, and cemented stems reduced and diffused these high stresses generated in the non-bonded stem. However, the proximally bonded stem generated more interface stresses near the distal end of the porous coated areas compared to the totally bonded and cemented stems.

Figure 3b shows the von Mises stresses in five regions along the cortical bone for the four models. In the distal region there were few differences between these models. However, the non-bonded stem transferred more stresses to the proximal medial cortical bones than the other stems. Increasing the bonding areas of the stems resulted in reducing the proximal bone stresses.

The compressive interface stresses of the two models with the different stem shapes (Omnifit and Omniflex) are shown in Fig. 4a. When made of the CoCrMo alloy, the more flexible shape (Omniflex stem) generated higher proximal stresses, but reduced and diffused distal stresses compared to the stiffer shape (Omnifit stem). With the Ti alloy stems, similar trends were found. With the stem shape of the Omnifit type, the maximum proximal stresses for the more flexible material (Ti alloy) were about 20% higher than for stiffer material (CoCrMo alloy), while the maximum distal stresses were about 50% lower. Similar stress patterns were found with the Omniflex shape.

The load transfer in these stems of varying stiffness affected the cortical bone stresses, as shown in Fig. 4b. There were few differences between the four

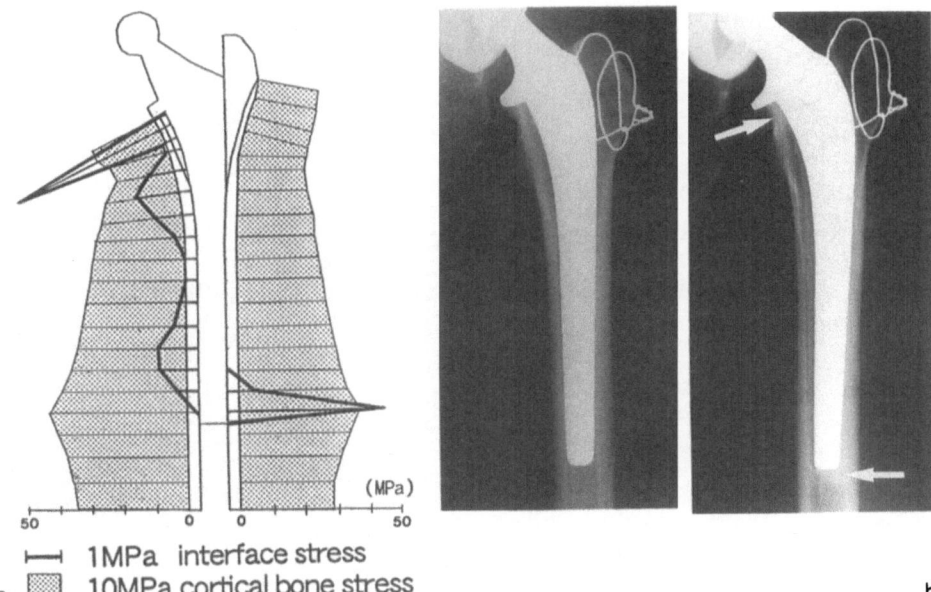

├──┤ 1MPa interface stress

a ▨ 10MPa cortical bone stress   b

Fig. 5. a Interface stresses and cortical bone stresses in the HS1 stem. **b** Immediate postoperative (*left*) and 3-year postoperative (*right*) radiographs of a patient with a HS1 stem. Endosteal bone formation has occurred, corresponding to the location of concentrated interface stresses (*arrows*)

models in the distal region. However, in the calcar region, the most flexible stem (Omniflex, Ti alloy) transferred more stresses to the cortical bones than the stiffer stems. The stiffest stem (Omnifit, CoCrMo alloy) reduced the load transfer to the calcar region by almost 25% compared to the most flexible stem.

Figures 5–7 show the results of FEM analyses and radiographs of the follow-up patients for three different stem types (the HS1, Omnifit, and Omniflex types).

The HS1 stem generated high interface stresses, 11.0 MPa in the proximal medial region and 10.5 MPa in the distal lateral region (Fig. 5a). Corresponding to these regions, endosteal bone formation appeared in about 80% of the radiographs of the HS1 patients (Fig. 5b). Cortical bone stresses tended to decrease towards the proximal region, but increased slightly in the calcar region (Fig. 5a). This phenomenon appeared as calcar sclerosis in about 95% of the radiographs (Fig. 5b).

At the distal end of the porous coated areas of the Omnifit stem, where maximum interface stresses of 5.8 MPa on the medial side were generated, endosteal bone formation occurred in 61% of the radiographs (Fig. 6a,b). Cortical bone stresses in the calcar region were very low, being only 15% of

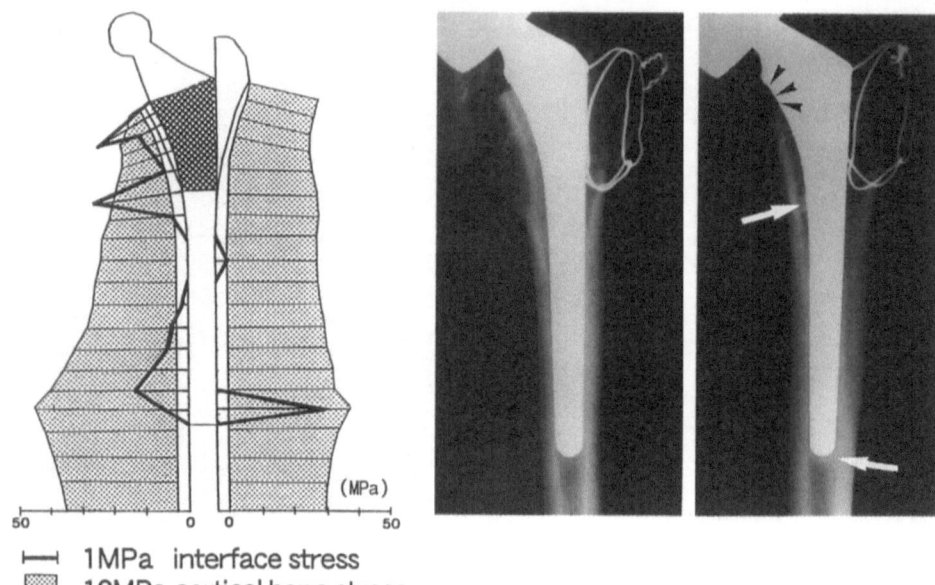

⊢──┤ 1MPa interface stress

a ▨ 10MPa cortical bone stress   b

Fig. 6. a Interface stresses and cortical bone stresses in the Omnifit stem. **b** Immediate postoperative (*left*) and 3-year postoperative (*right*) radiographs of a patient with an Omnifit stem. Calcar resorption due to stress shielding is apparent (*black arrows*). Endosteal bone formation, due to the localized stress transfer, is found near the distal end of the porous coating and around the tip of the stem (*white arrows*)

those in the tip of stem (Figs. 6a and 8a). Calcar resorption and atrophy occurred in 88% of the Omnifit cases radiographically (Fig. 6b).

The cortical stress patterns of the Omniflex stem were similar to those of the Omnifit stem (Fig. 7a). However, the calcar stresses were 22% of those in the tip of the stem, and they were about 7% higher than those of the Omnifit stem (Fig. 8a). The radiographic incidence of calcar resorption and atrophy was 53% in the Omniflex patients, and this figure was significantly lower than that in the Omnifit patients ($P < 0.05$). The medial interface stress in the distal end of the porous coating was 6.1 MPa, almost the same value as that in the Omnifit patients. Endosteal bone formation in this region occurred in 55% of the radiographs.

The maximum interface stresses of the three different stem types are summarized in Fig. 9a. The maximum interface stress of the HS1 stem was about two-fold higher than that of the Omnifit and Omniflex stems.

In the radiographic study, fixation was evaluated as stable in 99 of the total 110 patients (90%) and in 11 patients (10%) it was regarded as unstable (Fig. 9b). Ten of the 11 (90.9%) patients evaluated as unstable had had HS1 stems inserted, and 1 (9.1%) had had an Omnifit stem. Twelve of the 22 patients

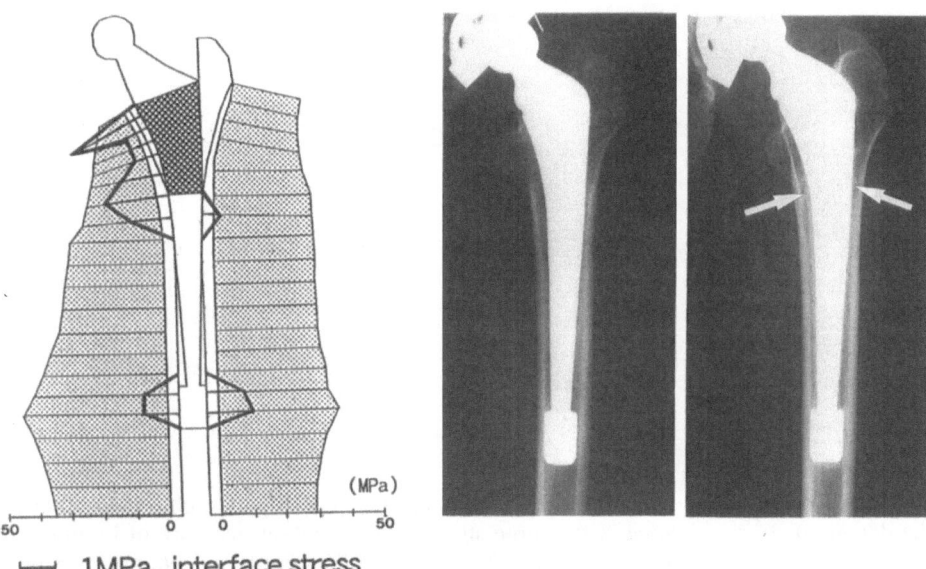

a   ⊢——⊣ 1MPa interface stress
           ▨ 10MPa cortical bone stress   b

Fig. 7. a Interface stresses and cortical bone stresses of the Omniflex stem. **b** Immediate postoperative (*left*) and 3-year postoperative (*right*) radiographs of a patient with an Omniflex stem. Endosteal bone formation is found at the distal end of the porous coating, as shown in Fig. 6b (*arrows*). Calcar atrophy and resorption is not as extensive as that shown in Fig. 6b

(55%) with a HS1 stem, 65 of the 66 patients (98%) with an Omnifit stem, and all 22 patients with an Omniflex stem were evaluated as having stable stem fixation.

The cortical bone stresses for the three different stem types, as determined by FEM analyses, and by the radiographic evaluation of the degree of stress shielding, are illustrated in Fig. 8a,b.

The Omniflex stem transferred more load to the cortex, except for the calcar region, than the other stem types (Fig. 8a). Calcar stresses were highest in the HS1 stem, in which they were about two-fold higher than in the other stem types.

In the radiographic study, it was found that the stem type affected the incidence of stress shielding, as shown in Fig. 8b. The first and second degree of stress shielding occurred in 27% of the HS1 patients, 65% of the Omnifit patients, and 94% of the Omniflex patients. The third and fourth degree occurred in 72% of the HS1 patients, 35% of the Omnifit patients, and 6% of the Omniflex patients. The HS1 patients had the highest incidence of severe stress shielding, while those with the Omniflex stem had the lowest incidence.

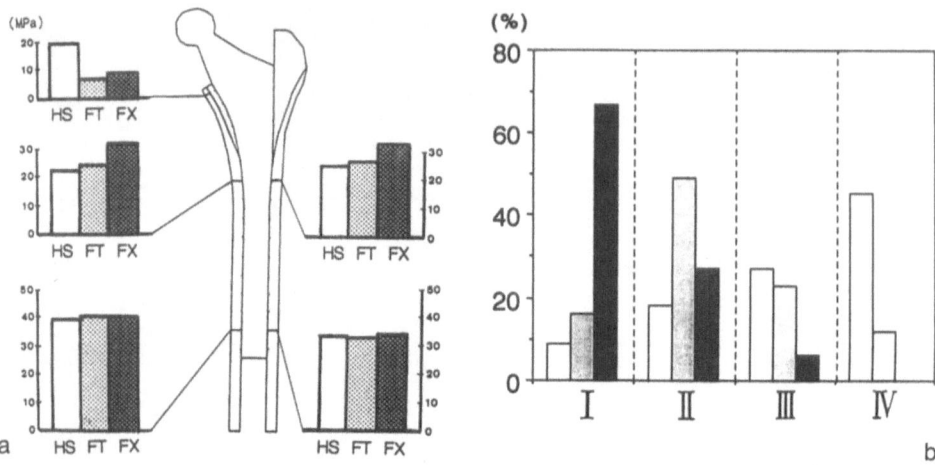

Fig. 8. a Cortical bone stresses in the three stem types. **b** Radiographic evaluation of the degree of stress shielding in the three stem types, based on the work of Engh et al. [13]. *HS*, HS1 stem (*open columns, n = 22*); *FT*, Omnifit stem (*gray columns, n = 66*); *FX*, Omniflex stem (*black columns, n = 22*)

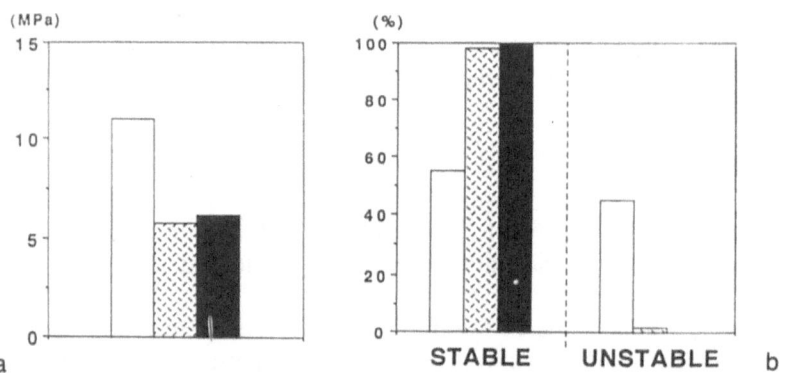

Fig. 9. a Maximum stem-bone interface stresses in the three stem types. **b** Radiographic evaluation of stem fixation in the three stem types. *HS*, HS1 stem (*open columns, n = 22*) *FT*, Omnifit stem (*patterned columns, n = 66*); *FX*, Omniflex stem (*black columns, n = 22*)

Discussion

Non-Linear Finite Element Method in This Study

The finite element method (FEM) is very useful for analyzing inner stresses of the femur and implant after total hip arthroplasty, particularly at the bone-implant interface. However, the limitations of this study must be understood in terms of the various assumed conditions, since these FEM models simplify the complex states that exist in the living body.

Firstly, this study used two-dimensional (2D) models, which could be analyzed more rapidly and more easily than the realistic three-dimensional (3D) models. However, it has been noted that the simple 2D model, as a "sandwich" construction, shows unrealistic mechanical behavior in terms of separating the cortices of the medial and lateral sides when a non-bonded condition is assumed at the stem-bone interface [14]. Therefore, to overcome this problem, we connected a side-plate, to simulate three-dimensional cortial bone continuity [8,9], to the simple 2D model (front-plate).

Secondly, these FEM models have limitations related to the boundary conditions at the stem-bone interface. As the original FEM is a linear analysis, the principle of which is to divide a continuous object into a finite number of small elements and then to calculate these elements respectively, the boundary condition must be assumed to be fixed. However, the bonded condition around stems can be simulated only in the case of cemented and bone ingrown cementless stems. Simulation of a smooth surfaced stem that cannot be bonded with bone after surgery requires non-linear conditions to represent the non-bonded interface. Several investigators [9,15] have emphasized the importance of non-linear FEM in the stress analysis of total hip arthroplasty. Two mathematical techniques for non-linear FEM are as follows: One is to put the condition of slip and friction in the nodal points of the stem-bone interface [15,16], and the other is to put these conditions in the interface elements [10,17]. This study used the latter method, inserting non-linear elements at the stem-bone interface to transfer only compressive stresses in the normal direction, but allowing for slip completely in the tangential direction [12]. Therefore, this method is unrealistic in the sense of not taking shear stresses into account. However, the friction coefficient of the shear stresses on the stem-bone interface remains unknown at present, and therefore the non-bonded condition of this study was assumed, as stated above.

Thirdly, these FEM models used a static loading condition, in which the parametric study of the load transfer mechanism was more easily interpreted than the realistic dynamic load.

Bonding Characteristics of Stems and Load Transfer

The results of this study showed that non-bonded stems, simulating smooth surfaced stems, generated higher concentrated interface stresses than other stems. If these stresses represent stem micromotion [18], this result suggests that smooth surfaced stems produce more micromotion than porous coated and cemented stems. Walker et al. [2] showed experimentally that cemented stems had significantly less micromotion than cementless press-fit stems. In our comparative study with the radiographic evaluation, HS1 stems, which have a smooth surface, generated the highest maximum interface stress of all the stem types, and were almost all evaluated as having unstable stem fixation. These findings suggest that a porous coated surface, which stimulates bone ingrowth, has a biomechanical advantage in terms of stable stem fixation.

With regard to the extent of the porous coated areas, the results of this study showed that a totally coated stem reduced load transfer to the proximal bone compared to the proximally coated stem. This phenomenon predicts that increasing the porous coated region would produce severe stress shielding of the proximal femur. Turner et al. [7] have confirmed this phenomenon experimentally, and Engh and Bobyn [5] confirmed it clinically.

In addition to this, the more interesting results obtained from this FEM analysis were that proximally coated stems, such as the Omnifit and Omniflex, generated relatively higher interface stresses at the distal end of the coating. Our clinical study of the Omnifit and Omniflex patients showed that endosteal bone formation at the corresponding areas was found in about 60% of those examined. These radiographic findings have been reported in the literature [5,19]. Geesink [19] found these bone changes at a higher incidence, of more than 90%, in a 2-year follow-up study of proximally hydroxyapatite-coated stems; these stems formed stronger bonds with living bone than did porous metal coating. Thus, these findings appear to be characteristic of proximally coated stems. These stems may prevent the stress shielding effect due to "stress bypass" of this area to the proximal femur.

Porous coated surfaces to stimulate bone ingrowth thus play an important role in stem fixation; however, limiting such coated surfaces to the proximal region is more advantageous in preventing bone loss in the proximal femur in long-term stem fixation. Almost all of these results above were obtained from a non-linear rather than from a linear FEM analysis, indicating that the non-linear method is of value for examining the effect of bonding on the mechanism responsible for load transfer.

Stem Stiffness and Load Transfer

Bobyn et al. [20] reported that stem stiffness was governed by the material modulus and geometric factors, and that the latter were related to the second moment of inertia of the stem. Therefore, increasing the stem diameter results in increasing stem stiffness. According to this rule, the Omnifit stem shape is stiffer than the Omniflex. The FEM results of this study showed that the Omniflex stem diffused and reduced distal interface stresses and increased load transfer to the proximal femur compared to the Omnifit stem. If the stem shape was the same, the Ti alloy had the same stress patterns as the CoCrMo alloy. This study confirmed the phenomenon already reported in the literature [4,6,21], namely, that reducing stem stiffness prevented the occurrence of proximal stress shielding of the femur.

In addition to this, the FEM results of this study showed that more flexible stems tended to generate greater proximal interface stresses than stiffer stems. In this respect, Huiskes et al. [21] pointed out that isoelastic stems (made of the same material as the cortex) tended to increase micromotion due to the high proximal interface stresses, and thus tended to cause stem loosening. Engh and Bobyn [5] emphasized the importance of individual bone quality, showing that stress shielding was influenced predominantly by stem stiffness relative to the femur, rather than by stem stiffness itself.

Determination of the level at which stem stiffness is reduced to prevent stress shielding and to keep stem micromotion to an acceptable level will require comparative studies of FEM analyses and clinical findings.

Significance of These FEM Analyses in our Clinical Cases

To interpret and evaluate the results of the FEM study, a comparative study of clinical findings is required. Therefore, in this study, we examined both radiographic evaluation and FEM analyses for the three kinds of stems that had been used in cementless total hip arthroplasties in our clinic. The FEM models in this comparative study were created by taking into account the characteristic features of each of the stems, namely, the stem design, extent of porous coating on the stem surface, and the stem material.

The results of the FEM analyses showed that the HS1 stem generated more concentrated interface stresses and higher maximum stresses than the other stems. The radiographic study showed that about 90% of the unstable stems were the HS1 type. When stem-bone interface stresses, representing the micromotion of the stem [18] were compared, the FEM results were found to be consistent with the results of the clinical study. As stated above, endosteal bone formation was shown to be occurred at a high incidence in the radiographs, corresponding to those locations where concentrated interface stresses were found, i.e., the distal end of the porous coated areas, the proximal medial region under the collar, and the distal end of the stem. These radiographic findings suggested that stem-bone interface stresses stimulated reactive bone changes around the stem.

With regard to the stress shielding of the femur, radiographic evaluation showed that the Omniflex stem had a lower incidence of severe stress shielding than the Omnifit stem; this could be understood to be due to the effect of stem stiffness, as shown by the FEM results. Radiographically, the HS1 stem had the highest incidence of severe stress shielding, despite showing higher calcar stresses than the other stems in the FEM analyses. These FEM results we considered to be due mainly to the differences in stem design, as well as to the non-bonding effect of the smooth surface. Collared stems, such as the HS1, transferred more compressive stresses through the collar directly to the calcar region, compared to collarless stems such as the Omnifit and Omniflex, resulting in the generation of higher calcar stresses in the HS1 stem. This load transfer mechanism has been confirmed by several experimental studies [22,23]. The high incidence of the severe stress shielding seen in the HS1 stem radiographically is probably related to the unstable fixation of this stem. Stem migration is thought to cause such biological reactions as disuse atrophy and thinning of the cortex, which features result in the radiographical evaluation of severe stress shielding. In other words, the biological bone reaction with unstable stems is considered to be the cause of the differences between the FEM results and the radiographic evaluation with regard to the stress shielding of the femur.

In conclusion, these FEM analyses, using non-linear elements, were of great value for the prediction of stem fixation and endosteal bone change around stems. However, with these analyses, predictions of the level at which interface stress causes endosteal bone change and stem migration, and predictions of the amount of femoral bone loss due to stem loosening could not be made. Overcoming these problems will require quantitative clinical studies of bone remodeling around stems and FEM studies that take the time factor into account.

References

1. Pilliar RM, Lee JM, Maniatopoulos C (1986) Observations on the effect of movement on bone ingrowth into porous-surfaced implants. Clin Orthop 208:108–133
2. Walker PS et al. (1987) Strains and micromotions of press-fit femoral stem prostheses. J Biomech 20:693–702
3. Perren SM (1984) The induction of bone resorption by prosthetic loosening. The cementless fixation of hip endoprostheses. Springer, Berlin Heidelberg, pp 39–41
4. Bobyn JD, Glassman AH, Goto H, Krygier JJ, Miller JE, Brooks CE (1990) The effect of stem stiffness on femoral bone resorption after canine porous-coated total hip arthroplasty. Clin Orthop 261:196–213
5. Engh CA, Bobyn JD (1987) The influence of stem size and extent of porous coating on femoral bone resorption after primary cementless hip arthroplasty. Clin Orthop 231:7–28
6. Maistrelli GL, Fornasier V, Binnington A, McKenzie K, Sessa V, Harrington I (1991) Effect of stem modulus in a total hip arthroplasty model. J Bone Joint Surg [Br] 73-B:43–46
7. Turner TM, Sumner DR, Urban RM (1986) A comparative study of porous coatings in a weight-bearing total hip arthroplasty model. J Bone Joint Surg [Am] 68-A:1396–1409
8. Huiskes R, Chao EYS (1983) A survey of finite element analysis orthopedic biomechanics: The first decade. J Biomech 16:385–409
9. Svensson NL, Valliappan S, Wood RD (1977) Stress analysis of human femur with Charnley prosthesis. J Biomech 10:581–588
10. Yettram AL (1989) Effect of interface conditions on the behaviour of a Freeman hip endoprosthesis. Sci Technical Record 11:520–524
11. Scholten R, Rohrle H, Scollbach W (1978) Analysis of stress distribution in natural and artificial hip joints using the finite element method. S Afr Mech Eng 28:220–225
12. Oomori H, Imura S, Gesso H (1992) Stress analysis of femoral stems in cementless total hip arthroplasty by two-dimensional finite element method using boundary friction layer. J Jpn Orthop Assoc 66:240–252
13. Engh CA, Bobyn JD, Glassman AH (1987) Porous-coated hip replacement. J Bone Joint Surg [Br] 69-B:45–55

14. Oonishi H, Kawaguchi A (1985) Pseudo-three-dimensional model formation for a two-dimensional FEM of a tubular bone. Proceedings of 1985 Annual meeting of the Japanese Society for Orthopaedic Biomechanics 10:79–84
15. Hampton SJ, Andriacchi TP, Draganich LF, Galante JO (1981) Stress following stem cement bond failure in femoral total hip implants. In: Hayes WC (ed) Transactions of the 27th annual meeting of the Orthopaedic Research Society. Orthopedic Research Society, Park Ridge, p 144
16. Azuma T (1985) Preparation of the acetabulum to correct severe acetabular deficiency for total hip replacement. J Jpn Orthop Assoc 59:269–283
17. Brown TD, DiGioia A III (1984) A contact-coupled finite element analysis of the natural adult hip. J Biomech 17:437–448
18. Huiskes R (1986) Biomechanics of bone-implant interactions. Frontiers in Biomechanics. Springer, New York Berlin Heidelberg Tokyo, pp 245–262
19. Geesink RGT (1990) Hydroxyapatite-coated total hip prostheses. Clin Orthop 261:39–58
20. Bobyn JD, Mortimer ES, Glassman AH, Engh CA, Miller JM, Brooks CE (1991) Producing and avoiding stress shielding. Clin Orthop 274:79–96
21. Huiskes R, Weinans H, Rietbergen B (1991) The relationship between stress shielding and bone resorption around total hip stems and the effects of flexible materials. Clin Orthop 274:124–134
22. Crowninshield RD, Brand RA, Johnston RC, Pedersen DR (1980) Analysis of collar function and the use of titanium in femoral prostheses. Clin Orthop 158:270–277
23. Markolf KL, Amstutz HC, Hirshowitz DL (1980) The effect of collar contact on femoral component micromovement. J Bone Joint Surg [Am] 62-A:1315–1323

24

Load-Bearing Mechanisms of Natural and Artificial Joints

Masanori Oka[1]

Summary. The transmission of forces through the hip joint was analyzed, using a new quasi three-dimensional finite element method program with gap elements, which enabled contact and sliding movements at articular surfaces. In the human hip joint model, stresses were evenly distributed toward the distal femur. In the Charnley total hip replacement model, a stress shielding phenomenon was found in the proximal femur. In a surface hip arthroplasty using a new artificial articular cartilage, it was confirmed that, by changing the design of the prosthesis, the stress pattern could be made to resemble that of the human hip joint.

Key words: Hip joint—Stress distribution—FEM—New artificial joint—Stress shielding

Introduction

It has been reported previously that the trabecular structure of the subchondral cancellous bone plays an important role in the joint load-bearing mechanism, particularly in regard to shock-absorbing capacity. In current artificial joint designs, however, the cartilage and the subchondral cancellous bone, with their capacity for recovery, are removed and replaced by artificial materials with completely different material properties [1–3]. We are currently developing a new artificial joint which will replace only the surface of the joint with artificial cartilage and will retain as much healthy cancellous bone as possible [4–6].

When designing a new artificial joint, we believe it is necessary to give due consideration to the biological responses to loading stress, i.e., remodeling.

With this in mind, we carried out stress analysis on a polyvinyl alcohol hydrogel (PVA-H) surface-replacement artificial hip, a Charnley-type artificial hip, and a healthy biological hip joint. To perform the stress analyses, we

[1]Department of Artificial Locomotive Systems, Research Center for Biomedical Engineering, Kyoto University, Shogoin, Sakyo-ku, Kyoto, 606 Japan

utilized quasi three-dimensional finite element modeling (FEM) with a bridging-plate. PVA-H is an artificial articular cartilage material that we are currently developing; it has excellent lubrication and shock-absorbing properties.

Methods

The program used to do the quasi three-dimensional FEM analyses was COSMOS/M from SRAC (Structural Research and Analysis, Santa Monica, Calif.). Hardware utilized was the PC-9801RA from NEC (Tokyo, Japan) [7].

Model Structure and Materials

To be able to take into account muscle forces and weight-bearing [8], a three-dimensional model of the femur and the pelvis (ilium, sacrum, pubis, and ischium) was constructed by extruding the front view two-dimensional model into space. This is seen in Fig. 1.

In the case of the biological hip joint and the PVA-H surface replacement artificial hip, the femur head was modeled as a sphere of 50-mm diameter and the acetabulum was modeled as a spherical indentation with a diameter of 52 mm.

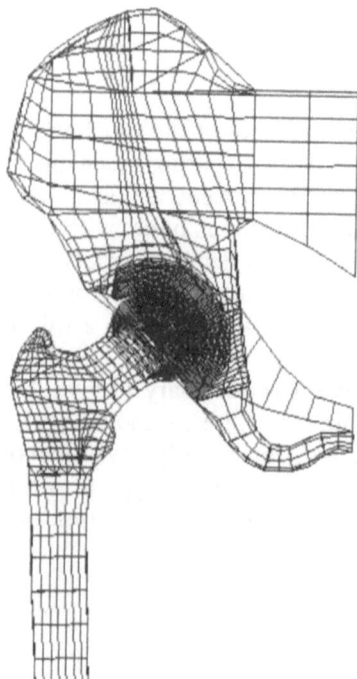

Fig. 1. Finite element model (*FEM*) mesh for human hip joint model, including proximal femur and pelvic bones

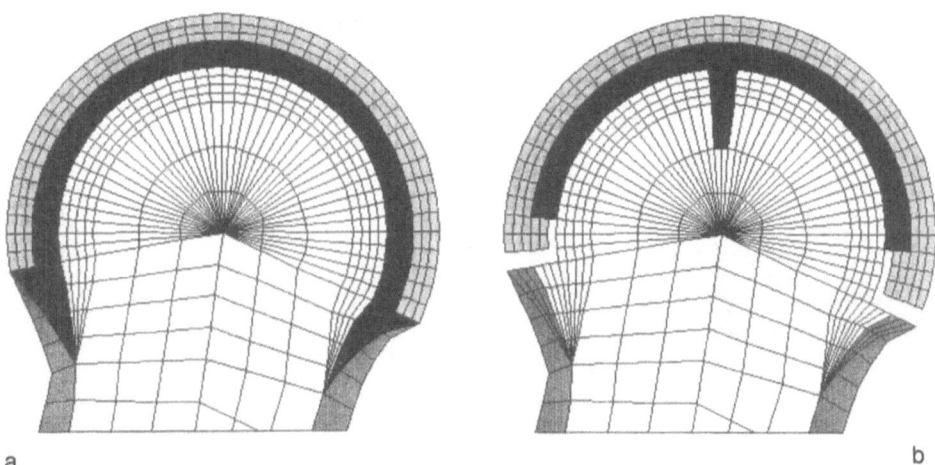

a   b

Fig. 2a,b. Surface hip arthroplasty. **a** Type 1, **b** type 2

For the material distribution around the joint in the biological model, a 3-mm-thick layer of cartilage was placed next to the joint surface, and underneath that, a 1-mm-thick layer of subchondral bone was placed.

In the PVA-H surface replacement artificial hip, we fixed the artificial cartilage to the femur head by injecting pure PVA liquid into the interior of a titanium fiber mesh porous artificial bone, creating an artificial osteochondral composite by bonding the two at the gel stage; this composite was then fixed to the head. Thus, the material distribution around the joint in the PVA model was modeled as 3-mm-thick PVA-H from the joint surface and, below that, as 3-mm-thick titanium mesh. In these experiments, we analyzed two models with differing composite configurations, in order to determine the effect of composite design on stress distribution (Fig. 2). The type-1 configuration seen in Fig. 2 replaces the subchondral bone in the femur with the titanium mesh and completely surrounds the cancellous bone of the femoral head. In the type-2 configuration, to ease surgery, the titanium mesh was made as a hemispherical shell and an oblong anchor was attached to the shell to ensure strong fixation. The remaining portions in both the biological and the PVA models were modeled as cortical bone, cancellous bone, and bone marrow.

For the Charnley artificial hip, the stem head diameter was set as 22 mm and the acetabulum was modeled as a hemispherical socket with a diameter of 22.88 mm (same curvature as the biological model, but with a 4% difference in radius). The three-dimensional model, constructed by extruding the two-dimensional model into space, similarly to the biological model, is illustrated in Fig. 3.

For material distribution, the Charnley model was modeled with a ceramic femoral head on a stainless steel stem, fixed with bone cement, and the

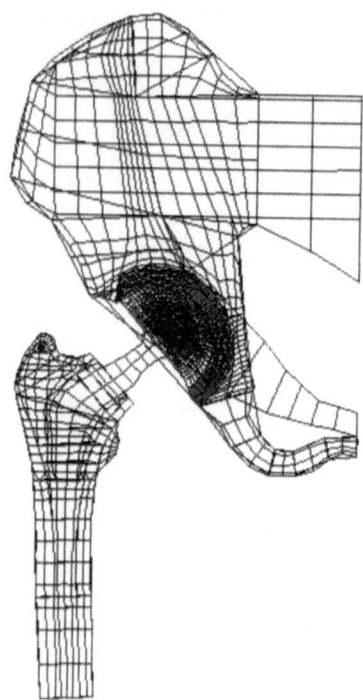

Fig. 3. FEM mesh for total hip replacement (*THR*) model, including proximal femur and pelvic bones

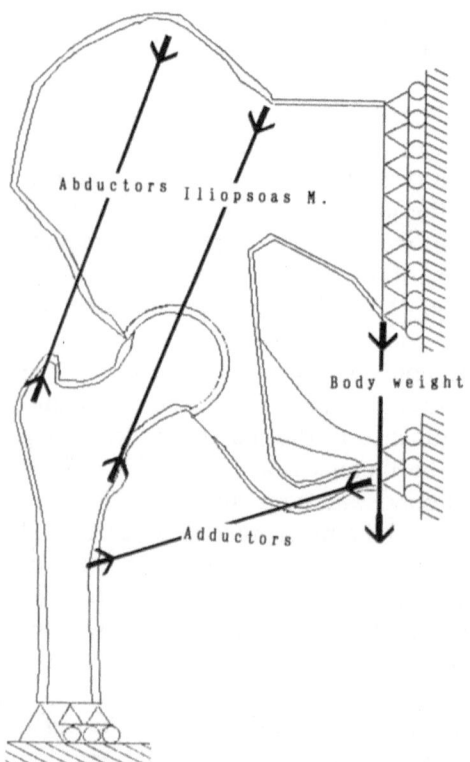

Abductors Iliopsoas M.

Body weight

Adductors

Fig. 4. Loading and boundary conditions

Table 1. Material property values.

Material	Young's modalus (MPa)	Poisson ratio
Cortical bone	7 000	0.25
Articular cartilage	10	0.49
Subchondral bone	3 000	0.25
Cancellous bone	1 000	0.3
PVA-H	100	0.4
Titanium fiber	30 000	0.3
UHMWPE	500	0.4
Bone cement	2 000	0.3
Stainless steel	100 000	0.3
Alumina	350 000	0.19
Spring	0.01	0.3

PVA-H, Polyvinyl alcohol Hydrogel; *UHMWPE*, ultra high molecular weight polyethylene

acetabulum was modeled as an 8-mm-thick high-density polyethylene (HDP) socket with 3 mm of bone cement underneath.

The constants used in calculations are listed in Table 1.

Parameters of the Analysis

Considering the abductor muscle, the iliopsoas muscle, the adductor muscle, and body weight, the loading conditions were set to 1800 N of resultant force located at the center of the femoral head, 16° from normal (Fig. 4). The model was constructed such that the distal end of the femur was fixed, with the rest of the distal portion restricted to horizontal motion. The proximal end of the femur (near the midline of the acetabulum) was restricted to vertical motion.

Also, as part of our analysis, we incorporated a 1-mm-gap in the biological model of the joint (0.44-mm-gap in the Charnley model) such that, initially, the femur and the pelvis were two separate bodies moving freely. When force was applied, the two bodies approached and forces were exerted between them upon contact. By using gap elements in our analysis, we can take into account the effects of interaction between soft compounds.

Results and Discussion

After FEM analysis was completed, the regions were divided into 15 color gradations based on the type of stress; these were displayed as stress distribution maps overlaid onto post-deformation cross sections.

When we compared the absolute values for the minimum and maximum principal stresses, we discovered that the minimum principal stress was one magnitude larger; the comparisons between the different models were therefore primarily based upon the minimum principal stress.

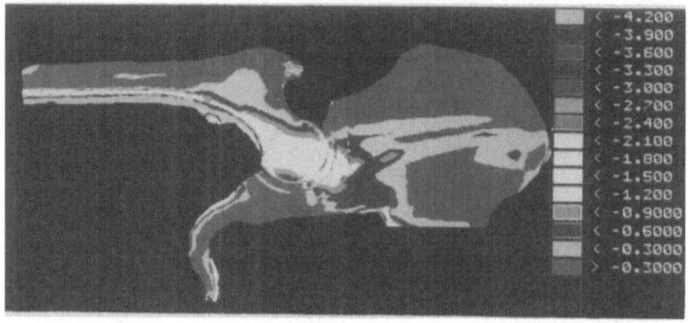

Fig. 5. Minimum principal stress distribution in human hip joint model, (−) values indicate compressive and (+) values, tensile stress

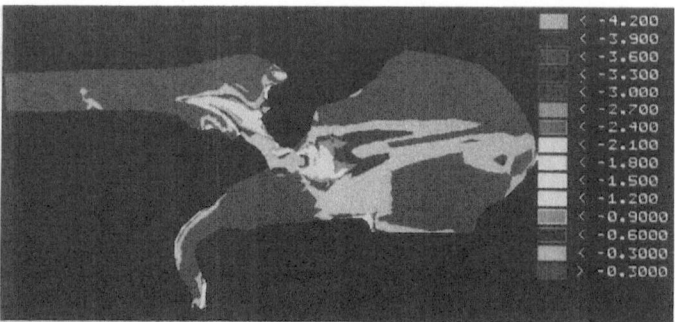

Fig. 6. Minimum principal stress distribution in artificial hip joint composed of stainless steel stem and ultra high molecular weight polyethylene (*UHMWPE*) socket

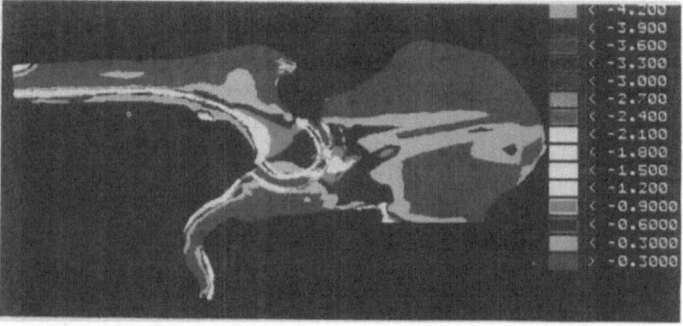

Fig. 7. Minimum principal stress distribution in surface arthroplasty created with artificial osteochondral composite materials composed of polyvinyl alcohol hydrogel (*PVA-H*) and titanium fiber mesh

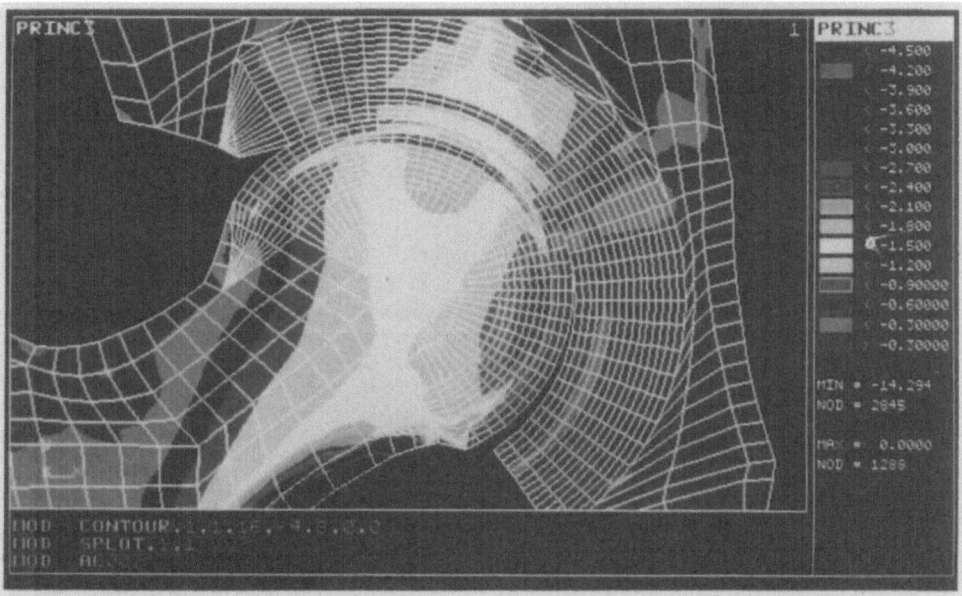

Fig. 8. Minimum principal stress distribution in human hip joint

In the biological model (Fig. 5), the transmission of pressure followed the trabecular distribution almost exactly and was spread over a large area in the joint surfaces. Also, a fairly large compressive stress, approximately −9.0 MPa, was seen within the inner cortex.

In the Charnley model (Fig. 6), very little compressive stress could be seen within the inner cortex; it is believed that this might be a cause of loosening due to stress shielding. Also, the spread of force in the acetabulum was small and demonstrated a different force distribution compared to that in the original hip. This would not be desirable from a remodeling standpoint.

In comparison to this, the minimum principal stress distribution map for the PVA model (Fig. 7) showed a similar distribution to the biological model, except within the femoral head.

With regard to the stress distribution within the femoral head in the biological model (Fig. 8), the pressure transmitted after contact spread through the cancellous bone to the cortical bone. In this model, the minimum principal stress of the cancellous bone in the femoral head was an average of, −1.39 MPa.

In comparison, the type-1 PVA model is designed such that the titanium mesh completely surrounds the cancellous bone in a shell. The pressure transmitted after contact passes through the titanium mesh to the cortical bone and does not pass to the cancellous bone. Thus, compared to the biological model, the average minimum principal stress of the cancellous bone in the femoral head in this model is −0.46 MPa and it shows stress shielding to the cancellous

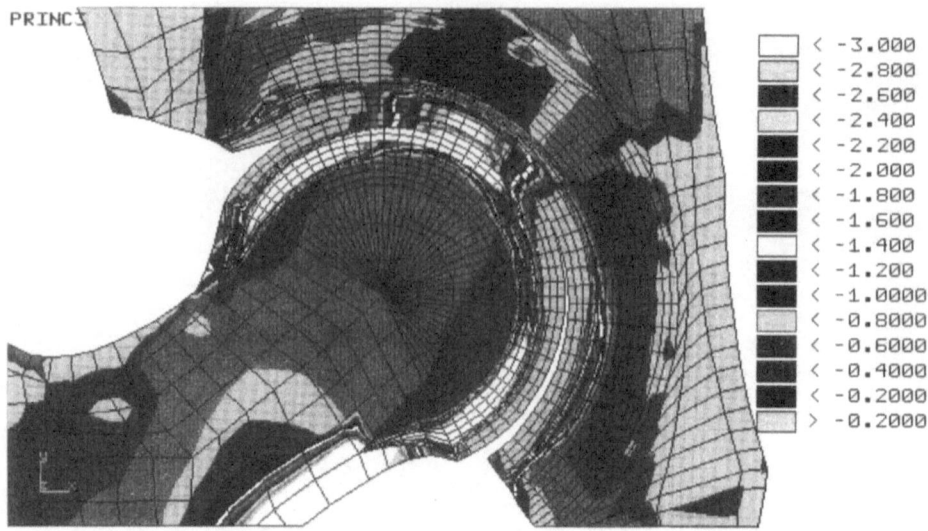

Fig. 9. Minimum principal stress distribution in type 1 surface hip arthroplasty

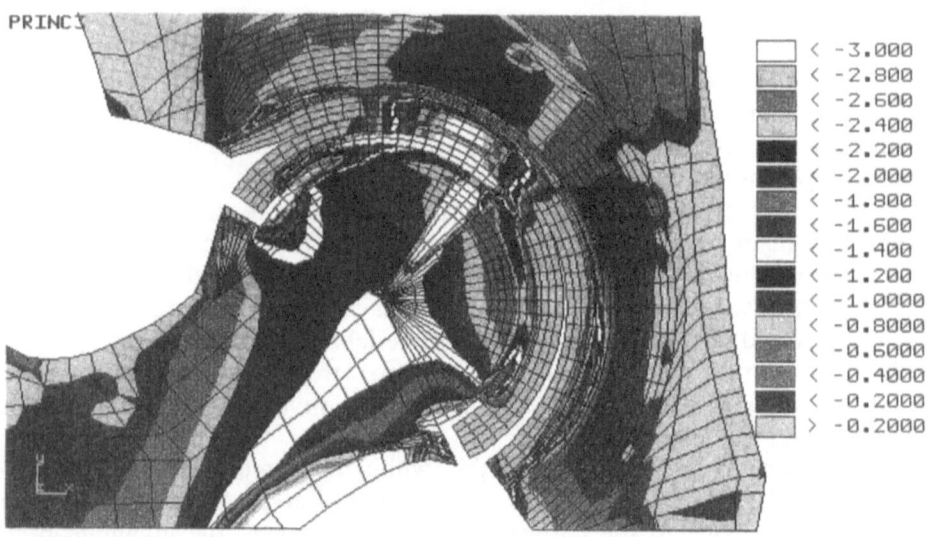

Fig. 10. Minimum principal stress distribution in type 2 surface hip arthroplasty

bone of the femoral head (Fig. 9). This is a possible cause of loosening of artificial osteochondral composite materials.

On the other hand, in the type-2 PVA model (Fig. 10), the pressure passes through the anchor to the cancellous bone and then to the cortical bone. The average minimum principal stress of the cancellous bone in the femoral head is −1.31 MPa; this can be seen to approach the value for that of the biological model. In other words, this model is desirable from a remodeling standpoint.

Finally, after the shear stress that occurred between the PVA-H and the titanium mesh was examined in the PVA model, our analysis showed a maximum shear stress of 0.85 MPa. Testing showed that the maximum sustainable shear stress was 2.2 MPa, so there would be little danger of the PVA-H separating from the titanium mesh.

Thus, by changing the shape of the titanium mesh in the PVA-H surface-replacement artificial hip, we showed that it was possible to approach the biologically transmitted stress. Once more, this finding emphasizes the importance of design in artificial hips.

Conclusions

1. We investigated loading transmission mechanisms in biological and artificial hip joints, using a FEM technique that allows deformation upon contact of two objects separated by a gap, and subsequent stress analysis.
2. In artificial joints, stress shielding occurred in the inner cortex, while in a biological hip joint the stresses were distributed evenly.
3. By changing the shape of the titanium mesh in a new surface-replacement artificial hip made with an artificial osteochondral composite material, it was possible to obtain stress transmission similar to that in the original hip.

References

1. Oka M, Ikeuchi K, Kotoura Y, et al (1988) Deformation characteristics of articular cartilage and prosthetic materials (in Japanese). Orthopedic Biomechanics 10:71–75
2. Oka M (1989) Biorheology of articular cartilage (in Japanese). J Jpn Rheol Assoc 3:152–163
3. Oka M (1990) Material properties of artificial joints (in Japanese). Joint Surg 9:41–52
4. Oka M, et al (1988) Present and future of artificial joints. Rheumatism 28:236–273
5. Oka M, Noguchi T, Kumar P, et al (1990) Development of an artificial articular cartilage. Clin Mat 6:361–381
6. Oka M (1990) Development of an artificial articular cartilage (in Japanese). Med Philosophica 10:52–58
7. Oka M, Tonomura K, Tsutsumi S, et al (1991) Load bearing mechanism of natural and artificial joints (fourth report) (in Japanese). Orthopedic Biomechanics 12:93–97
8. Oka M (1989) Theory of osteotomy, viewed from biomechanics of hip joint (in Japanese). Joint Surg 8:949–958

25

Biomechanical Features of Various Cementless Total Hip Replacement Systems

Takatoshi Ide, Rikio Amano, and Noriya Akamatsu[1]

Summary. The selection of an implant and the surgical placement of a prosthetic component in total hip replacement are based on limited radiographic and clinical information. To optimize surgical outcome and anticipate intraoperative problems, computer-aided simulation/analysis software was developed based on the rigid body spring model (RBSM), using plane radiographs (two-dimensional [2-D] analysis) to obtain implant/bone interface stress, ligamentous tension, and relative displacement for bony segments and implants. In the study of our total hip replacement (THR) system which features three spikes on the socket, the main assessment criterion used was an implant fixation index based on implant/bone interface stress and relative displacement of the prosthesis in reference to bone. The computer simulation was consistent with the clinical results for our cementless THR system and it showed that implant design and placement significantly affected prosthesis/bone interface stresses. Therefore, this method can be expected to provide useful information in the preoperative planning of difficult reconstructive cases involving the hip.

Key words: Cementless THR—Stress analysis—Computer simulation—Rigid-Bogy Spring Model

Introduction

Of the many different types of cementless total hip replacement (THR) systems now available it is almost impossible to select one over another. One important factor to be heeded when assessing their relative merits is the interface between the stem and the bone. In this study, our objectives were to estimate the force distribution in the prosthesis/bone interface, using a computer simulation technique, and to predict the prognosis of cementless THR systems biomechanically.

[1] Department of Orthopedic Surgery, Yamanashi Medical University, 1110 Shimokato Tamaho-cho, Nakakoma-gun, Yamanashi, 409-38 Japan

266 T. Ide et al.

Methods

Computer Simulation Model of the Cementless THR

Stress analysis was performed using Kawai's rigid body spring model (RBSM) [1]. A frontal section at the hip joint, consisting of the pelvis and femur, artificial acetabular component and the femoral component, was used as a two-dimensional and four-element computer simulation model (Fig. 1). This model assumes that bone segments and implant components are rigid bodies connected by distributed compressive and shear spring elements at the articulating surface and the implant/bone interfaces. The system will then be investigated, using equilibrium analysis based on energy principles, in an iterative manner.

The virtual springs which resist the compressive and shear force were positioned 1 mm apart along the implant/bone interface. The virtual muscle structure springs, which resist only tensile force, were positioned between the pelvis and the great trochanter of the femur as the abduction muscle. An external load of 50 kgf, as the body weight, was loaded at the center of the fifth lumbar spine, and the femur element was fixed simultaneously. The center of gravity of the pelvis was positioned at the center of the fifth lumbar spine, and the femoral center of gravity was positioned at the mid-point of the femoral shaft.

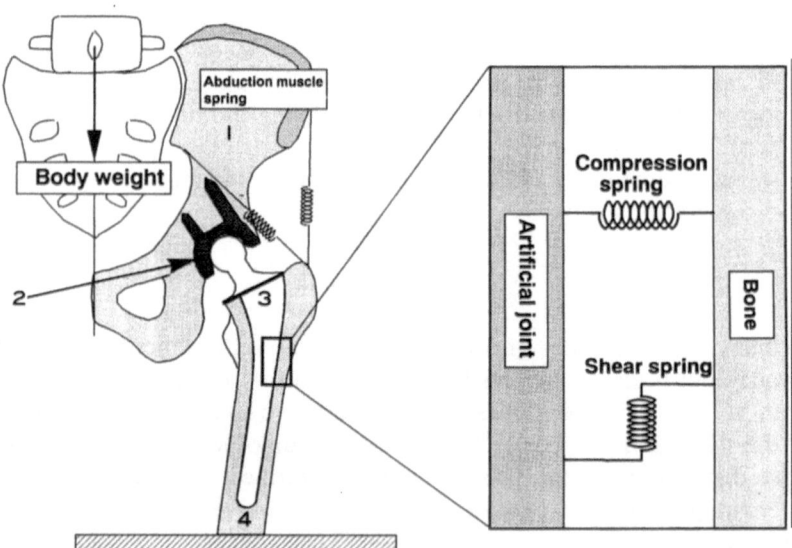

Fig. 1. Computer simulation model for stress analysis using the rigid body spring model (*RBSM*). Bone segments and implants are assumed to be rigid bodies connected by distributed virtual springs at the bony/implant interfaces. *Element 1*, Pelvis; *2*, socket; *3*, stem; *4*, femur

$$\text{Fixation index} = \frac{100}{10} \sum \left[\begin{array}{cc} \dfrac{\text{Mean interface stress of femoral head A}}{\text{Mean interface stress of femoral head R}} & \dfrac{\text{Max interface stress of socket A}}{\text{Max interface stress of socket R}} \\[2mm] \dfrac{\text{Virtual displacement of socket A}}{\text{Virtual displacement of socket R}} & \dfrac{\text{Max interface stress of stem A}}{\text{Max interface stress of stem R}} \\[2mm] \dfrac{\text{Max interface stress of femoral head A}}{\text{Max interface stress of femoral head R}} & \dfrac{\text{Rotation of socket A}}{\text{Rotation of socket R}} \\[2mm] \dfrac{\text{Mean interface stress of socket A}}{\text{Mean interface stress of socket R}} & \dfrac{\text{Virtual displacement of stem A}}{\text{Virtual displacement of stem R}} \\[2mm] \dfrac{\text{Mean interface stress of stem A}}{\text{Mean interface stress of stem R}} & \dfrac{\text{Rotation of stem A}}{\text{Rotation of stem R}} \end{array} \right]$$

Socket & stem A : The implant to be evaluated
Socket & stem R : The reference implant to be compared with

Fig. 2. Fixation stiffness index in total hip replacement: An implant fixation index was used as the main assessment criterion, and the model-Y2 was defined

From the analysis using the RBSM, the stress distribution on the implant/bone interface was estimated in the one leg standing situation.

Assessment Criteria of Mechanical Stability in THR System

In this study, the major assessment criterion used was an implant fixation index, based on the implant/bone interface stress and relative displacement of the prosthesis in reference to bone. Stress concentration at the implant/bone interface is an important indicator of the biomechanical stability of THR systems (Fig. 2). The maximum compressive stress on the bone/implant interface was estimated, via analysis using the RBSM, as an indicator of stress concentration. It was also possible to estimate the rotational and X-Y plane deviation in this study, since each of the elements in the RBSM was assumed to be a rigid body. Since the rotational deviation of the perpendicular line on the center of gravity of the element is equivalent to the medial and lateral shift of the stem, or to the opening of the socket, we selected rotational deviation as an indicator for the shifting of the stem in this study.

Objectives

Various THR Systems for Stress Analysis

Different types of cementless THR systems, both those used in the past and those now in use, were analyzed. The JIAT (Jikei Itami, Akamatsu, and Tomita) is a first generation cementless THR system originated in Japan, and the model-Y and model-Y2 [2] were developed at the Department of Orthopaedic Surgery, Yamanashi Medical University (Fig. 3). Stress analysis of various other cementless THR systems on the market was also performed using the computer simulation technique with the RBSM. A cemented type THR and the model-Y2 collarless type, a hypothetical computer simulation

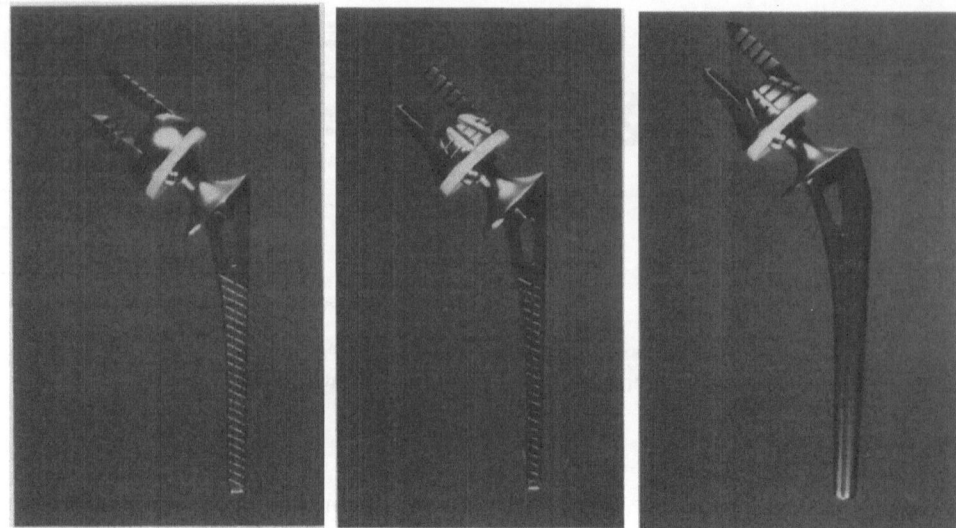

Cases : JIAT : 6   Model–Y : 20   Model–Y2 : 27

SEX	: M 7, F 46
AGE (Yrs. Old)	: 63.2
FOLLOW UP (Yrs.)	: 3.3
DISEASES	: OA 35, RDC 8, AN 8, RA 2

Fig. 3. Our cementless total hip replacement (*THR*) system with three spikes on the socket. *OA*, Osteoarthritis; *RDC*, rapidly destructive coxarthritis; *AN*, aseptic necrosis; *RA*, rheumatoid arthritis

model, were also analyzed and compared with other cementless THR systems (Table 1).

Stress Analysis in Clinical Cases and Clinical Evaluation

A total of 53 patients who had received our cementless THR (which has three spikes on the socket) were analyzed using the computer simulation technique with the RBSM. To identify the key landmarks for the origin and insertion points of the abduction muscle on the great trochanter, the inner and lateral edge of the pelvis and the center of the fifth lumbar spine were selected. To determine these key landmarks, a computer graphics X-ray image analyzing system was used. The radiographic image obtained immediately after the THR operation was scanned and entered into a personal computer through a video scanner, and key landmarks were digitized on the cathode ray tube (CRT) screen. The geometric data for each implant type were prepared by digitizing implant templates, and were stored in a computer. The insertion angle of the

Table 1. Various total hip replacement systems subjected to stress analysis using the RBSM.

Name	Type/design
JIAT	Original socket with three spikes and stem, 45° collar angle
Model-Y	Y-socket and JIAT stem
Model-Y2	Y-socket and new stem, 20-mm elongated and 35° collar angle
Model Y2-xx	Simulated hypothetical collarless model-Y2
PCA	Three-dimensional interlock, collarless and porous coated
BIOMET	Anatomical press fit
H/G Porous	Porous coated fiber mesh
LORD	Gerald LORD fluted retention prosthesis (22-mm head, 48-mm socket)
RING	First generation, old type
AML	Porous coated stem
Kyoto Ceramic	With cement type THR

RBSM, Rigid body spring model; *THR*, total hip replacement

acetabular component, spike migrations of the socket into the inner pelvic space, size of bone transplantation for the acetabular hypoplasia, stem size, and press-fit conditions of the stem were also taken into consideration for the stress analysis using the RBSM.

The Clinical results of these patients, with an average 3.3 years follow-up, were evaluated and classified into three categories based on X-ray findings. A finding of a clear zone of less than 2 mm around the socket and stem, without shifting of either component, was evaluated as excellent. Less than 2 mm central migration or less than 3° socket opening, less than 3° medial or lateral shifting of the stem, and a clear zone of less than 2 mm around both components was evaluated as good. A clear zone of more than 2 mm with more than 3° shifting or with central migration of the socket of more than 2 mm was defined as fair.

Results

Stress Distribution of Cementless THR

The stress distribution of the model-Y2 is shown in Fig. 4. The arrows around both components show compressive stresses over the artificial joint component; compressive stress was found on the lateral part of the socket and spikes. On the stem site, a large compressive stress was recognized at the site of the calcar collar (created to prevent the stem sinking), and a compressive stress pattern was also noted on the distal lateral part, the distal stem tip, and the curved proximal medial area. However, there was no compressive stress on the distal medial and proximal lateral areas of the stem. For the artificial femoral head,

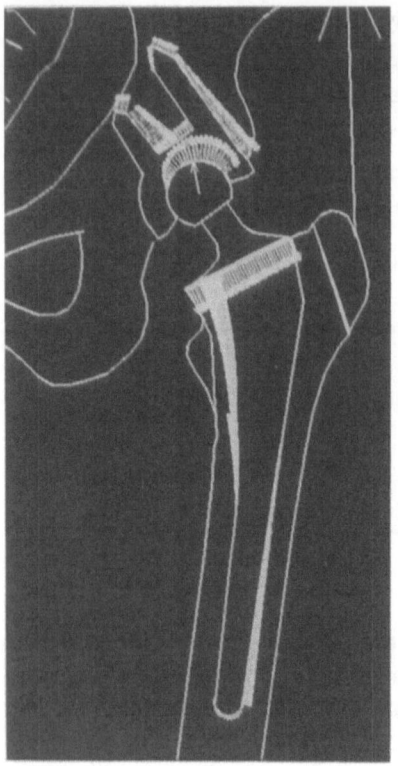

Fig. 4. Estimated stress distribution of the model-Y2 cementless THR system. *Arrow* shows compressive stress on bone/implant interface

compressive stress distribution was recognized on the outer 2/3 of the joint contact area.

From our analysis of various types of THR systems (Figs. 5–7), it was generally observed that the acetabular component was shifted toward the opening of the socket, whereas the femoral component shift was unsettled. The pattern of stress distribution on the femoral component was affected by the presence of a collar, by its design, and by its angle. Especially in the case of the Harris-Galanty (H/G) type, there was more stress over the proximal third of the stem at its medial side at the site of the collar (Fig. 6). Comparing the cemented type Kyoto Ceramics THR with our original THR, we found that the stress distribution was almost the same over the medial and proximal sides of the stem and the curved medial side of the collar (Fig. 7).

Maximum Stress Value

Stress concentration at the bone interface is an important indicator of the biomechanical stability of a THR system. The results of our analysis of the acetabular component showed that the least stress, 1.9 kg/mm, occurred in model-Y, followed by the model-Y2 collarless type, while the (PCA) showed a

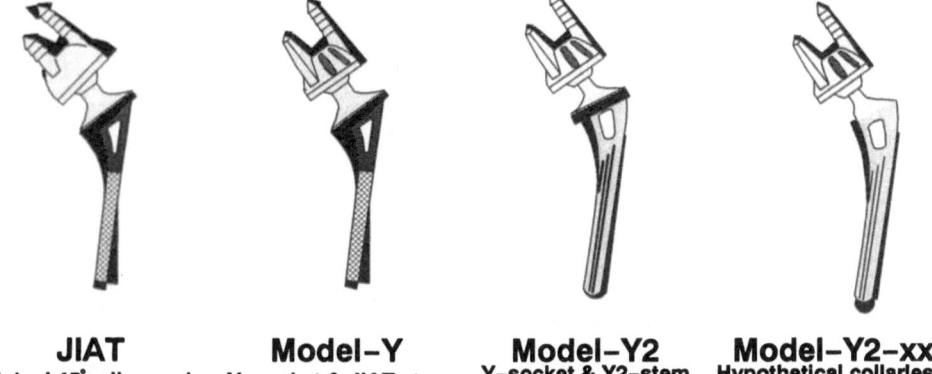

JIAT
Original, 45°collar angle

Model–Y
Y–socket & JIAT stem

Model–Y2
Y–socket & Y2–stem
(35°collar angle etc.)

Model–Y2–xx
Hypothetical collarless
simulation model

Fig. 5. Estimated stress distribution in our cementless THR systems with three spikes on the socket

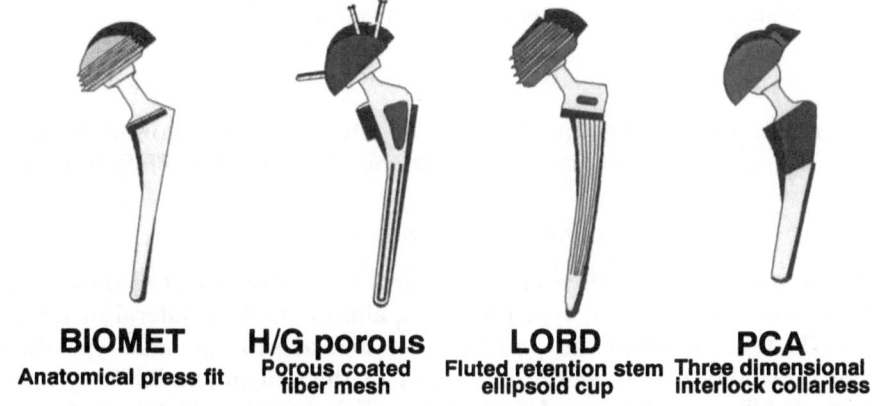

BIOMET
Anatomical press fit

H/G porous
Porous coated
fiber mesh

LORD
Fluted retention stem
ellipsoid cup

PCA
Three dimensional
interlock collarless

Fig. 6. Estimated stress distribution of various THR systems. The stress on the collar of the Harris-Galanty (*H/G*) type was greater than that in other systems

Fig. 7. Estimated stress distribution in various THR systems

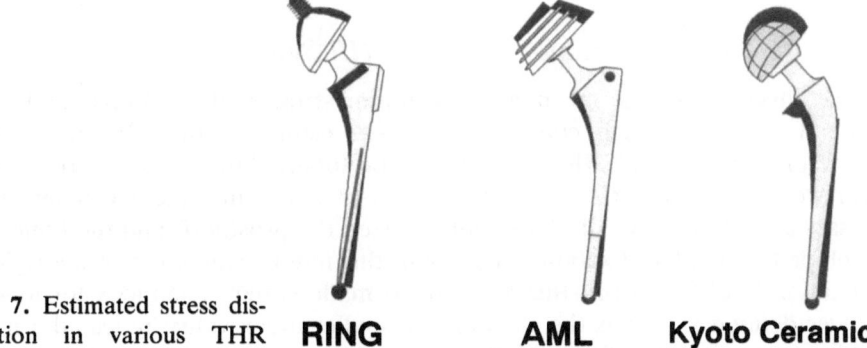

RING
Old type

AML
Porous coated stem

Kyoto Ceramic
With cement

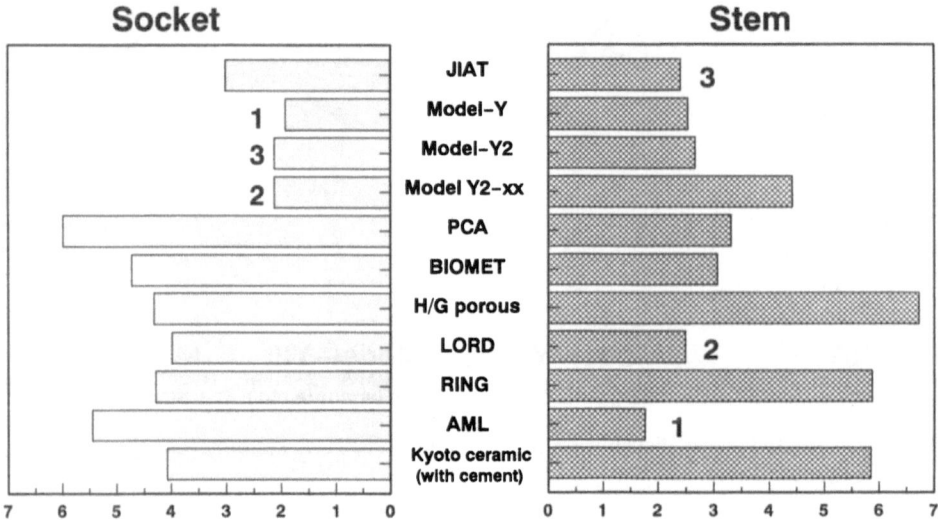

Fig. 8. Maximum compressive stress on bone/implant interface

maximum stress of 5.98 kg/mm. In the stem site, the AML showed the least stress, 1.78 kg/mm, followed by the LORD and JIAT types of THR (Fig. 8).

Rotational Deviation of the Stem

Results of analysis of the stem are shown in Fig. 9. Rotational deviation was categorized into three types, type one being almost stable in lateral or medial shift, while type two had the possibility of medial shifting, and type three had the possibility of lateral shifting. Most of the cementless THR systems, including JIAT, model-Y, model-Y2, BIOMET, LORD, RING, and AML were classified as type three. Only the Harris-Galanty THR system was classified as type two and the collarless model-Y2 and PCA were classified as type one.

Fixation Index of THR System

In the investigation of mean and maximum stress and rotational and X-Y plane deviation of each component, it was estimated that the mechanical characteristics of each THR system would be totally different when socket and stem were combined. Combining the ratio of maximum and mean interface stresses and the relative displacement between the prosthesis and the bone, we developed an implant fixation index. With this index, evaluation of the rigidity of fixation in different prosthesis designs is made easier. The index for model-Y2 was defined as 100 as the standard; a smaller index value indicated a more stable mechanical condition of the THR system. This analysis showed that

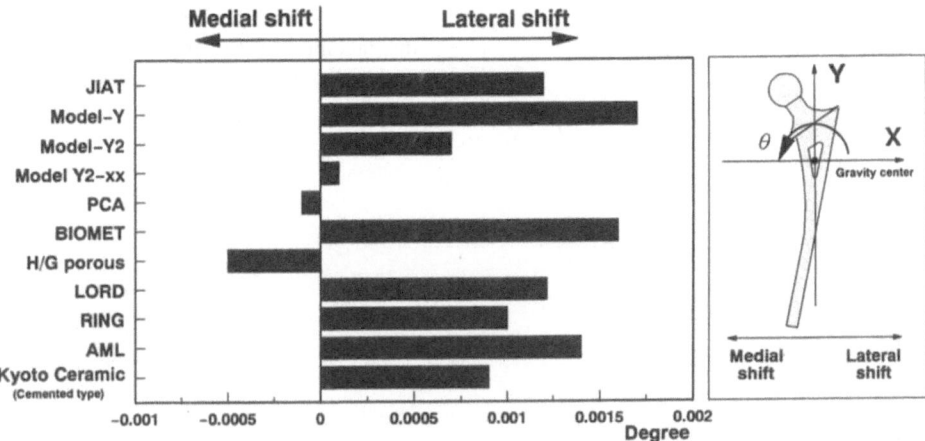

Fig. 9. Rotation of the femoral component. It was possible to categorize calculated rotational deviation as being of three types, neutral, lateral, and medial shifting

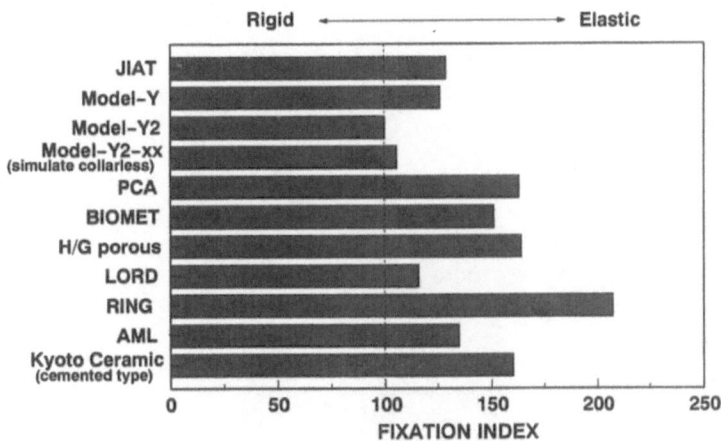

Fig. 10. Fixation indices of various THR systems

model-Y2 was the most stable, followed by the model-Y2 collarless and LORD types of THR (Fig. 10).

Comparison of Calculated and Clinical Results

The stress distribution and deviation schema for the JIAT type stem (Fig. 11) indicated that the stem tended to sink and shift laterally. This schema was very similar to the X-ray finding for the clinical case, wherein stem sinking and lateral shifting was observed. (Fig. 12).

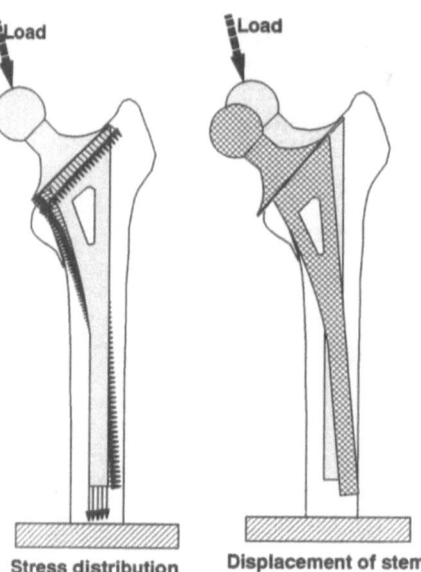

Fig. 11. Stress distribution and virtual displacement of the JIAT stem. Sinking and lateral shifting were observed

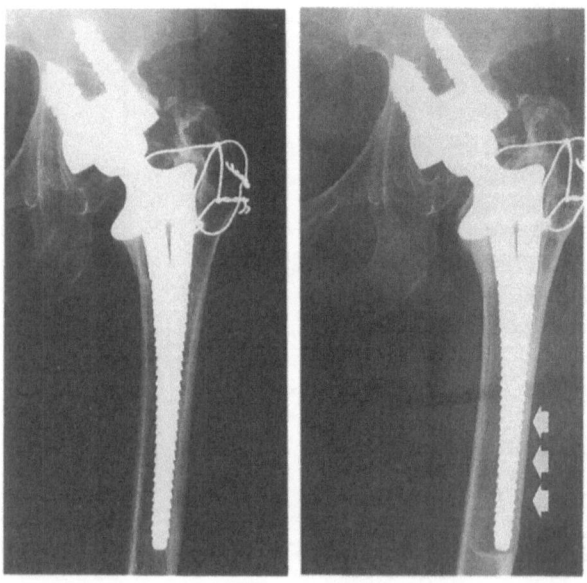

Fig. 12. Sinking and lateral shifting of the JIAT stem. The X-ray finding in this clinical case was similar to the computer simulation using the RBSM

The fixation index was calculated for each clinical case, using the RBSM, and was compared with clinical evaluations. The mean fixation index in the excellent clinical evaluation group was 139.6, the mean value of the fixation index in the good group was 161.0 and that in the fair group was 236.8 (Fig. 13). Regarding THR types, all JIAT were evaluated as fair, model-Y was mainly evaluated as good, and model-Y2 was evaluated as excellent.

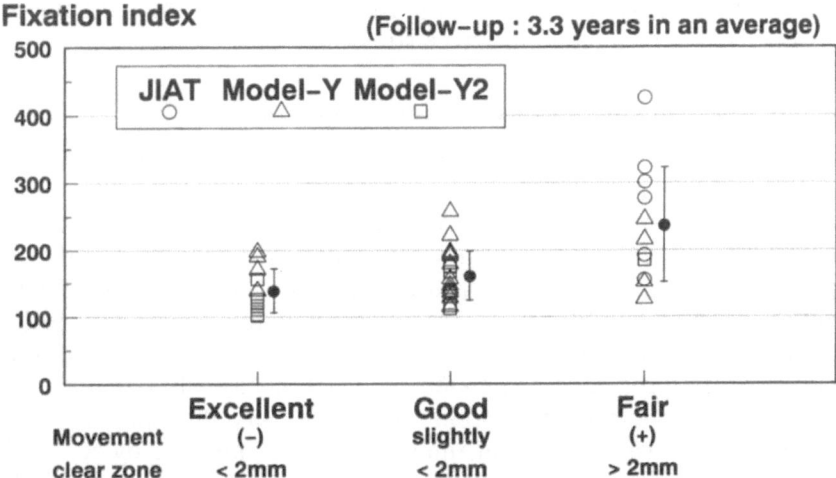

Fig. 13. Fixation index in clinical cases. The model-Y2 was evaluated as clinically excellent; further, the calculated fixation index was lower than that of the other models shown here. *Open circles*, JIAT; *open triangles*, model-Y; *open squares*, model Y-2

Discussion

Biomechanical Problems in Cementless THR and Stress Analysis

In the past, the biomechanical problems of THR have been discussed mainly in terms of stem fractures or abrasions of the high density polyethylene. Problems at the bone/implant interface have been the major focus of attention in discussions of how to fix implants of the cemented type, and how to create press-fitting in the cementless type. Traditional bond/implant stress interface analysis has been performed using either two-dimensional or three-dimensional finite element models (FEM). However, such techniques are extremely time-consuming and expensive, and require a large computer and special personnel, which features make parametric analysis nonfeasible. For preoperative planning and for the study of critical implant/bone interface stresses, a simpler analysis technique and simulation model based on the RBSM technique was developed. The limitation of such an analysis is that the internal stresses in the bone and in the prosthetic stem are not determined.

Utilizing the RBSM analysis technique, we studied various designs of prosthetic implants to determine their effects on interface stress patterns and distribution. The presence of a collar and the collar angle, and design significantly affected interface stresses, a wide collar covering the femoral cortices in the medial, anterior, and posterior aspects being the most effective in reducing stem/bone interface stresses. Stress concentration was observed in a collar covering only the medial cortex in the Harris-Galanty type stem. We considered that the design of the calcar collar was one of the most important factors involved in stem/bone interface stresses.

Clinical and Computer Simulation Results

Many factors affect clinical results, e.g., surgical technique, size of implant, position, insertion angle of the socket, bone defect, body weight, grade of osteoporosis, discrepancy of lower extremity length, bilateral or unilateral lesion, and so on. In this study, however, we found that the results of the computer simulation were consistent with the clinical results obtained with our cementless THR system [3–5], since it was possible to estimate prognosis with the implant fixation index. The selection of the implant and the surgical placement of the prosthetic component in THR are usually based on limited radiographic and clinical information, since it is very difficult technically to analyze biomechanical behavior at the implant/bone interface by conventional analytic methods. We consider that this simplified analysis using the RBSM is far more efficient and practical than the FEM technique. We believe that a stress analysis and computer simulation based on the RBSM could provide useful information for the preoperative planning of difficult reconstructive cases involving the hip. Through an operator interactive mode, results in graphic form could be obtained instantaneously to facilitate the selection of the ideal size and placement of prosthetic components. In cases when there is a bone defect or unusual angular deformity, the need for bone grafting or a custom prosthesis could be identified.

Conclusions

1. Results of stress analysis using the RBSM were consistent with the clinical results of our cementless THR.
2. To acquire suitable initial fixation of prosthetic components, implant design is of great importance.

References

1. Kawai T (1977) A new discrete model for analysis of solid mechanics problems (in Japanese). Seisan Kenkyu 29:208–210
2. Akamatsu N, Hamada Y, Nakajima I, et al (1989) Basic research on the design of cementless total hip system. Jpn J Artif Organs 18:1626–1635
3. Ide T, Hamada Y, Nakajima I, et al (1989) Non-linear stress analysis of various kinds of cementless total hip systems. Jpn J Artif Organs 18:348–351
4. Ide T, Amano R (1992) Biomechanical features of various cementless total hip systems. J Joint Surg 11:827–834
5. Yamamoto Y (1990) Clinical and basic research on the design of cementless total hip system. Jpn J Artif Organs 19:1476–1486

26

—Overview—
Application of Thermoelastic Stress Analysis Method to Joint Biomechanics

Koji Hyodo and Tetsuya Tateishi[1]

Summary. The thermoelastic stress analysis method utilizes the thermoelastic qualities of hard tissues such as bones and teeth, and enables one to measure and image surface stress distribution easily. Using this approach, we performed stress analyses of human femurs and tibias in simulated loading conditions. The femur and the tibia were fixed to loading devices. A 1-Hz sinusoidal load was applied with a hydraulic oscillator, and the emitted infrared radiation was measured. The oscillatory amplitude was 2.8 kN (max 2.9 kN, min 0.1 kN). The surface stress distribution was then obtained by calculating the differential value. Thermoelastic stress images were obtained for the anterior, posterior, medial, and lateral aspects of the femur. In the intact femur at standard position, the compressive stress was measured primarily in the medial aspect. The greatest compressive stress occurred around the femur neck and the proximal part of the femur. The compressive stress increased both at 15° adduction and at 15° abduction. At 15° abduction, in particular, the compressive stress was increased in the medial shaft of the femur. In comparison, in the femur with a total hip joint, a remarkable reduction of compressive stress was observed around the medial proximal part of the femur. At 15° abduction, the compressive stress was greater in the medial distal part of the femur than at standard position. Surface total hip replacement or reduction of stem rigidity may be a method of obtaining better biomechanical compatibility.

For the tibial experiment, in normal loading the greatest stress image was obtained on the posterior aspect, indicating that the majority of the stress was concentrated in the posterior part of the tibia. The epiphysis of the proximal tibia, which consists predominantly of cancellous bone, showed compressive stress in every plane. These results indicated that the tensile stress in the epiphyseal surface indirectly reflected the impact-absorbing properties of cancellous bone. In 2.5° varus loading, the compressive stress

---

[1] Biomechanics Division, Mechanical Engineering Laboratory, Agency of Industrial Science and Technology, MITI, Namiki 1-2, Tsukuba, Ibaraki, 305 Japan

in the medial aspect increased, while that found in the lateral aspect decreased. In this setting, the main compressive stress had shifted to the posteromedial aspect of the tibia. In 5° varus loading, the load transmission shifted to the medial aspect, and therefore the compressive stress in the medial aspect increased further. Also, in this setting, the tibia was deflected to bend to the medial side. These results demonstrated that small changes in the alignment of the knee joint have a great influence on the stress pattern of the proximal tibia.

Key words: Biomechanics—Hip joint—Artificial hip joint—Experimental stress analysis—Thermoelastic stress analysis

Introduction

Stress analysis, which is indispensable for the understanding of the biomechanical function of hard tissues such as bones and teeth, is also of great importance for the biomechanical evaluation of prosthetic systems. In this research field, the finite element method (FEM) in addition to various experimental methods, plays a very important role in giving us very useful biomechanical information. A crucial requirement for practical FEM analysis is the creation of the most suitable FEM model to include complex biological tissues and surface interface conditions.

Data from experimental, as well as biomechanical research, are necessary for the creation and evaluation of such models. Many researchers have used strain gauges, photoelastic modelling, and laser holography methods for analyzing stress and strain.

The thermoelastic stress analysis method utilizes the thermoelastic properties of materials. Compared with the former methods it enables one to easily measure and image the distribution of the sum of principal surface stresses [1–4]. It is a non-contact *whole stress field* analysis method, and is free from some of the difficulties inherent in many other methods. For example, it requires no strain gauges, no two-dimensional plastic modelling, and no precision optical systems for making holograms. We applied this method to stress distribution imaging in the field of biomechanics. Stress distribution changes in the femur, depending on varus-valgus angles and the influence of the artificial hip joint, were demonstrated. Experimental stress analyses of the tibia were also performed.

Materials and Methods

Materials

Hip Joint (Femur) Experiment

A pair of dried human femurs and an artificial hip joint made of Co-Cr alloy with a ϕ24-mm ceramic hip ball were used for the biomechanical evaluation of

Fig. 1. Intact femur (*left leg*) and total hip replacement femur (*right leg*)

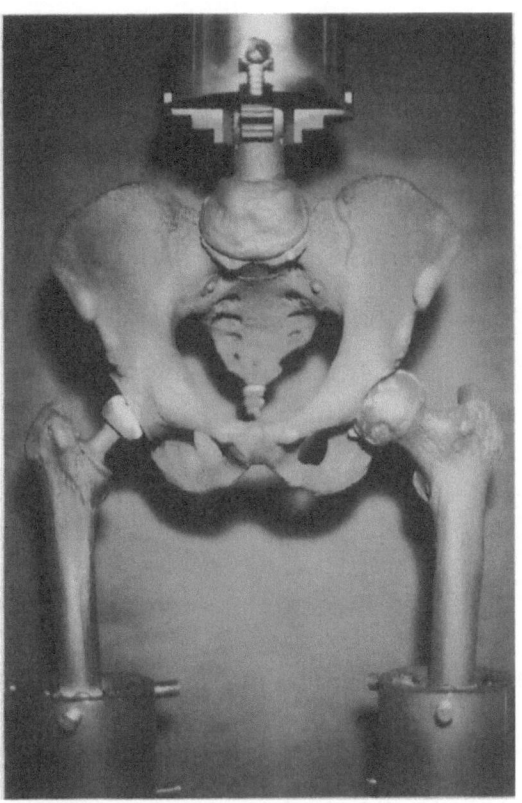

the artificial joint system (Fig. 1). The distal part of the femur was fixed in a custom-made device (Fig. 1) which reproduced biomechanical loading conditions via the use of an oscillator unit. During the experiment, three kinds of femoral position, standard, 15° abduction, and 15° adduction were reproduced. The femurs were painted frosted black to suppress heat reflection.

Tibia Experiment

Three fresh human cadaveric legs were used. All extraneous soft tissues, other than the extensor mechanism and capsular, and ligamentous structures and fibula were removed to expose bony structure surface. The proximal part of the femur and the distal part of the tibia were fixed in the custom-made device outlined above (Fig. 2).

Thermoelasticity

The thermoelastic effect is the change in temperature that accompanies the adiabatic elastic deformation of a body and is governed by the following relations:

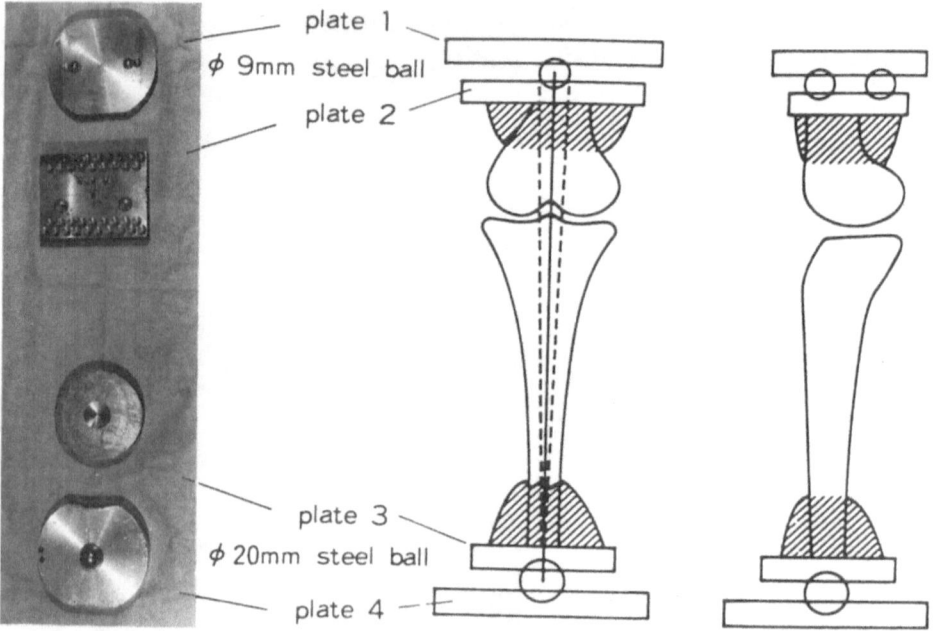

Fig. 2. Femur and tibia fixed in a custom-made device

$$\Delta T = -k \cdot \Delta \varepsilon$$

where T is the absolute temperature, k, the thermoelastic constant, and ε, the volumetric strain.

The thermoelastic constant, k, is given by

$$k = K \cdot \beta \cdot T \cdot V/Cv$$

where K is the isothermal bulk modulus, β, the coefficient of cubic expansion, T, the absolute temperature, V, the volume, and Cv, the specific heat at constant volume.

There is a linear relationship between the change in the surface principal stress changes and the produced temperature changes, ΔT, which is given by

$$\Delta(\sigma x + \sigma y) = -1/k \cdot E/(1 - 2v) \cdot \Delta T$$

where σ, is the principal stress, v, is Poisson's ratio, and E, is Young's modulus.

The local temperature changes measured from the emitted infrared radiation provide detailed information on the associated stresses.

Methods

The measurement apparatus consisted of two units (Fig. 3a–c), the first being the oscillator unit (Tokyo Shikenki) and the second the thermoelastic stress

Fig.   3a–c. Thermoelastic stress measurement apparatus. **a** Hardware blockchart, *VTR*, video tape recorder. **b** oscillator unit, **c** thermoelastic stress analysis unit

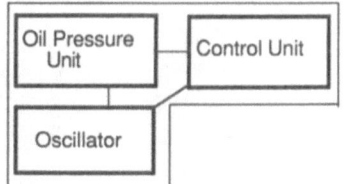

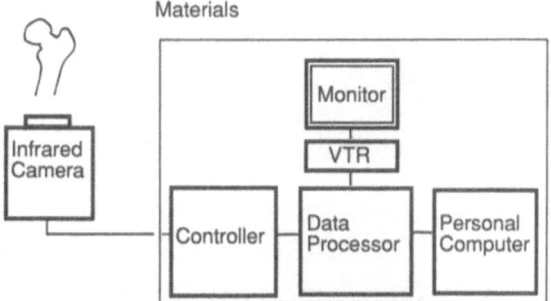

Materials

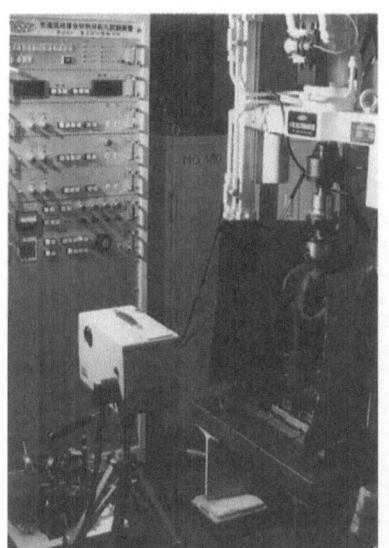

The Thermoelastic Stress Analysis Unit **a**

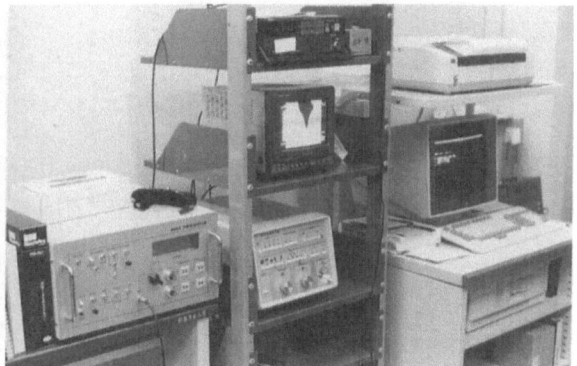

b   c

analysis unit (Fujitsu, Tokyo). The oscillator unit applied 1-Hz sinusoidal loads to the materials. The oscillatory amplitude was 2.8 kN (max 2.9 kN, min 0.1kN). Thermal radiation from the surface of the materials was detected with an infrared camera (Fujitsu, Tokyo, Japan; HgCdTe infrared sensor; 10-μm peak) with 60-Hz scanning; 0.01°C was detectable when integrated 16 times). Data collection was synchronized with the load cycle (upper peak and lower peak) to obtain the temperature difference. The data were analyzed to relative stress changes (ΔT or $\Delta(\sigma x + \sigma y)$) and imaged (224 × 224 dots) by the thermoelastic stress analysis unit. The stress imaging was displayed on a TV monitor, with stress levels represented by 25 colors. The stress data were also stored on floppy disks in a personal computer (Fujitsu) (Fig. 3a,c).

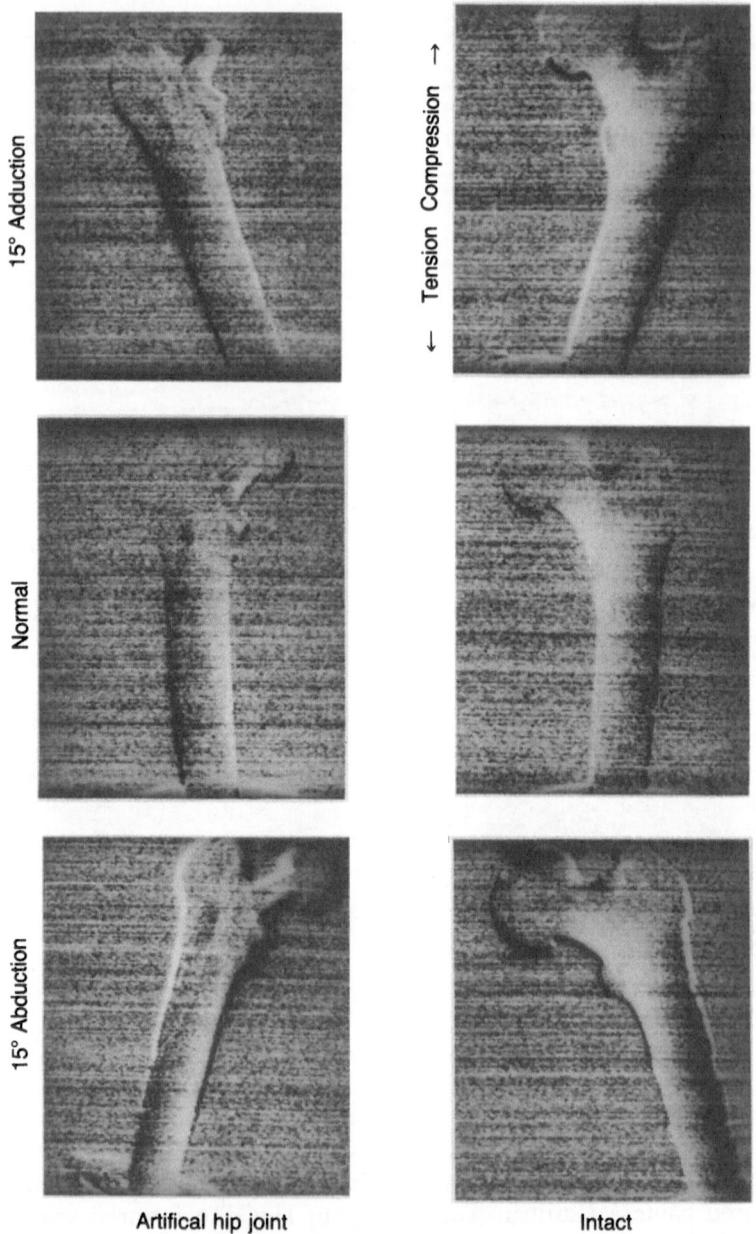

Fig. 4. Thermoelastic stress image of the femur (anterior view; sinusoidal load 1.5 ± 1.4 kN, 1 Hz)

Results and Discussion

Hip joint (Femur) Experiment

Thermoelastic stress images were obtained in the anterior (Fig. 4), posterior, medial, and lateral aspects of the femur. In the intact femur at standard position, the compressive stress was measured predominantly in the medial aspect. The greatest compressive stress was found to occur around the femur neck and the proximal part of the femur, the stress being increased both at 15° adduction and at 15° abduction. At 15° abduction, in particular, the compressive stress was increased in the medial shaft of the femur.

In comparison, in the femur with a total hip joint, a remarkable reduction of compressive stress was observed around the medial proximal part of the femur. At 15° abduction, the compressive stress was greater in the medial distal part than at standard position.

These experimental results support the FEM analysis, showing that total hip replacement produces a reduction of cortical bone stress in the medio-proximal part of the femur. From the clinical point of view, this phenomeon reduces bone density around the area and is one of the reasons for stem loosening [5].

In this experiment, we used a cementless artificial hip joint stem with a peg and without surface coating. Stress distribution changes depending on the shape and material design of the stem and on the interface betweem the stem and the bone. However the shape and the material design of the stem may be varied, if we select the same method, i.e., insertion of a rigid stem into the femur, the mechanical imbalance will remain. This imbalance is similar to that noted above (in the second paragraph of the discussion of the results of this experiment). To obtain better biomechanical compatibility, surface total hip replacement or reduction in stem rigidity may be a solution. It is clear that fatigue test and wear test data must be taken into consideration with this kind of stress analysis.

Tibial Experiment

Thermoelastic stress images were demonstrated in the medial, lateral, and posterior aspects of the tibia (Fig. 5) and little stress image was noted on the anterior side. The greatest stress image was obtained in the posterior aspect, indicating that under these testing conditions, the major stress was concentrated on the posterior surface of the tibia.

The bony surface of the epiphysis and the metaphysis in the tibia, which consist of cancellous bone predominantly, showed tensile stress, and that of the diaphysis, which consists of compact bone, showed compressive stresses in every plane.

In varus loading, with a simulated 2.5° deformity between the tibia and femur (gray lines in Fig. 5), the compressive stresses in the medial aspect of the diaphysis increased, while the compressive stresses in the lateral aspect decreased, indicating that little stress existed on this side. In this testing

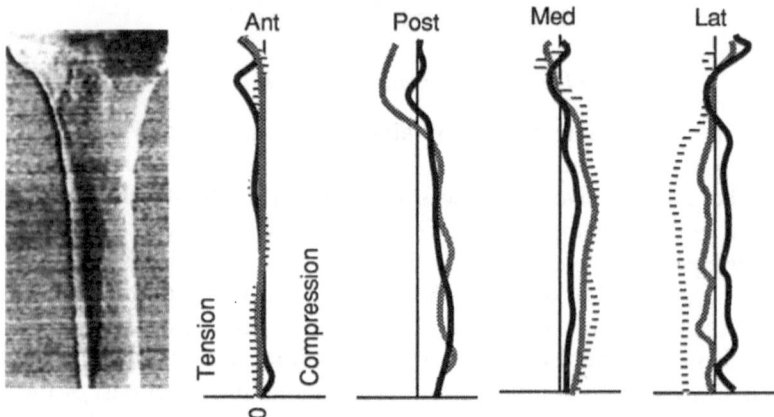

Fig. 5. Stress distribution along the bone axis of the tibia at various aspects. *Black lines*, Normal; *gray lines*, 2.5° varus; *dashed lines*, 5° varus

condition, the high stress image had shifted medially and the medial and posterior aspects of the tibia showed similar stress concentration. In 5° varus loading (dashed lines in Fig. 5), the compressive stress was further increased in the medial aspect of the diaphysis, while tensile stress was demonstrated in the lateral aspect of the tibia.

These results indicate that the tibia is loaded equally between the medial and the lateral condyle and that the load is concentrated posteriorly in normal alignment. In moderate varus loading, with 2.5° varus deformity, the major compressive stress had shifted to the posteromedial aspect of the tibia. In marked varus loading, with 5° varus deformity, the load transmission had shifted to the medial aspect.

Valgus loading showed that, with 2.5° valgus deformity, the load was concentrated on the posterolateral aspect of the diaphysis of the tibia, while with 5° valgus deformity, the load was concentrated almost entirely on the lateral aspect. These results were similar to those observed in varus loading, except that the medial and lateral aspects were reversed.

These results demonstrated that small changes in the alignment of the knee joint have great influence on the stress pattern of the proximal tibia.

Conclusions

The thermoelastic stress analysis method was very effective for performing stress analysis of the femur and the tibia. We readily detected the influence of differences in abduction and adduction and varus and valgus angles, as well as the effects of an artificial joint. It is clear that total hip replacement reduces the stress on the cortical bone of the anterial proximal part of the femur. Surface total hip replacement or reduction in the rigidity of the artificial joint stem may provide better mechanical conditions. It was shown that small changes of alignment in the knee joint had great influence on the stress pattern of the proximal tibia.

The thermoelastic stress analysis method detects the sum of the major surface stresses ($\sigma x + \sigma y$). Combining this with photoelastic stress analysis, which detects ($\sigma x - \sigma y$), would produce interesting results.

References

1. Duncan JL, Cummings WM (1984) Thermoelastic stress analysis of fresh bone. Abstracts of the first international conference of stress analysis by thermoelastic techniques. Sira, London, Section 10, pp 1–5
2. Hyodo K (1986) Application of the thermoelastic stress analysis method to biomechanics. Mech Eng Lab News 9:5–8
3. Tateishi T, Hyodo K, Homma K, Yamada M (1990) Visualization methods in biomechanics. In: Heimke G, Soltész U, Lee AJC (eds) Clinical implant materials. Advances in biomaterials, vol 9. Elsevier, Amsterdam, pp 651–656
4. Yamada M, Kurosaka M, Hirohata K, Tateishi K (1990) Thermoelastic stress analysis of the human tibia. Clin Rheumatol 3:192–200
5. Saejong S, Hirano S, Granholm JW, Walker PS (1987) Strains and micromotions of press-fit femoral stem prostheses. J of Biomech 20(7):693–702

V

Design and Fixation of Femoral Components in Total Hip Arthroplasty

27

Basic Study of Hip Prosthesis Design: Analysis of Shape of the Femoral Medullary Canal in Japanese Subjects by Computed Tomographic Scanning

Shigeru Yanagimoto and Toyonori Sakamaki[1]

Summary. Japanese patients with osteoarthritis of the hip have a characteristic femoral shape. Predominantly these patients develop secondary osteoarthritis, the primary cause of which is congenital dislocation of the hip. Our aim in this study was to determine a stem shape suitable for an artificial hip joint in Japanese patients. To date, it has been difficult to carry out precise measurements of the femur, due to the complexity of its shape, the major difficulty being the determination of a standard basal central axis. We have overcome this problem with a novel method, using computer graphics. We took computed tomographic (CT) scans of 52 Japanese patients with osteoarthritis of the hip. From the CT scan, we assessed the optimal extent of the tapered reamer occupation of the medullary canal in each patient. We then took the axis of the tapered-reamer to be the central basal axis of the proximal femur and we measured various parameters. By the use of computer graphics and our new method, we were able to make morphological measurements of the proximal medullary canal of the femur. We then analyzed our data statistically. We found that the femoral diameter was positively correlated with the height, the femoral length, the femoral transverse diameter, and the medullary transverse diameter. We found regular proportionality in the femurs of these patients. The femoral head offset was not correlated with any measurement data. The medullary canal transverse diameter of the diaphyseal portion, as well as the ratio of the medullary transverse diameter to the femoral transverse diameter, increased with age. The taper angle, i.e., the degree of tapering of the medullary canal, of the proximal medullary canal had a mean value of 3.8°. It was positively correlated only with the medullary transverse diameter. In the light of these results, we concluded that hip prostheses suitable for Japanese patients must have these characteristics: whether small or large, the stems should be similarly shaped. The length and width of the stems should increase in an almost proportional way. A modular system would be desirable, so that the

---

[1] Department of Orthopaedic Surgery, School of Medicine, Keio University, 35 Shinanomachi, Shinjuku-ku, Tokyo, 160 Japan

preferred offset length could be selected. In those patients whose medullary canal is enlarged by aging, special stems should be prepared. The taper angle of the stems should be almost 4°.

Key words: Artificial hip joint—Anatomy of proximal femur—Femoral stem design—CT measurement—Osteoarthritis of hip joint—Japanese people—Computer graphic method

Introduction

Total hip arthroplasty (THA) has spread worldwide following Charnley's 1979 report on low friction arthroplasty [1]. This procedure provides reliable, excellent short-term results, but aseptic loosening of components and implant-related wear debris in the long-term are of great concern. To prevent cement-induced disease and problems with cement removal during revision surgery, many surgeons prefer cementless THA [2,3]. Whether the stem is fixed with or without cement, it is important for it to be properly positioned in the medullary canal. Thus, there is even greater need to ascertain the precise anatomical shape of the femoral medullary canal.

The shape of the femur itself is quite complex, in that there is, for example, anteversion and antetorsion in the proximal portion of the femoral neck and anterior bowing and tapering in the middle of the medullary canal [4,5]. It is thus crucial to determine the basal point when taking measurements for fitting. However, it is difficult to precisely determine the basal central axis in the medullary canal. Until now, we have not been able to precisely analyze the anatomical shape of the femur.

Japanese patients predominantly develop secondary osteoarthritis, the primary cause of which is congenital dislocation of the hip (CDH). Since difficulties are occasionally encountered when imported hip prostheses are used in Japanese patients, there is a need for suitable new hip prostheses for these patients. We used computed tomographic (CT) scans to measure and analyze the anatomical shape of the proximal femoral medullary canal in Japanese patients, with the aim of determining the parameters required to create an artificial hip joint stem suitable for use in Japanese patients.

Materials and Methods

We analyzed 52 Japanese patients (6 males and 46 females) with osteoarthritis of the hip. In all cases the condition was secondary osteoarthritis caused by CDH and hip dysplasia. The average age of the patients was 50.8 years (range, 18–86 years). Before taking CT measurements, we evaluated all patients on plain X-rays. The mean center-edge (CE) angle was 5.7° and the Sharp angle was 46.6°. Most patients had severe osteoarthritis and were admitted to our hospital for the purpose of undergoing THA or pelvic osteotomy.

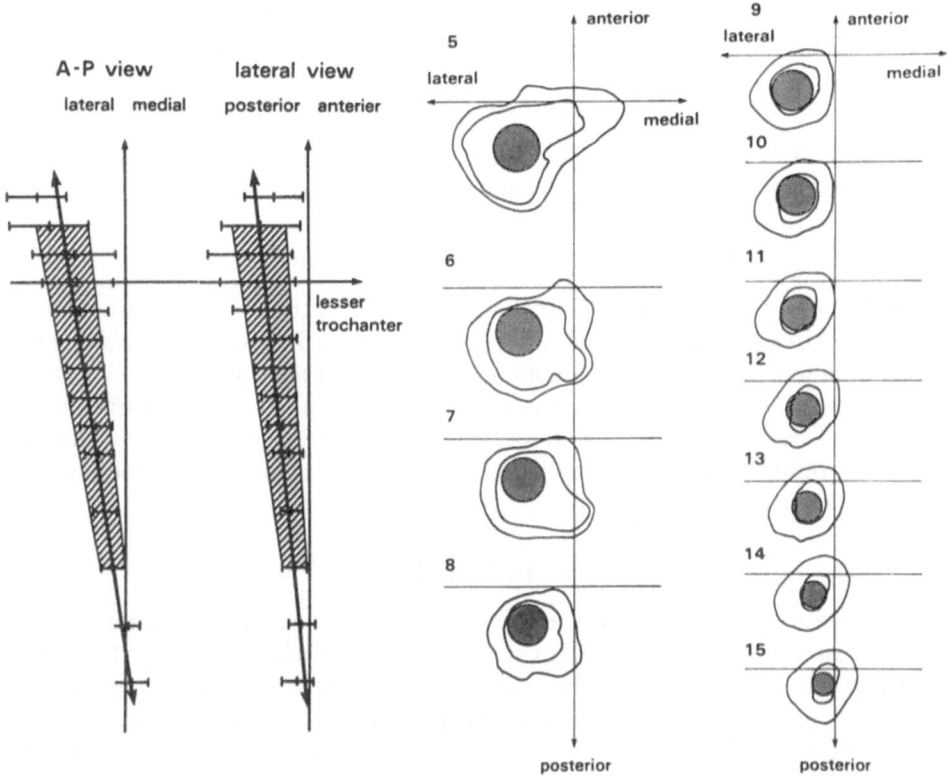

Fig. 1. Optimal tapered reamer occupation of the medullary canal; anteroposterior (*A-P*) lateral view, cross-section view

In each case, we took CT scans (2-mm-thick slices) of the area from the femoral head to the midpoint of the femur. In each case, one additional slice was taken at the height of the femoral condyles to measure the femoral anteversion angle. The scanning interval was 1 cm in the proximal portion and 2 cm in the distal portion. Scanning distance was generally 20 cm. In general, the CT scan window level was +100 and window 500.

The anteroposterior (A-P) view revealed that the femoral canal was almost symmetrical, while the lateral view revealed asymmetry (owing to anterior bowing in the middle). It was thus difficult to determine the basal central straight axis for measurement. We first determined this axis in the medullary canal. When we placed a cementless press-fit stem in the medullary canal, we usually used a tapered reamer to ream the canal. When a suitable sized tapered reamer is used, we can feel that it is already fixed in the medullary canal. In other words, when a tapered reamer is used, the central axis of the stem has already been determined. Using a computer graphics method, we determined the final tapered reamer for each patient from the CT scan measurement data (Fig. 1). We then took this axis of the tapered reamer to be the basal central

---

axis of the proximal femoral medullary canal. Using this axis, we measured several parameters on the CT scan and analyzed them.

Femoral Length

From the CT scan, we measured the femoral length (from the top of the femoral head to the bottom of the femoral medial condyle).

Femoral Head Diameter (Horizontal Plane)

From the cross-section of the CT scan at the height of the femoral head center, we drew a circle overlapping the outline of the femoral head. From this circle, we determined the center and diameter of the femoral head. In many cases, the femoral head was severely deformed, making it sometimes difficult to determine whether the circles were overlapping. In these cases, we determined the circles based on the data for the lateral shape in the horizontal plane or on the A-P view in the X-ray.

Anteversion Angle of Femoral Neck

The direction from the central axis of the proximal medullary canal to the center of the femoral head reveals right anteversion of the femoral neck. Thus, the right anteversion angle of the femoral neck is the angle that is created by the following two straight lines: one that passes through the central axis and the center of the femoral head, and the other going through the posterior surfaces of the femoral medial and lateral condyles.

Femoral Head Offset (Fig. 2)

The distance from the central axis of the proximal medullary canal to the center of the femoral head is the femoral head offset. We can easily and correctly measure this distance.

Taper Angle of the Proximal Femoral Medullary Canal (Fig. 2)

From the point of the lesser trochanter to the point of the isthmus, the femoral medullary canal has a relatively regular tapered shape. We took the taper angle of the final tapered reamer (estimated based on computer-graphics) as the taper angle of the proximal medullary canal.

Femoral and Medullary Transverse Diameters (Fig. 3)

On the CT scan cross-section, we drew four kinds of circles: maximum inscribed and minimum circumscribed circles along the outline of the femur and the same along the outline of the medullary canal. We calculated the average diameter of the inscribed and circumscribed circles along the outline of femur, and took this to be the femoral transverse diameter. Using this method,

Fig. 2. Taper angle and femoral head offset

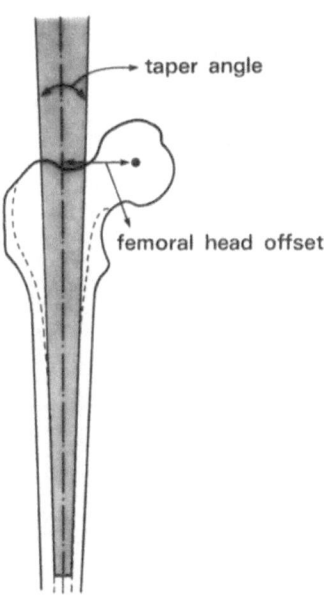

Fig. 3. Measurement of femoral and medullary transverse diameters. F_E, circumscribed circle of femur; F_I, inscribed circle of femur; C_E, circumscribed circle of medullary canal; C_I, inscribed circle of medullary canal. Femoral transverse diameter, $\dfrac{F_E + F_I}{2}$; medullary transverse diameter, $\dfrac{C_E + C_I}{2}$

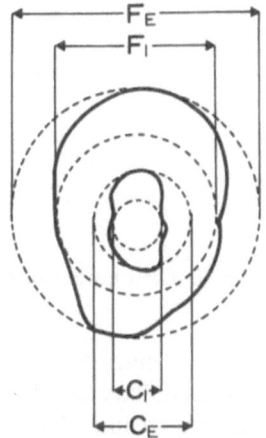

we also calculated the medullary transverse diameter. We calculated these parameters at a height of 4 cm distal from the medial tip of the lesser trochanter. We also calculated the ratio of the medullary to the femoral transverse diameter (medullary transverse diameter/femoral transverse diameter).

We measured the femoral and transverse diameters at a height of 4 cm distal from the medial tip of the lesser trochanter. According to Dai [6] and Noble [7], the results of their anatomical analyses of the femur, showed the point just below (about 2 cm) the lesser trochanter to be the most predictable dimension

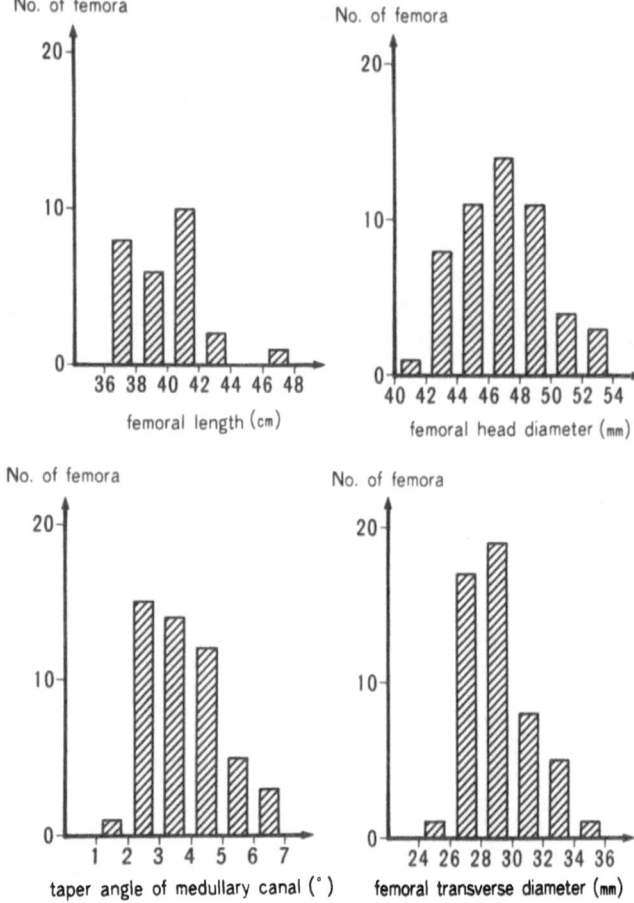

Fig. 4. Distribution of parameters by computed tomographic (*CT*) measurements

of the femur. As we have noted in our earlier reports [8,9], the cross section of the medullary canal at the point just below the lesser trochanter is circular. The tapered reamer has good contact with the inner cortex all around the circle. We believe that the size of the cementless press-fit stem should be based on the diameter of the medullary canal at this point. We therefore measured the diameter at a height just below the lesser trochanter.

We also applied the values for age, height, and CT scan parameters in each patient to statistical analysis.

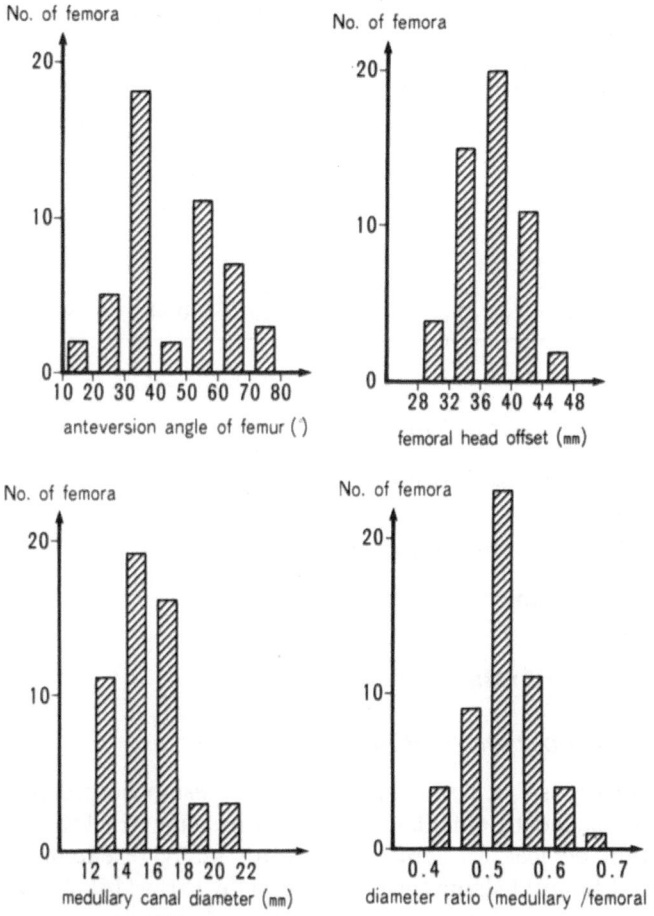

Fig. 4. *Continued*

Results

Reliability

We input the figures for the femoral transverse shape on the CT scan slices for each patient. Using computer graphics, we determined the best tapered reamer for each patient (Fig. 1). These tapered reamers had good contact with the inner side of the femoral cortex just below the lesser trochanter to the midpoint of the femoral shaft (proximal from the isthmus). However, proximally from the lesser trochanter, the tapered reamer does not contact the cortex and does not ream; thus the use of a rasp is required. We believe that, since the tapered reamer determined via computer graphics is suitable for the tapered reamer used at operation, the central axis determined from the tapered reamer can be taken to be the central axis of the proximal femur.

Table 1. Parameters determined by computed tomographic (*CT*) measurements.

	No. of femora	Min.	Max.	Av.	SD
Femoral length (cm)	27	36.1	46.5	39.5	2.4
Femoral head diameter (mm)	52	41	52	46.0	2.8
Femoral head offset (mm)	52	28	45.5	37.0	3.9
Anteversion of femur (°)	48	17	72	44.9	14.9
Taper angle of medullary canal (°)	52	1.9	6.4	3.8	1.2
Femoral transverse diameter (mm)	52	25.8	34.0	29.1	2.0
Medullary canal diameter (mm)	52	12.0	21.8	15.7	2.2
Diameter ratio (medullary/femoral)	52	0.44	0.68	0.54	0.05

Basic Anatomical Data

The parameters determined by CT are shown in Table 1 and Fig. 4.

Correlations Among Parameters

These correlations are shown in Table 2.

Discussion

Purpose and Methods

To produce the novel hip prosthesis, it was very important to establish precise anatomical measurement data and to elucidate the femoral shape. Following advances in internal fixation devices for femoral fracture, many researchers have measured and analyzed femoral shape by analyzing plain X-rays [4]. Very few authors have taken three-dimensional measurements of femoral shape, particularly of the femoral medullary canal. For this reason, until now the exact, three-dimensional taper angle of the proximal femoral medullary canal has not been determined. We cannot obtain the exact taper angle from analyzing plain X-rays. As patients with severe osteoarthritis of the hip have hip joint contracture, we cannot even obtain a precise A-P and lateral view of the femoral X-ray. Because of the rotational contracture, we recommend CT scanning [10]. We have used CT scanning methods and have devised original methods. As outlined above, we first determined the final tapered reamer, which occupied a large portion of the medullary canal, from the CT scan slices of each patient. We then took the central axis of the tapered reamer to be the central axis of the proximal femoral medullary canal, and we measured various parameters. We are now hoping to create a new hip prosthesis that occupies a large portion of the medullary canal.

Whether we use a cemented or cementless prosthesis, clarifying the characteristic femoral shape in Japanese patients with osteoarthritis of the hip is of great importance. Because this condition in a majority of Japanese patients

Table 2. Correlations of parameters.

	Age	Height	Femoral length	Femoral head diameter	Femoral head offset	Anteversion angle	Taper angle	Femoral transverse diamter	Medullary canal diameter	Canal/femur ratio
Age		$P < 0.01$ (−)				$P < 0.01$ (−)			$P < 0.01$ (+)	$P < 0.001$ (+)
Height	$P < 0.01$ (−)									
Femoral length		$P < 0.001$ (+)								
Femoral head diameter	$P < 0.001$ (+)	$P < 0.001$ (+)	$P < 0.001$ (+)					$P < 0.01$ (+)	$P < 0.01$ (+)	$P < 0.05$ (+)
Femoral head offset										
Anteversion angle	$P < 0.01$ (−)									
Taper angle									$P < 0.05$ (+)	
Femoral transverse diameter	$P < 0.01$ (+)			$P < 0.01$ (+)					$P < 0.001$ (+)	$P < 0.05$ (+)
Medullary canal diameter	$P < 0.01$ (+)			$P < 0.01$ (+)			$P < 0.05$ (+)	$P < 0.001$ (+)		$P < 0.001$ (+)
Canal/femur ratio	$P < 0.001$ (+)			$P < 0.05$ (+)				$P < 0.05$ (+)	$P < 0.001$ (+)	

(+) Positive correlation between the two; (−) negative correlation between the two

results from CDH, their femora are distinctive from those of patients in other countries. Thus, imported hip prostheses do not ensure a precise fit when used in Japanese patients. To produce a new hip prosthesis suitable for Japanese patients, we took precise measurements of the anatomical shape of the femoral canal; the three-dimensional and direct measurement data were then analyzed statistically. Our CT measurement data clearly revealed the basal data required for producing a hip prosthesis (e.g., femoral head offset, taper angle, and medullary transverse diameter). We believe our original method will be very valuable when applied to stem design.

Anatomical Data

CT measurement parameters are divided into two categories [7,10]: external form, i.e., femoral length, femoral head diameter, femoral head offset, and femoral transverse diameter; and internal form, i.e., medullary canal shape, which includes the taper angle of the medullary canal and medullary transverse diameter. External form parameters tend to be correlated with other parameters in the category. After a person has stopped growing, the external form is not greatly changed by the aging process [6,7]. In contrast, the internal form parameters (medullary canal shape) are affected by age, sex, and walking ability. Medullary canal diameter is increased with age [11,12]. Precise clarification of internal form (the medullary canal) is necessary to achieve optimal medullary canal occupation by the stem. External form must be clarified to determine the proximal portion of the stem (femoral head; femoral neck).

External Form

The femoral head diameter is correlated with height, femoral length, femoral transverse diameter, and medullary transverse diameter. However, the femoral head offset is not correlated with any of these parameters. These results indicate that the external femoral shape in Japanese patients with osteoarthritis of the hip has parameters with regular proportionality (excluding femoral head offset). We therefore believe that stems, regardless of their size, should be similar in shape. From small to large, the shape of the stem (including length and width) should increase in an almost proportional way (excluding head offset). Noble reported a statistical correlation of external femoral shape with external (periosteal) form, including femoral head offset [7]. However, our results indicate that femoral head offset is not correlated with other parameters. This appears to be due to the severe hypoplasia of the femoral neck that occurs in Japanese patients as a result of CDH-induced osteoarthritis of the hip. Thus, femoral head offset often varies and has no relationship with other parameters for external form.

Internal Form

Until now, no authors have taken three-dimensional measurements of the taper angle of the proximal femoral medullary canal. We have definitely

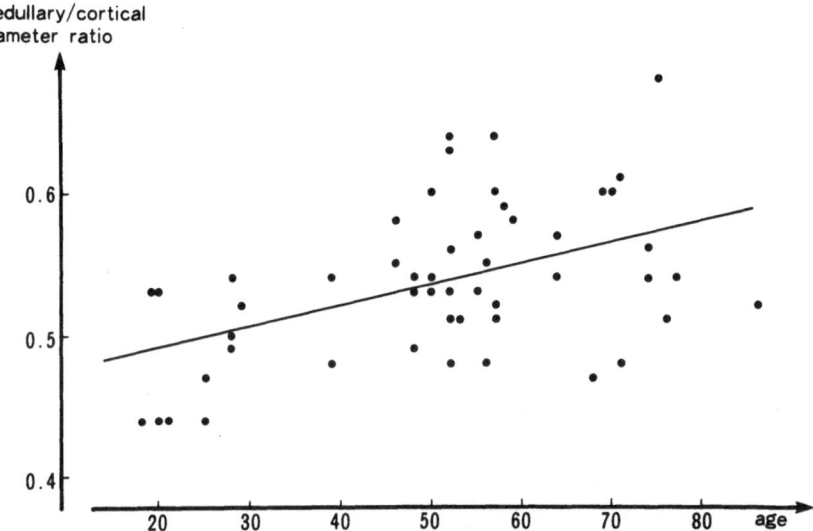

Fig. 5. Correlation between age and transverse diameter ratio (medullary/femoral)

accomplished this; we found this angle to be 3.8° on average. We know that the tapered reamer has good contact with the inner cortex from just below the lesser trochanter to the isthmus of the medullary canal. In this portion only, the femoral medullary canal has 3.8° tapering. This is the portion in which the shaft of the stem is set. We found that deviations of the taper angle were not substantial; however, the taper angle correlated only with medullary canal diameter. Femurs with large medullary canal diameters had relatively large taper angles.

The medullary canal diameter was strongly correlated with age. That the medullary canal diameter increases gradually with age was shown by the results, i.e., the diameter ratio (medullary canal diameter/femoral transverse diameter) had a strong positive correlation with age (Fig. 5). Different stems may be suitable depending on the patient's age at the time THA surgery is performed. If the patient is not so old, a smaller stem may be selected, as the patient ages, a larger stem would be necessary for the larger medullary canal. But the proximal portion of the femur, femoral neck length, for example, may be the same. Thus, if we produce hip prostheses to be used for different age groups, many types of stem (not one standard proportion) may be required.

Anteversion Angle

We cannot say whether anteversion angle is internal or external form. The femoral anteversion angle we derived from our data (44.9 + 14.9°) is greater than that derived from the standard data of others (average: about 20°) [13]. The reason for this may be that our cases were secondary osteoarthritis caused by CDH. Many authors have used CT scans to measure the anteversion angle

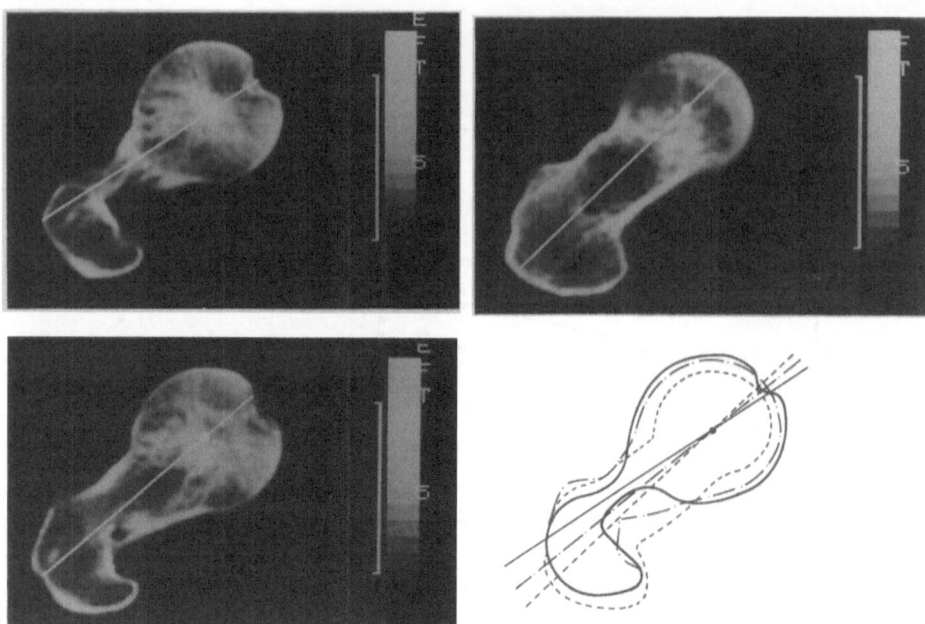

Fig. 6. Femoral anteversion angles on CT scans; differing results due to measuring at different heights

[5,14], but the measurements have been made at various heights; data thus cannot be standardized (Fig. 6). If the central axis of the proximal femur is not determined, exact measurements cannot be taken. Our methods allow exact anteversion to be easily obtained.

If a cementless press-fit stem that has good contact with the inner cortex is implanted in Japanese secondary osteoarthritis patients excessive anteversion of the stem results. We do not know whether an implanted stem with excessive anteversion should be corrected (e.g., the stem may show retroversion on the neck). Such a conclusion can only be drawn following a biomechanical investigation.

Conclusions

We suggest the following as optimal parameters for hip prosthesis stems suitable for CDH-induced secondary osteoarthritis in Japanese patients.

Taper Angle of Proximal Femoral Medullary Canal

The taper angle is not correlated with height, femoral head offset, femoral head diameter, or femoral transverse diameter. Thus, regardless of the stem size, the taper angle of the stem shaft is almost always 3.8°. However, for elderly patients who have a large medullary canal, a stem with a larger taper angle (perhaps 4.5°) is recommended. For younger patients who have narrow medullary canals, a stem with a smaller taper angle (perhaps 3.5°) is recommended.

Proportionality of Stem

If various size stems are to be produced, they should be of similar proportions; from small to large, the length and transverse diameter of the stem should increase gradually in a proportional way. However, the standard neck offset of the stem should remain constant and various neck lengths are to be used (modular type).

Age Related Problems

Since the medullary canal increases in size with age, stems in standard proportions do not suit all age groups. If a cementless stem that will perfectly occupy the medullary canal is to be prepared for all age groups, various types of stems will be needed. For elderly patients especially, stems in which the distal portion are larger than the proximal portion are needed. However, the question of whether cemented or cementless prostheses are better remains to be resolved. We recommend cemented stems for elderly patients, especially for those who have large medullary canals.

References

1. Charnley J (1979) Low friction arthroplasty of the hip. Springer Berlin Heidelberg New York, pp 185–298
2. Morscher EW (1983) Cementless total hip arthroplasty. Clin Orthop 181:76–91
3. Walker PS, Robertson DD (1988) Design and fabrication of cementless hip stems. Clin Orthop 235:25–34
4. Harper MC, Carson WL (1987) Curvature of the femur and the proximal entry point for an intramedullary rod. Clin Orthop 220:155–161
5. Murphy PC (1987) Femoral anteversion. J Bone Joint Surg [Am] 69-A:1159–1176
6. Dai KP, An KN (1985) Geometric and biomechanical analysis of the human femur. Trans Orthop Res Soc 10:99
7. Noble PC, Jerry WA (1988) The anatomic basis of femoral component design. Clin Orthop 235:148–165
8. Yanagimoto S, Sakamaki T (1987) Basic studies of cementless hip prosthesis design (3rd report): Offset of human femur (in Japanese). Cent Jpn J Orthop Trauma Surg 30:304–308
9. Yanagimoto S (1991) Basic study of cementless hip prosthesis design: Analysis of the proximal femur in Japanese patients with osteoarthritis of the hip (in Japanese). J Jpn Orthop Assoc 65:731–744
10. Rubin PJ, Leyvraz PF (1992) The morphology of the proximal femur. J Bone Joint Surg [Br] 74-B:28–32

11. Smith RW, Walker RR (1964) Femoral expansion in aging women: Implications for osteoporosis and fractures. Science 145:156–157
12. Trotter M, Peterson RR (1967) Transverse diameter of the femur: On roentgenograms and on bones. Clin Orthop 52:233–239
13. Ikeda K (1977) Femoral anteversion (in Japanese). Rinsho Seikei-geka 12:1033–1040
14. Peterson HA, Klassen RA (1981) The use of computerized tomography in dislocation of the hip and femoral neck anteversion in children. J Bone Joint Surg [Br] 63-B:198–208

28

Analysis of the Endosteal Geometry of the Proximal Femur in Japanese Patients with Osteoarthritic Hips: Use in Femoral Stem Design

Yasumasa Matsuda, Kazuhiko Sawai, Tomokazu Hattori, and Shigeo Niwa[1]

Summary. A study was undertaken to establish a basic design for a straight femoral stem to be used in total hip arthroplasty (THA). This stem would be fitted to the endosteal geometry of femurs in Japanese patients with advanced osteoarthritis of the hip due to congenital displacement of the hip (CDH). Using our three-dimensional computer-aided design (3D-CAD) system, we investigated the geometry of the proximal femur, based on computed tomography (CT) slices obtained from 30 femurs in patients with advanced osteoarthritis and 20 femurs in healthy controls. We found that almost 90% of the femurs in the patients with advanced osteoarthritis had a straight section in the marrow cavity, between 40 and 60 mm below the tip of the lesser trochanter. An adequate basic axis, as a standard for measuring and as a guide for reaming the marrow cavity of the femur, could be obtained by extending the connecting line of the inner center of each slice in this specific straight section. Reaming along our basic axis, simulated by the 3D-CAD system, cut off less cortical bone (79 mm^3) than reaming along a conventional axis determined by connecting the center of two cross sections, one at the entry point of the marrow cavity and the other at the maximum depth of reaming (203 mm^3). Excessive variations were observed in the cross-sectional geometry of the upper part of the canal of the proximal femur. It was found that, in THA, to obtain a standard straight marrow cavity with no excessive variation and under-cut, the femoral neck should be cut off less than 15 mm above the tip of the lesser trochanter, with a cutting angle of less than 20°. Two radiuses of curvature (R1, 100 mm; R2, 150 mm) were proposed to simplify the curve on the medial flare of the femoral canal between the levels of 15 mm above and 40 mm below the tip of the lesser trochanter. Two taper angles (T1, 2°; T2, 1°) were proposed, to simplify the inclination of the medial femoral canal between 40 and 80 mm below the tip of the lesser trochanter. Correlations were found between patient's canal

---

[1] Department of Orthopedic Surgery, Aichi Medical University, 21 Yazako Karimata, Nagakute, Aichi, 480-11, Japan

flare indices (CFIs) and canal widths 60 mm below the tip of the lesser trochanter in both the frontal ($r = -0.6697$) and sagittal ($r = -0.7961$) dimensions. No correlation, however, was revealed between frontal and sagittal CFIs.

Key words: Femoral stem—Femoral canal—Computerization

Introduction

The aim of this study was to establish a fundamental design for a straight femoral stem to be used in total hip arthroplasties (THA) carried out with bone cement. This design was to be applied to the endosteal geometry of the femur in Japanese patients with advanced osteoarthritis of the hips. In such patients, the geometry of the proximal femur varies greatly, primarily due to congenital displacement of the hip (CDH).

Thus, the two fundamental problems to be solved in designing a femoral stem for these osteoarthritic femurs were: (1) How to establish a basic axis in the proximal cavity of the femur and (2) how to deal with the excessive variations in the proximal cavity of the femur. The following studies were carried out to solve these problems.

Material and Methods

Initially, plain X-ray lateral views of 42 femurs in Japanese patients with advanced osteoarthritis of the hip were investigated to locate a straight section in the proximal marrow cavity. Anterior and posterior endosteal outlines were traced on the plain X-ray lateral view by using a digitizer; the outlines were then compressed, using a three-dimensional computer-aided design (3D-CAD) system to enhance their curvature toward the longitudinal axis of the femur. A transparent grid was then placed over these outlines to measure deviation, and the endosteal outline was considered straight when the outline did not deviate by more than 0.5 mm from a standard straight line on the grid. The location of the straight section in the proximal marrow cavity was confirmed when two straight endosteal lines were found to coexist opposite each other on the plain X-ray lateral view.

Computed tomographic (CT) slices of 30 femurs in 22 Japanese patients with advanced osteoarthritis of the hip and 20 femurs in 10 healthy Japanese controls were used in the study. Data for these were processed with the 3D-CAD system. A circle was inscribed within the outline of the marrow canal to determine the inner center of the femoral canal on CT slices made at 10-mm intervals. The inner centers of the femoral canal on CT slices made at intervals of 20 mm were connected by a straight line in the proximal marrow cavity. The perpendicular distance between the connecting line and the inner center of a CT slice midway between the connected CT slices was measured; the location

Fig. 1. Perpendicular distance was measured to check
alignment of inner centers

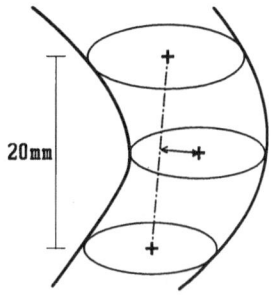

of these inner centers in the straight section detected on the lateral X-ray was
regarded as straight when deviation from the straight line was less than 0.5 mm
(Fig. 1). A basic axis through the marrow cavity was assumed by extending the
line which connected the inner centers of each slice in this specific straight
section of the femoral canal. The perpendicular distance between the inner
center and this axis was measured at 13 levels. We considered this basic axis to
be a suitable standard for measuring dimensions of endosteal geometry when
these distances were less than 1 mm. The use of this basic axis as a guide for
reaming the femoral canal was also investigated, by measuring the volumetric
loss of cortical bone in ten femurs after reaming along the basic axis by 3D-
CAD simulation and comparing the results to reaming along the conventional
axis.

In order to access various dimensions (e.g., the dimension perpendicular to
the femoral shaft or neck), wire-frame models of the marrow cavity were
generated by interpolating these accumulated CT slices. These models were
processed so they could be expressed in standard 3D coordinates. The basic
axis was defined as the z-axis. The x-axis was defined as the line which
connected the intersecting point of the basic axis and the apex of the medial
projection of the marrow cavity at the neck level; this was at right angles to the
z-axis. The y-axis was then set up at right angles to both of these axes. Using
dimensions on the x–y plane, the distance between the z-axis and the endosteal
surface was measured every 22.5° around the z-axis in the individual models.
These distances in each osteoarthritic femoral cavity were compared statistically to those in the healthy controls to investigate variations in the femoral
neck.

An adequate and suitable cutting line on the femoral neck was carefully
determined, one which would remove any excessive variations of the proximal
femur and would produce a straight marrow cavity with no under-cut. Various
cutting lines were thus considered, related to (a) cutting levels on the femoral
neck (15 mm and 20 mm above the tip of the lesser trochanter) and (b) cutting
angles on the femoral neck (0°, 20°, 30°, and 45° to the horizontal).

The marrow cavity was simplified to facilitate the design of a prototype stem.
A medial flare, represented by a simple arc, was established by averaging the

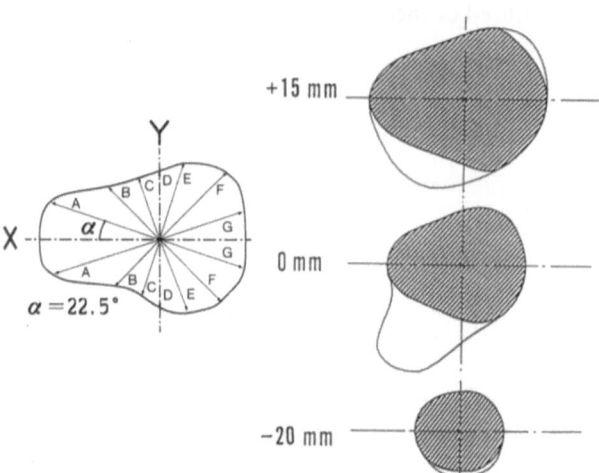

Fig. 2. Simplified cross sections established by symmetrizing along the x-axis

distance between the z-axis and the endosteal medial surface of the femur of all patients at each level from 15 mm above to 40 mm below the tip of the lesser trochanter. Simplification was considered to have been achieved when this simple arc passed within ±1 mm of the average distance at each level.

A straight endosteal medial line was established below the flare to determine the taper angle. Simplification was achieved when this line passed within ±1 mm of the average distance at each level.

Simplified cross sections were established by symmetrizing along the x-axis. The outlines were limited to the length of the shorter of each of two corresponding distances from the x-axis to the anterior and posterior endosteal outline at each level (Fig. 2).

In order to facilitate stem design, canal proportionality was determined by using a geometric parameter called the canal flare index (CFI), defined as the ratio of the width of the femoral canal 15 mm above to that 60 mm below the tip of the lesser trochanter.

Results

In about 90% of the patients, a specific straight section was observed in the marrow cavity of the femurs between 40 and 60 mm below the tip of the lesser trochanter (Fig. 3).

In the CT slices of the femoral canal, two inner centers of the inscribed circles in CT slices taken at intervals of 20 mm were connected by a straight line through this section, and the deviation was 0.35 mm or less from the inner center of a CT slice midway between the connected CT slices. Extended through the section 20 mm above and 100 mm below the tip of the lesser trochanter of the femoral canal, this straight line ran through near the inner

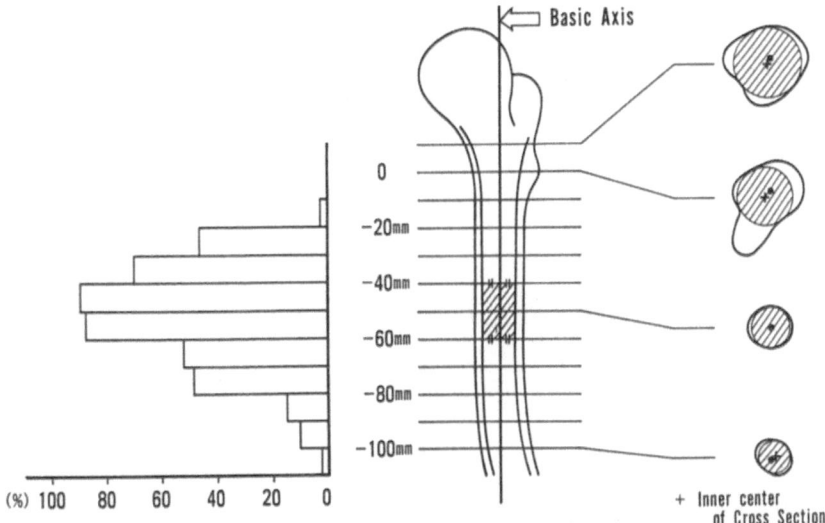

Fig. 3. In about 90% of the patients, a specific straight section was observed in the marrow cavity of the femurs between 40 and 60 mm below the tip of the lesser trochanter

center of every slice, with a mean deviation of 0.85 mm. The maximum distance between this axis and the inner center was 2 mm toward the great trochanter at the level of 20 mm above the tip of the lesser trochanter. In the simulation study, the volumetric loss of anterior cortical bone was only 79 mm³ near the lowest point of the reamed cavity when the basic axis was used as a reamer guide. On the other hand, when the conventional axis served as the reamer guide, the volumetric loss of posterior cortical bone was 203 mm³ along the mid-portion of the reamed cavity. By reaming along the basic axis, the volumetric loss of cortical bone was reduced to less than 34% of that produced by reaming along the conventional axis (Fig. 4).

Excessive variations were observed in the cross-sectional geometry of the upper part of the femoral canal in the patients. Significant differences were found in the mean absolute deviation of distance from the z-axis to the endosteal outline between the patients and the healthy controls in the portion of the neck higher than 20 mm above the tip of the lesser trochanter (Fig. 5).

It was shown that, to obtain a standard straight marrow cavity with no under-cut, the femoral neck should be cut off less than 15 mm above the tip of the lesser trochanter, with a cutting angle of less than 20°. The cross-sectional geometry of the resulting proximal femoral cavities was almost the same in osteoarthritic patients as in healthy controls after adequate cutting of the femoral neck was performed (Fig. 6).

To simplify the medial flare, two radiuses of curvature (R1, 100 mm; R2, 150 mm) were proposed for the femoral canal between 15 mm above and 40 mm below the tip of the lesser trochanter.

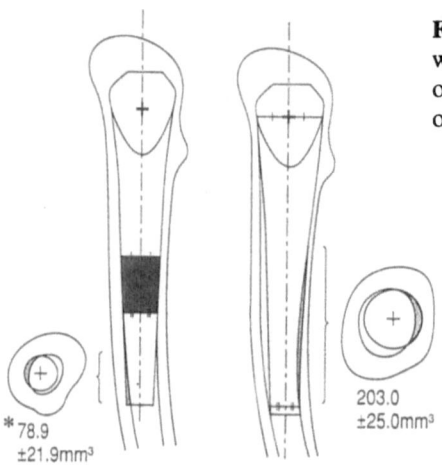

Fig. 4. Volumetric loss of cut cortical bone with reaming along our basic axis (*left*) was one-third of that lost by reaming along the other straight axis (*right*) (*n* = 10). *P < 0.01

203.0
±25.0mm³

*78.9
±21.9mm³

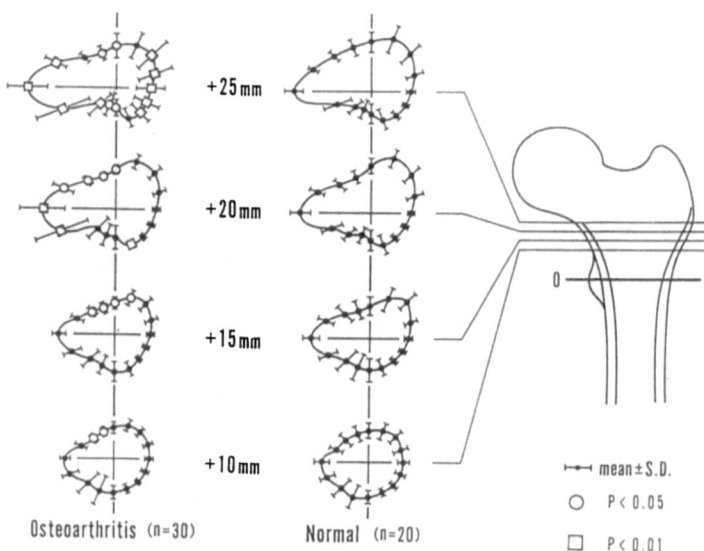

+25 mm

+20 mm

0

+15 mm

+10 mm

⊢——⊣ mean±S.D.

○   P < 0.05

Osteoarthritis (n=30)   Normal (n=20)   □   P < 0.01

Fig. 5. Statistical analysis of cross-sectional variations. *Solid circles* indicate mean ± SD; *open circles*, P < 0.05; *open squares*, P < 0.01

To simplify the endosteal medial outline, two taper angles (T1, 2°; T2, 1°) were determined, by establishing a straight line on the medial femoral canal between 40 and 80 mm below the tip of the lesser trochanter (Fig. 7).

The CFI was defined as the ratio of the canal widths at levels 15 mm above and 60 mm below the tip of the lesser trochanter on the x–z and the y–z planes. Correlations ($r = -0.6697$, x–z plane; $r = -0.7961$, y–z plane) were

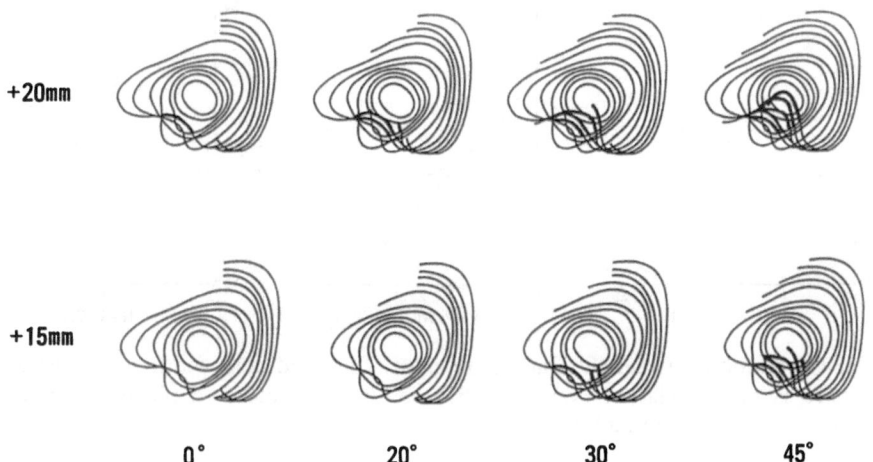

+20mm

+15mm

0°   20°   30°   45°

Fig. 6. Cross-sectional geometry of femoral cavity obtained through the various cutting lines and levels

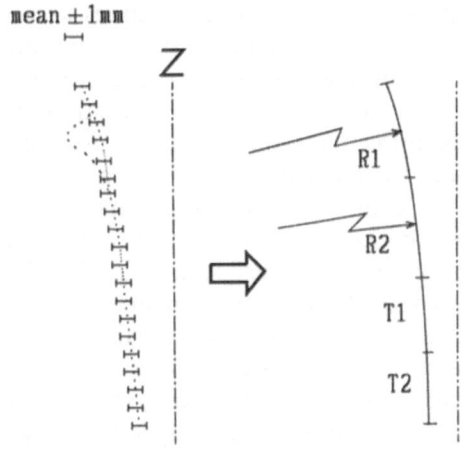

Fig. 7. Two radiuses of curvature (R1, 100 mm; R2, 150 mm) proposed for the medial flare, and two taper angles (T1, 2°; T2, 1°) proposed for the distal medial femoral canal

found between these CFIs and both canal widths at levels 60 mm bleow the tip of the lesser trochanter (Fig. 8).

Discussion

Plain X-ray lateral views of 42 femurs were observed in order to locate section with straight outlines in the geometry of the proximal marrow cavity on the sagittal plane. Nunn et al. [1], in their measurements, noted magnification, tracing, and inter-observer errors with regard to X-ray measurements. To

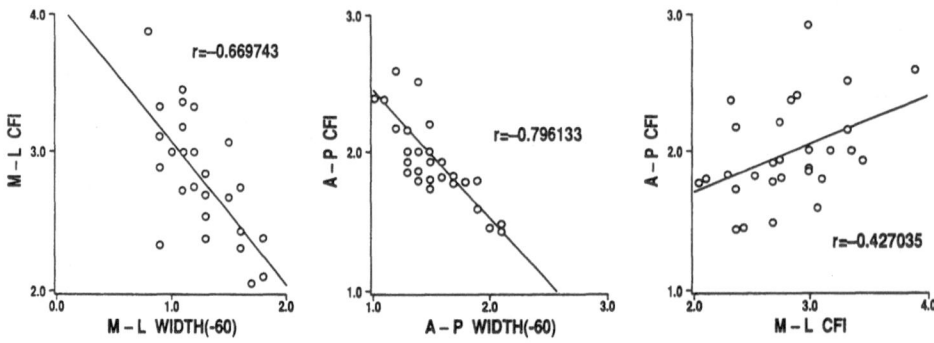

Fig. 8. Correlation between the Canal flare index (*CFI*) and canal width is demonstrated, but there is no correlation between CFIs on the x–z and y–z planes. *A-P*, anteroposterior; *M-L*, medio-lateral

correct for magnification error in that study, Nunn et al. multiplied values in X-ray views by a factor of 0.8. However, in our study, in order to establish a strict range of error, the magnification error was not corrected. There was no inter-observer error, since the outlines were traced by only one of the authors. Nunn et al. [1] noted errors of 0.5 mm in both the horizontal and vertical planes with regard to the tracing error. Accordingly, we took deviation distances of 0.5 mm to be the tracing error. The straight section we defined was confirmed on condition that two straight endosteal lines coexisted opposite each other on both the anterior and posterior surfaces of a plain X-ray lateral view [2,3].

To confirm, three-dimensionally, the presence of a straight section so defined, we checked the alignment of inner centers by using CT slices made at 10-mm intervals along each femoral cavity. The perpendicular distance was measured to check this alignment by drawing a connecting line between the inner centers at 20-mm intervals and measuring the extent to which the inner center of a CT slice midway between the connected CT slices deviated from this line. These mean distances of deviation were found to be 0.47, 0.23, and 0.15 mm at levels of 30, 40, and 50 mm, respectively, below the tip of the lesser trochanter.

Robertson and Huang [4] reported that errors in measuring the area of the medullary canal were 4%–11% of real value with regard to measuring errors in CTs. When these ratios of error were applied to the dimensions of the femoral cavities in our cases, the measuring errors were found to be 0.72–1.5 mm. Woolson et al. [5] noted that measuring errors in the medullary canal were 1–3 mm. In our study, the deviation distances were less than the sum of the errors, which was 2.0 mm: a tracing error of 0.5 mm plus measuring errors in the CT of 1.5 mm. Therefore, the section between 40 and 60 mm below the tip of the lesser trochanter was defined as straight for the alignment of the inner centers.

To test the suitability of the basic axis as a standard for measuring three-dimensional endosteal geometry, we measured perpendicular distances between the extended line connecting the inner centers of both ends of the straight section and the inner centers of the CT slices at each level. The maximum deviation distance at each level 20 mm above the tip of the lesser trochanter was 2.3 mm, and these distances were less than 1.9 mm in the portion lower than this level. These distances were less than the sum of the errors estimated as measuring errors in CT and tracing. This line was therefore regarded as useful for determining the basic axis of three-dimensional measurements in the portion less than 15 mm above the tip of the lesser trochanter.

Further, to test the suitability of this basic axis as a guide axis for surgical reaming or inserting a femoral stem, we measured volumetric loss of cortical bone when reaming was done along this basic axis as the guide. When a line (the conventional axis, so to say) connecting the inner centers both at the entry point of the marrow cavity and at the distal maximum depth of reaming was used with centralizer and/or aligning devices at both the proximal and distal portions of the femoral stem [6,7], the volumetric loss of cortical bone was three times greater than when our basic axis (i.e., the extended line connecting the inner centers of both ends of the straight section) was used as a guide for surgical reaming. This indicated that our basic axis was also more suitable as a guide for surgical reaming. According to our study [2] of the geometry of the femoral neck in Japanese patients with advanced osteoarthritis due to CDH, about one-half of the patients had nearly normal geometry, while the other half had severe varus or valgus deformities, with marked anteversion and/or antetorsion. Reikeras et al. [8] recorded similar findings. In these osteoarthritic patients, variations of geometry in the portion more than 20 mm above the lesser trochanter were greater than in the healthy controls, particularly on the sagittal plane. These results, indicate that it would be next to impossible to fit so-called curved stems, even those designed to fit the femoral canal on the sagittal plane.

In THAs in which bone cement is used, a straight stem is favorable for obtaining the necessary forcible anchoring and for the filling of the cavity with bone cement. Cutting of excessive variation of the proximal portion of the femoral neck at a level 15 mm above the tip of the lesser trochanter and removing it is the most effective method of fitting a straight stem in the proximal femoral canal; such a method is also used in some total knee arthroplasty (TKA) procedures. Further, in Japanese patients, any remaining projection of the posterior cortical bone of the femoral neck toward the anterior, with marked anteversion and/or antetorsion, causes malalignment in the setting of the stem in the sagittal plane and a lower rate of occuparcy of the femoral cavity by the stem. [9,10] Therefore, we determined a cutting line on the femoral neck which would produce a straight marrow cavity with no projection of the posterior cortical bone.

A minimum number of simple arcs and straight lines with taper angles on the x–z plane were employed to simplify the endosteal outline for the design of the prototype stem. The line which we selected and used to simplify the endosteal outline passed within ±1 mm of the average distances between the z-axis and the endosteal medial surface at each level of the femur in all patients, after we took into account the above-mentioned measuring errors in CT [3]. Canal proportionality is important in regard to standardization of the femoral cavity. Correlation between the CFI and canal width was demonstrated, but there was no correlation between CFIs on the x–z and y–z planes. Noble et al. [7,11] calculated a CFI from the medio-lateral width at a level 20 mm above the tip of the lesser trochanter and the isthmus, using X-ray anteroposterior (A-P) views of 200 femurs, but they observed no correlation between the medio-lateral width of the isthmus and the CFI. In contrast, we demonstrated a correlation between the width of the medullary canal at a level 60 mm below the tip of the lesser trochanter (i.e., the distal end of the straight section of the proximal femoral cavity) and our CFI. This discrepancy may be due to the characteristics of the geometry of the femoral canal in Japanese patients with advanced osteoarthritis due to CDH.

Unlike previous studies, which produced CFIs using two-dimensional measurements, the present study was based upon the three-dimensional geometry of the endosteal cavity. Our research indicates the superiority of using a three-dimensional CFI for determining variations in the shape of femoral stems.

References

1. Nunn D, Freeman MAR, Hill PF, et al (1989) The measurement of migration of the acetabular component of hip prostheses. J Bone Joint Surg [Br] 71-B:629–631
2. Matsuda Y, Sawai K, Yamazaki S, et al (1986) The geometry of the proximal femoral canal in the osteoarthritic hip. Cent Jpn J Orthop Traumat 29:601–606
3. Matsuda Y, et al (1987) The design of the femoral component. Cent Jpn J Orthop Traumat 30:308–312
4. Robertson DD, Huang HK (1986) Quantitative bone measurements using X-ray computed tomography with second-order correction. Med Phys 13:474–479
5. Woolson ST, Dev P, Fellingham LL, et al (1986) Three-dimensional image of bone from computerized tomography. Clin Orthop 202:239–248
6. Levy RN, Noble PC, Scheller A Jr, et al (1988) Prolonged fixation of cemented total hip replacement. Surg Rounds for Orthopaedics April:15–22
7. Noble PC, Tullos HS, Landon GC (1991) The optimum cement mantle for total hip replacement: Theory and practice. In: Eilert RE (ed) Advances in total hip reconstruction: Instructional course lectures. American Academy of Orthopaedic Surgeons, pp 145–150
8. Reikeras O, Bjerkreim I, Kolbenstredt A (1983) Anteversion of the acetabulum and femoral neck in normals and in patients with osteoarthritis of the hip. Acta Orthop Scand 54:18–23
9. Garg A, Deland J, Walker PS (1985) Design of intramedullary femoral stems using computer graphics. Eng Med 14:89–93
10. Hirose S, Yamazaki S, Sawai K, et al (1986) Radiological assessment of lateral view of the hip after total hip replacement. Cent Jpn J Orthop Traumat 29:611–613
11. Noble PC, Alekunder JW, Lindahl LJ, et al (1988) The anatomic basis of femoral component design. Clin Rothop 235:148–165

29

Torsional Fixation of the Femoral Component in Cementless Total Hip Arthroplasty: Newly Designed Funnel Shape Femoral Component

Hajime Sugiyama, Kagehisa Murota, Yoshitsugu Tomita, Takuya Ohtani[1], and Yakichi Higo[2]

Summary. Rotational stability of the femoral component is now regarded as the most important factor in the minimization loosening in artificial hip joints. In this study, we report the development of a new type of femoral component from which excellent rotational stability is expected. The design of the new Jikei total hip funnel shape femoral component is characterized by the curvature of the proximal part of the stem being the same on both the lateral and medial sides. By fitting the shape of the stem so that its greatest diameter fits snugly into the greatest diameter of the proximal femoral canal, excellent rotational stability is obtained. Using the finite element method (FEM), we compared the rotational stability of this component with that of other femoral component designs. In the newly designed Jikei femoral component stem, the rotational stresses were distributed to the cortical bone, with less stress concentration being transmitted to the cancellous bone in the middle portion of the stem and without the occurrence of micromotion. Superior rotational stability was found in this extended breadth funnel-shaped stem.

Key words: Total hip arthroplasty (THA)—Torsional fixation—Finite element technique method (FEM)—Micromotion—Stress distribution

Introduction

In cementless total hip arthroplasty, loosening of the implant is the most important determinant of operation results. The strength and direction of the dynamic stresses loaded on each component are closely connected with the

[1] Department of Orthopaedic Surgery, The Jikei University School of Medicine, 3-25-8 Nishishinbashi, Minato-ku, Tokyo, 105 Japan
[2] Tokyo Institute of Technology, Research Laboratory of Precision Machinery and Electronics, 4259 Nagatsuta, Midori-ku, Yokohama, 227 Japan

occurrence of loosening. Therefore, when implants are inserted, it is extremely important to determine whether each component can be stabilized sufficiently to tolerate such stresses. Various implant designs have recently been employed to minimize loosening in cementless total hip arthroplasty. The concept underlying the development of these implants is the improvement of stability. In this procedure in particular, the stability of the femoral component is very important. However, stability in terms of the direction of the stress has yet to be clarified. With regard to the stem in total hip arthroplasty, the fixation of a longer stem is not affected by the vertical stresses imparted during walking. However, since such stems have very low stability against rotational stresses, swaying of the stem in the direction of rotation is apt to occur when the patient stands up from a chair or climbs stairs. While most studies in the literature have evaluated stress distribution and micromotion with axial loading [1–5], only a few have also evaluated the problem of torsional loading. This limited number of studies has indicated that the cause of loosening is insufficient rotatory stability of the femoral component [6–8]. More recently, fundamental studies have shown that the rotational stability of the stem in cementless total hip femoral components is much lower than that in cement-type prostheses [9,10]. In the light of these findings, various methods have recently been developed to improve the rotational stability of cementless total hip femoral components. These methods include fixation of the stem to the bone with screws [11] and the use of materials with the same elasticity as the femur [12]. There is also a procedure to fit the stem in the peripheral area by a press-fit technique following under reaming of the distal femur [13]. Moreover, a so-called retaining neck type femoral component has been developed, the rotational stability of which component is improved by retaining sufficient neck through minimizing neck resection, thus making the lever arm longer [14]. All of these methods are actually effective in improving stability against rotation. However, in some cases where these techniques are applied, cracking of the femur or wearing or breaking of the implant has occurred. Difficulties are also experienced with insertion of the stem for retaining neck type femoral components in patients who have secondary osteoarthritis secondary to congenital dislocation of the hip, a feature often seen in Japanese patients. In view of these facts, we developed a new type of stem from which we expect excellent rotational stability. Here we describe the stem and report on a comparison of its stability with that of femoral components of other designs, using a finite element method (FEM).

Materials and Methods

New Design of Femoral Component in Jikei Total Hip Arthroplasty

The design for the newly developed femoral component in the Jikei total hip arthroplasty (Fig. 1) is characterized by the proximal part of the stem being shaped similarly on both the lateral and medial sides, so as to better fit the

Fig. 1. The newly developed femoral component of the Jikei total hip arthroplasty. The proximal part of the stem has a similar shape on both the lateral and medial sides

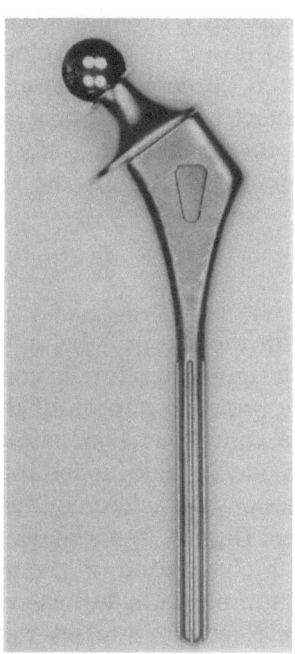

shape of the medullary cavity in the proximal part of the femur. That is, the component is funnel-shaped. Thus, vertical stresses can be distributed evenly from the stem to the lateral and medial sides of the proximal part of the femur. This component was designed to obtain excellent rotational stability by fitting the shape of the stem so that its greatest diameter fits snugly into the greatest diameter of the proximal femoral canal. In addition, so as to fit the stem, in terms of cross-sectional shape, to the medullary cavity in the proximal part of the femur, the thickness is tapered in a modified wedge shape so that the lateral side is of greater thickness than the medial side. Moreover, to avoid stress concentration, each corner is rounded.

Rotational Stability of the New Funnel Shaped Component: Evaluation of Stability by Computed Simulation Using FEM

We used bleached human femoral bones, taking the axis connecting the center of the diaphysis to the center of the lesser trochanter; a cross section parallel to the longer axis was prepared by computed tomographic (CT) scan. The condition of this cross section after the insertion of a total hip femoral component was evaluated by computed simulation using FEM. For FEM, the elasticity constants were 20 GPa for cortical bone, 2 GPa for cancellous bone, and 200 GPa for the femoral component. Two fundamental designs were compared; one being the non-lateral type as used in the old-type Jikei total hip prosthesis [15] and in Moore's endoprosthesis; the other being the lateral type used in our newly-developed femoral component. For both these types, both square-cornered and round-cornered variations were simulated. The area in contact with the cortical bone was regarded as well fixed to the cortical bone. As rotational stress, only torque was given, with the center of the stem in the femoral diaphysis as the axis.

Results

With the stem of the conventional femoral component, which has no lateral side and is therefore in contact with the cortical bone only on the medial side in the proximal portion of the femur, stresses were remarkably concentrated on the cancellous bone around the stem, as well as on the cortical bone of the medial side under rotational stresses. Thus, the stem was torsioned and considerable micromotion arose (Figs. 2 and 3). With the square-cornered type stem (Fig. 2) in particular, stresses were concentrated at the corners.

In comparison, in the new extended breadth stem, the lateral side similarly to the medial side, was in contact with the cortical bone. In this case, even the rotational stresses were distributed to the cortical bone, with less stress concentration on the cancellous bone in the middle portion of the stem, and without the occurrence of micromotion (Figs. 4 and 5). In particular, when a stem of the rounded-corner type was used, micromotion did not occur; as shown in Fig. 5, the stresses on the cortical bone were distributed, since there were no corners. Figure 6 shows the stress loaded on the cortical bone more clearly. The stress was distributed over the entire cortical bone. The stem of the round-cornered type of component, which comes into contact with the bone cortex on both the medial and lateral sides, showed excellent stability against rotational stresses.

Discussion

Many types of implants have recently been developed, and in each of these various methods are employed to improve stability. For example, the implant is designed to fit the shape of the bone, screws or spokes are used to fix the implant in contact with the hard portion of the bone to enhance stability, or materials with elasticity similar to that of the bone are used to minimize displacement and to increase stability. However, in our previous studies, performed to determine the rotational stability of cementless total hip femoral components, such implants were shown to have very low stability [10,13]. With the AML femoral component (De Puy, Warsaw, Ind.) in particular, considered to be insufficiently stable because most of the tissues surrounding the stem consisted of cancellous bones, remarkable subsidence was seen under rotational stresses [13]. With the Ortholoc femoral component (Dow Corning Wright,

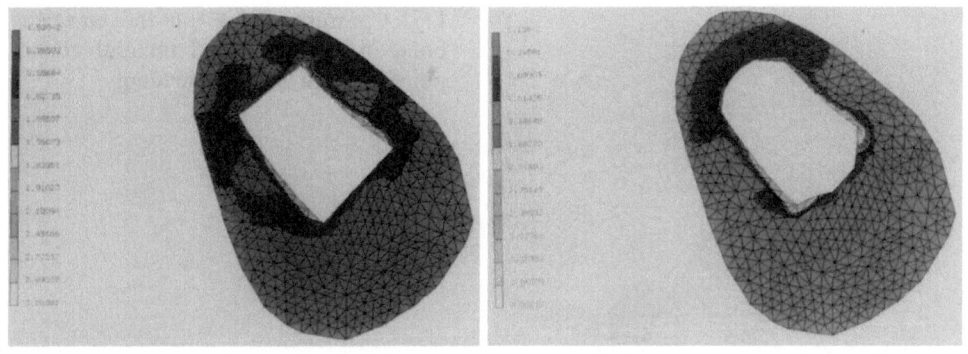

2   3

Fig. 2. Stress concentration and micromotion under torsional loading in conventional femoral component with square corners

Fig. 3. Stress concentration and micromotion under torsional loading in conventional femoral component with rounded corners

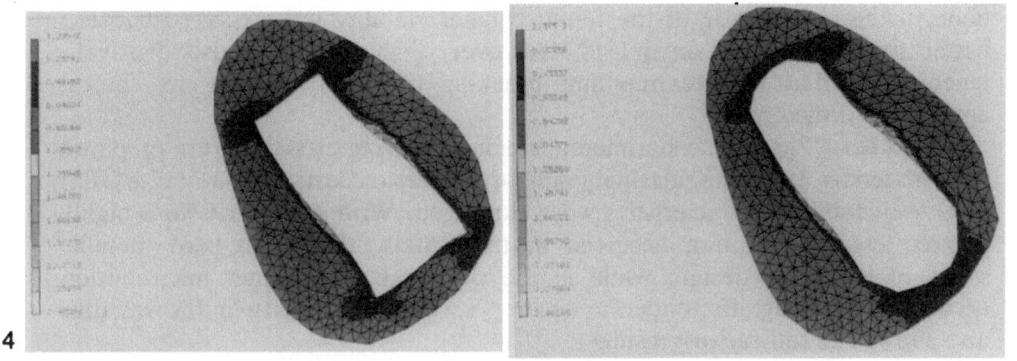

4   5

Fig. 4. Stress concentration and micromotion under torsional loading in extended breadth femoral component with square corners

Fig. 5. Stress concentration and micromotion under torsional loading in extended breadth femoral component with rounded corners

Arlington, Tenn.), which has a large A/P diameter, some degree of fixation was obtained, since the stem was fixed to the cortical bone on both the anterior and the posterior sides [10]. However, its subsidence exceeded 100 μm under a torsional load of 20 N-m, which is the level of stress that occurs in daily activities. On the other hand, with an AML stem that was press-fit in the distal portion of the femur, the stability was much better than that of the AML stem with press-fit only in the central part [13], indicating that the distal press-fit

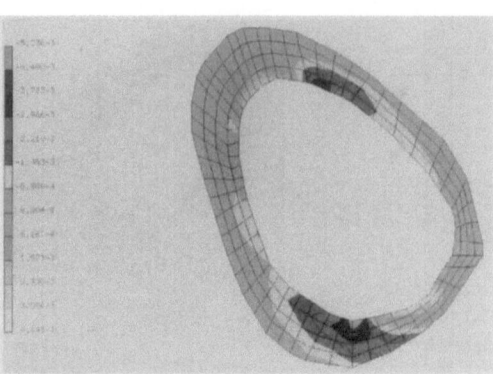

Fig. 6. Stress loaded on the cortical bone in the new Jikei femoral component under torsional loading

technique is more effective for fixation, since the distal part of the femur is harder and has less variation in shape than the central part. With this technique also, subsidence of more than 50 μm was seen under a torsional load of 20 N-m [16] which corresponds to the stress level encountered in daily activities. For the Ortholoc total hip femoral component fixed with screws to the cortical bone in the medial side of the femur, higher stability was expected because better fixation was obtained [11]. However, this method is also disadvantagenous in that the wearing and breaking of implants make its clinical application difficult.

In another [12] study, experimental stems of varying elasticity were prepared with materials exhibiting elasticity similar to that of cortical bone (17.6 GPa) and their stability characteristics were examined. With stems that were highly elastic, less micromotion occurred in the distally than with hard metallic stems of the same design, while in the proximal part greater micromotion occurred; loosening also tended to occur. Varying the elasticity in this way thus also showed unsatisfactory results.

Recently, so-called retaining neck total hip femoral components, such as the Freeman [14] and the new Whiteside femoral components, have been developed. For such components, the neck of the femur is retained so as to fix the stem; improved rotational stability can thus be expected. It is considered that stability is greatly affected by the shape of the neck area.

In osteoarthritis deformans of the hip joint secondary to congenital dislocation of the hip, a common disorder in Japanese patients, not only is fixation of the stem difficult but even insertion is difficult to achieve. With these aspects in mind, we prepared a new type of femoral component using titanium alloy. This type of stem was designed so that the lateral side was expanded to be similar in shape to the medial side in the proximal part, and it has been applied clinically. Differing from conventional methods, this stem was developed to improve rotational stability, based on a quite new idea, namely, fitting the stem to the medullary cavity in the portion with the largest diameter. As noted above, various other methods have been developed to improve rotational

Fig. 7. Anteroposterior roentgenogram of new Jikei total hip arthroplasty (1 year after surgery)

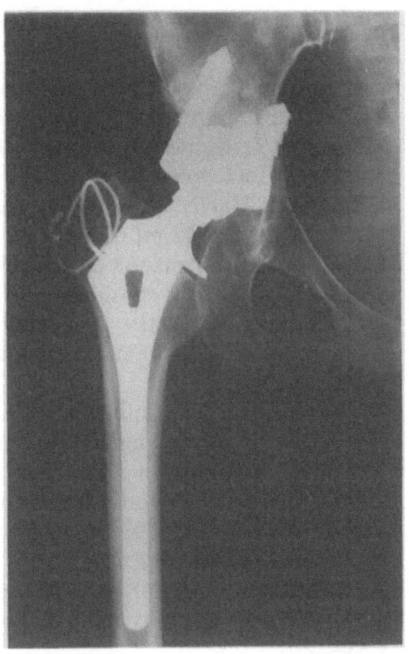

stability, but, since conventionally designed total hip femoral components are used without any modification, satisfactory results are not always obtained in terms of rotational stability. Most of the currently available cementless total hip femoral components are based on Moore's endoprosthesis, with the AML total hip femoral component as the model. With these components little attention is paid to the fit on the lateral side. However, the curve on the lateral side of the femur in the contacted portion is similar to that on the medial side. A stem curved on the medial side alone is not satisfactory in terms of ensuring an even distribution of stress and rotational stability. Using the Charnley total hip femoral component, the greater trochanter will also be cemented if the operation is done correctly, and the stress will be distributed not only on the medial but also on the lateral side. With this method, stable fixation can be obtained.

The results of this study, in which rotational stability was determined by the use of the FEM, showed that with the new Jikei funnel-shaped total hip femoral component, there was far less stress concentration on cancellous bone, than with the non-lateral type component, and high stability was demonstrated. We consider that the stress is distributed evenly to the bone cortex and that favorable long-term stability can be obtained with this new component (Fig. 7).

References

1. Andriacchi TP, Galante JO, Belytschko TB, Hampton S (1976) A Stress analysis of the femoral stem in total hip prostheses. J Bone Joint Surg [Am] 58A:618–624
2. Charnley J, Kettlewell J (1965) The elimination of slip between prosthesis and femur. J Bone Joint Surg [Br] 47B:57–60
3. Oh, I and Harris WH (1978) Proximal strain distribution in the loaded femur. J Bone Joint Surg [Am] 60A:75–85
4. Whiteside LA, Amador D, Russell K (1988) The effects of the collar on total hip femoral component subsidence. Clin Orthop 231:120–126
5. Whiteside LA, Easley JC (1989) The effect of stem fit and collar seating on micromotion of the femoral component in uncemented total hip replacement. Clin Orthop 239:145–151.
6. Crowninshield RD, Johnston RC, Andrews JG, Brand RA (1978) A biomechanical investigation of the human hip. J Biomech 11:75–85
7. Mjoberg B, Hansson LI, Selvik G (1984) Instability of total hip prostheses at rotational stress. Acta Orthop Scand 55:504–506
8. Wroblewski BM (1979) The mechanism of fracture of the femoral prosthesis in total hip replacement. Int Orthop 3:137–139
9. Burke DW, O'Connor DO, Zalenski EB, Jasty M, Harris WH (1991) Micromotion of cemented and uncemented femoral components. J Bone Joint Surg [Br] 73B: 33–37
10. Sugiyama H, Whiteside LA, Kaiser AD (1989) Examination of rotational fixation of the femoral component in total hip arthroplasty. Clin Orthop 249:122–128
11. Martin JW, Sugiyama H, Kaiser AD, Hoech JV, Whiteside LA (1990) An analysis of screw fixation of the femoral component in cementless hip arthroplasty. J Arthroplasty 5:15–20
12. Ohtani T, Whiteside LA (1992) Failure of cementless fixation of the femoral component in total hip arthroplasty. Orthop Clin North Am 23:335–346
13. Sugiyama H, Whiteside LA, Engh CA (1992) Torsional fixation of the femoral component in total hip arthroplasty. Clin Orthop 275:187–193
14. Freeman MAR (1986) Why resect the neck? J Bone Joint Surg [Br] 68B:346–349
15. Itami Y, Akamatsu N, Tomita Y, Nagai M, Nakajima I (1983) A clinical study of the results of cementless total hip replacement. Arch Orthop Trauma Surg 102:1–10
16. Davy DT, Kotzar GM, Brown RH, Heiple KG, Goldberg VM, Heiple Jr, KG, Berilla J, Burstein AH (1988) Telemetric force measurements across the hip after total arthroplasty. J Bone Joint Surg [Am] 70A:45–50

30

—Overview—
Ideal Stem Design for Cementless Total Hip System

Yoshiki Hamada, Noriya Akamatsu, Takatoshi Ide, Shigeru Tatsugi, and Hiroshi Watanabe[1]

Summary. To improve cementless stem design to achieve long-term stability, simulations using graphic processing systems or cadaver specimens, and biomechanical analyses, using rigid body spring models (RBSM) or finite element models (FEM) have been used. Simulations using graphic processing systems or cadaver specimens are useful for investigating the shape or size of the stem which provides stem press fit into the medullary canal. These methods, however, have a disadvantage in that, when they are used, it is not possible to comprehensively evaluate biomechanical stability, in terms of evaluating such factors as the resection angle of the femoral neck, the stem length, and the absence or presence of calcar femorale, which greatly influence this stability. Biomechanical analysis of computer simulations using the RBSM, on the other hand, makes it possible to comprehensively evaluate biomechanical stability in terms of these factors. Further, the results obtained for such simulations are so reliable as to be consistent with clinical results. Against this background, we performed computer simulations, using the RBSM, to develop a new cementless stem, the model Y 2, which was used for 48 hips in 48 patients. This model has a 35° resection angle of the femoral neck and a stem length of 178 mm. The stem has a smooth surface with several longitudinal grooves. In the 29 hips of 29 patients who were followed-up for more than 2 years there was an excellent postoperative course, only 3 patients (10%) being affected by thigh pain. Roentgenographic findings showed no shifting or sinking of the stem and a clear zone less than 2-mm-wide was observed in only seven cases. Based both on our clinical results in patients who received model Y 2 replacements, and on the advantages gained in terms of comprehensively evaluating biomechanical stability, we believe that biomechanical analysis of computer simulations, using the RBSM, is a very useful method for the evaluation of the ideal design for a cementless stem.

[1] Department of Orthopaedic Surgery, Yamanashi Medical University, 1110 Shimokato, Tamaho-cho, Nakakomagun, Yamanashi, 409-38 Japan

Key words: Cementless total hip system—Biomechanics—Computer simulation—Press fit—Rigid body spring model (RBSM)—Surgical procedure—Hip joint

Introduction

In cementless total hip systems, certain problems arise with regard to the stems, e.g., thigh pain, sinking, varus and valgus shift, and loosening of the stem. To overcome these problems, and to obtain excellent initial fixation, followed by stable secondary fixation with new bone formation, various factors should be considered, i.e., the development of new materials, improvements of stem design for the achievement of favorable press fit, and improving surgical procedures and postoperative care. With regard to improving the design of the stem to make it more mechanically stable, means for investigating such improvements are: simulation, using graphic processing systems or cadaver specimens, and biomechanical analyses, using the rigid body spring model (RBSM) (Fig. 1) or the finite element model (FEM). Computer simulation with the RBSM has routinely been used in our department. We investigated certain aspects that should be considered when assessing stem designs for cementless total hip systems and the value of the RBSM.

Methods for Assessing Stem Design

Graphic Processing Stem Design

With this method computed tomographic (CT) scans of transverse sections of the femur are input into a computer to show the medullary canal three-

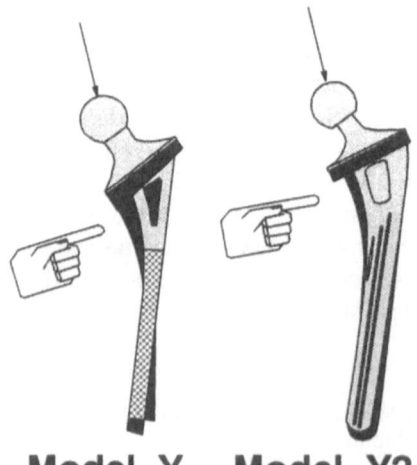

Model-Y   Model-Y2

Fig. 1. Biomechanical analysis of computer simulation using the rigid body spring model (*RBSM*). This shows one of the results obtained in the biomechanical analysis of the computer simulation using the RBSM. The stress distribution (*dark area*) around the model Y 2 stem is more uniform than that in the model Y stem

dimensionally, and the stem design which best fits the femoral medullary canal is determined. At present, this method is being used to assess designs that provide favorable press fit into the medullary canal. Although it is useful as a means of developing such stem designs, there are associated problems. It is difficult to measure and standardize the size and shape of the medullary canal since these features are very complex in patients with osteoarthritis of the hip. Further, the accuracy of reproduction of the shape of the medullary canal varies with the number of CT scan slices, and this may affect the results. With regard to the stability of cementless total hip systems, favorable press fit of the stem into the medullary canal is very important, but how well the shape of the stem fits into the medullary canal is not known. Akamatsu et al. in 1989 [1] and Yamamoto in 1990 [2] have emphasized the importance of achieving excellent press fit for the cementless stem and the need to have more uniform stress distribution between the bone and the stem for biomechanical stability. The biomechanical stability of the stem is greatly influenced by the resection angle of the femoral neck, by the length of the stem, and by the presence or absence of a collar. This method, i.e., the use of a graphic processing system, has disadvantages in that it cannot be used to comprehensively evaluate stability in terms of all these factors.

Investigation of Stem Design Using Cadaveric Femurs

Thigh pain experienced by patients with cementless total hip systems is thought to be caused by micromotion due to the reduced mechanical stability of the stem when exposed to the torsional load that occurs when the patient ascends and descends stairs and stands up from a chair [3]. In 1991, Burke et al. [4] developed a jig that simulates the position of the hip joint and the quantity and direction of the weight-bearing load on the femoral head at the time that a 52-kg person stands on one leg or ascends or descends stairs. They compared cemented and cementless stems inserted into the medullary canal by the press-fit technique in seven cadaver femurs and they reported that both stems provided almost the same stability against compression and shear force, but that the stability against torsional load in the cementless stem was less than that in the cemented stem. They also emphasized that innovations in stem designs to improve stability against torsional loads, as well as restriction of weight-bearing on the operated hip for 3 months after operation, were necessary to prevent micromotion. However, they did not describe any specific stem design.

Sugiyama et al. [5] discussed specific stem designs which had excellent biomechanical stability when exposed to torsional load using this method. They reported that the stem should fit into the widest diameter of the intramedullary canal in the proximal part of the femur. However, this design has an inherent problem in that the stem requires the reaming of a large amount of the medullary canal at the proximal part of the femur so that it can be supported by the cortex, which is thinner than the distal part of the femur. Also, it is doubtful whether the intended stability would be exhibited.

The biomechanical stability of cementless stems should be assessed comprehensively by including the resection angle and stem length, and so on, as described above. This method is thus also still problematic in this regard.

Biomechanical Analysis by Computer Simulation Using RBSM

In our department, we use a cementless total hip system that has a socket with three spikes. In 1983, Akamatsu et al. [6] reported varus and valgus shift, as well as sinking of the stem, in the original type of stem, the JIAT (Jikei, Itami, Akamatsu, and Tomita) in 12 (24%) and 19 (36%) out of 50 patients (50 hip joints) followed-up for more than 5 years after operation. They thus exposed a problem in the stability of this original stem. To develop a new type of stem which offers more biomechanical stability, we studied such stems by computer simulation, using the RBSM developed by Kawai [7].

Value of Computer Simulation

When investigating stem designs using computer simulations, it is most important that the results obtained such simulations are so reliable as to be consistent with the clinical results. It is also of the utmost importance that this method makes it possible to comprehensively evaluate the biomechanical stability of the stem, in terms of the resection angle of the femoral neck, the stem length, and the degree of press fit. The results of computer simulation using RBSM, which is being used in our department, coincided with the clinical findings, as reported by Ide et al. [8], Amano et al. [9], and Yamamoto [2]. Furthermore, using this system, the mechanical stability of the stem can be examined comprehensively, unlike simulations using graphic processing systems or cadaver specimens.

Method for Investigation of Biomechanical Stability of Stem in Computer Simulation Using RBSM

Prior to analysis, assuming that a person weighing 60 kg stands up on one leg, a 2000 N load is applied to the femoral head at an angle of 286°. The value of the rotational deviation, which reflects biomechanical stability against the varus and valgus shift of the stem, and the stress distribution around the collar and the stem, were evaluated under the following conditions (collodiaphyseal angle of the stem was 135°):

1. The amount of stem offset was kept constant. Osteotomy was performed immediately above the lesser trochanter with a 45° resection angle. From 45° with complete resection of the calcar femorale, the angle was changed to 30°, then to 20° with the remaining calcar femorale.
2. The stem was inserted so that it yielded favorable press fit with a 30° resection angle of the femoral neck. The stem length was then changed from 158 mm to 178 mm, then to 198 mm.

Results (Fig. 2)

When the value of the rotational deviation, 0.02218, with a constant offset and a 45° resection angle of the femoral neck was regarded as 100, its value was reduced as the resection angle of the femoral neck decreased, i.e., to 50.2% at 30° and to 27.8% at 20°, and stress distribution around the collar and the stem became more uniform as the resection angle of the femoral neck decreased.

When the value of the rotational deviation with a 45° resection angle of the femoral neck and a stem length of 158 mm was regarded as 100, the value was reduced to 62.2% and 38.2% when the stem was lengthened by 20 mm and 40 mm, respectively. The stress distribution around the stem also became equalized in proportion to the elongation of the stem.

Stem Design

The results of biomechanical analysis by computer simulation using RBSM showed that when the resection angle of the femoral neck was smaller, the length of the stem was longer at a collodiaphyseal angle of 135°, and the stem became more mechanically stable. Such a design, however, is not practical, since, as the resection angle of the femoral neck was reduced and the length of the intact calcar femorale, was increased, surgical procedures, such as stem insertion, reduction of the hip joint, and reattachment of the greater trochanter became more difficult, and there was the possibility of postoperative restriction of hip abduction. Therefore, the ideal stem, from the viewpoint of surgical procedure, was considered to have a 35° resection angle of the femoral neck, an 8-mm length of intact remaining calcar, and a 35° open socket angle, although biomechanical stability would thus be rather poor.

With regard to stem length, on the other hand, insertion of stems that are longer than 40 mm is difficult because the femur has an anterior curvature at the diaphysis. In this case, elongation by 20 mm may be ideal, taking these clinical restrictions into consideration.

Characteristics of Model Y 2 Cementless Total Hip System Using a Socket with Three Spikes and Innovations in Surgical Procedure (Fig. 3)

The above results of computer simulation using the RBSM, in addition to those gained in animal experiments, and our past clinical experience, led us to change the design of the original JIAT and model Y stems (the stem design of the model Y being the same as that of the JIAT); we developed the Y 2 cementless stem. This stem, which has a femoral neck resection angle of 35° degrees and is 178 mm in length, is made of titanium alloy. The surface of the stem is smooth, with several longitudinal grooves. There are three stem sizes in terms of width: small, standard, and large. The width of each of the model Y 2 stems is 1 to 3 mm thicker than the model Y stem. In surgery for the original model Y, the inner cortex is reamed to sufficiently enlarge the medullary canal,

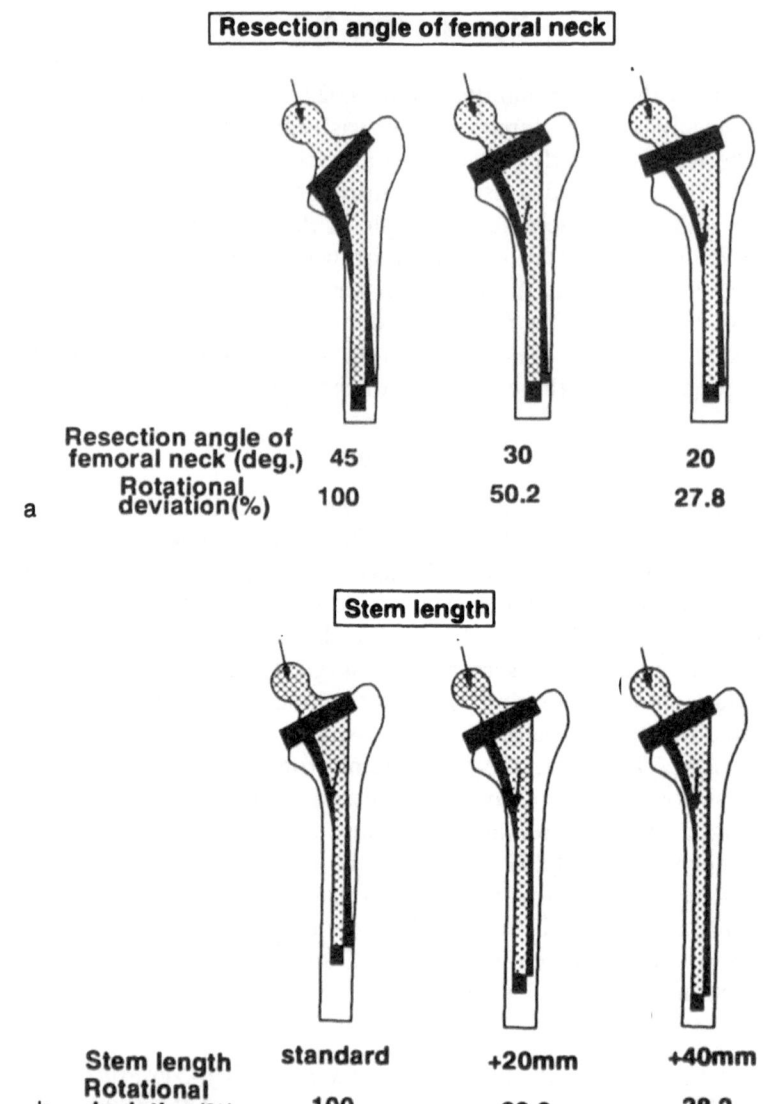

Fig. 2.a,b. Results of Computer Simulation using RBSM. **a** The resection angle of the femoral neck was changed from 45° to 30°, then to 20°, and then to 10°. Value for the rotational deviation was reduced and the stress distribution (*dark area*) became more uniform as the resection angle of the femoral neck decreased. **b** The stem length was changed from the standard 158 mm to 178 mm (+20 mm), and then to 198 mm (+40 mm) when it was inserted, so that it yielded a favorable press fit with a 30° resection angle of the femoral neck. Value for the rotational deviation was reduced and the stress distribution was equalized in proportion to the elongation of the stem

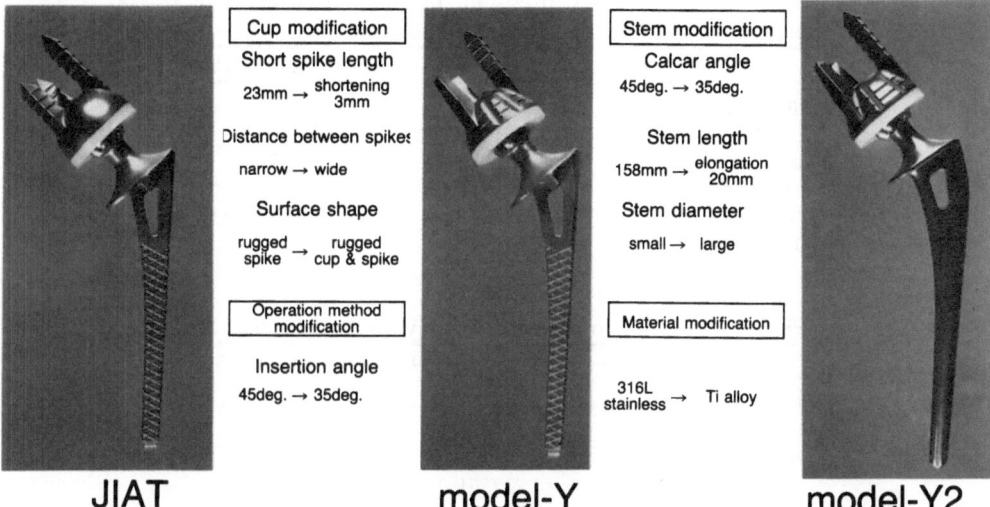

Fig. 3. Improvement of cementless total hip replacement (*THR*) using a socket with three spikes. The stems of the Jikei, Itami, Akamatsu, and Tomita (*JIAT*) THR and the model Y have a 45° resection angle of the femoral neck, are 158 mm in length, and have a notched surface. The stem of the model Y 2, however, has a 35° resection angle of the femoral neck, is 178 mm in length, and has a smooth surface with a fluted design. Currently, we use the model Y 2

so that the entire stem can be inserted manually. In contrast, for the model Y 2, the medullary canal of the femur is rasped to the minimal extent necessary, and the stem can then be manually inserted until the distance between the collar of the stem and the resected surface of the femoral neck is 2–3 cm.

In the computer simulation, using RBSM, when the stress distribution around the model Y stem was compared with model Y 2 stem under optimal press-fit conditions, stress distribution with the latter was more uniform than with the former (Fig. 1).

Clinical Results (Tables 1, 2)

The model Y 2 cementless stem has already been used for 48 joints in 48 patients. The age at operation ranged from 40 to 81 years, with the average being 65 years. The diseases for which treatment was indicated were: osteo-arthritis, in 37 patients, including 16 bilateral cases; rapidly destructive cox-arthritis, in 6 patients, and aseptic necrosis of the femoral head in 5 patients. The clinical and roentgenographic results were investigated in the 29 (29 joints) of these patients who were followed-up for more than 2 years. The longest follow-up was 4 years and 2 months, with the average being 2 years and 6 months after operation.

Table 1. Evaluation of press fit in patients who were implanted by model Y2 or model Y replacements.

	Model Y2 $n = 29$ (%)	Model Y $n = 26$ (%)
Exellent (≥70%)	24 (83)	7 (27)
Fair (50%–70%)	5 (17)	8 (31)
Poor (<50%)	0	11 (42)

More patients with model Y2 had a favorable press fit than pitients with model Y

Table 2. Results in patients with favorable press fit.

	Model Y2 $n = 29$ (%)	Model Y $n = 15$ (%)
Thigh pain	3 (10)	5 (33)
Clear zone		
Typel 1 (none)	22 (76)	6 (40)
Type2 (<2 mm)	7 (24)	3 (20)
Type3 (≥2 mm)	0	6 (40)
Shift and sinking	0	6 (40)

The model Y2 stem showed better results than the model Y, both clinically and roentgenographically

Evaluation of Press Fit

On the roentgenographic anteroposterior (AP) view obtained immediately after operation, we calculated the contact ratio, that is, the percentage of the area where the space between the inner cortex and the stem surface was less than 1 mm, at the lateral and medial cortex, from the lower base of the lesser trochanter to the distal one-third of the stem. A value of 70% or higher was regarded as excellent press fit, 50%–69% as fair, and lower than 49% as poor. Twenty-four patients (83%) showed excellent press fit, and 5 (17%) showed fair press fit. No patients showed poor press fit.

Roentgenographic Findings Around the Stem

With regard to a clear zone around the stem, results were classified into three types; absence of clear zone being type 1, presence of a clear zone less than 2-mm in width as type 2, and the presence of a clear zone 2 mm in width or more as type 3. Twenty-two patients (76%) showed type 1, and 7 (24%) showed type 2. No patients showed type 3.

Shifting and Sinking of the Stem

No shifting or sinking of the stem was seen in any patients.

Thigh Pain

Thigh pain affected only three patients (10%), two of whom had excellent press fit; the third patient had fair press fit.

Cases

Case 1

Case 1 was a 63-year-old female who underwent total hip replacement with a model Y 2 due to aseptic necrosis of the femoral head on the right side. On roentgenographic AP findings obtained immediately after surgery, the stem had a favorable press fit with an excellent contact ratio. Over the 4 years and 2 months of the postoperative course, there was no pain or limping; roentgenographically, no shifting or sinking of the stem was seen, and there was no clear zone around the stem (type 1) (Fig. 4).

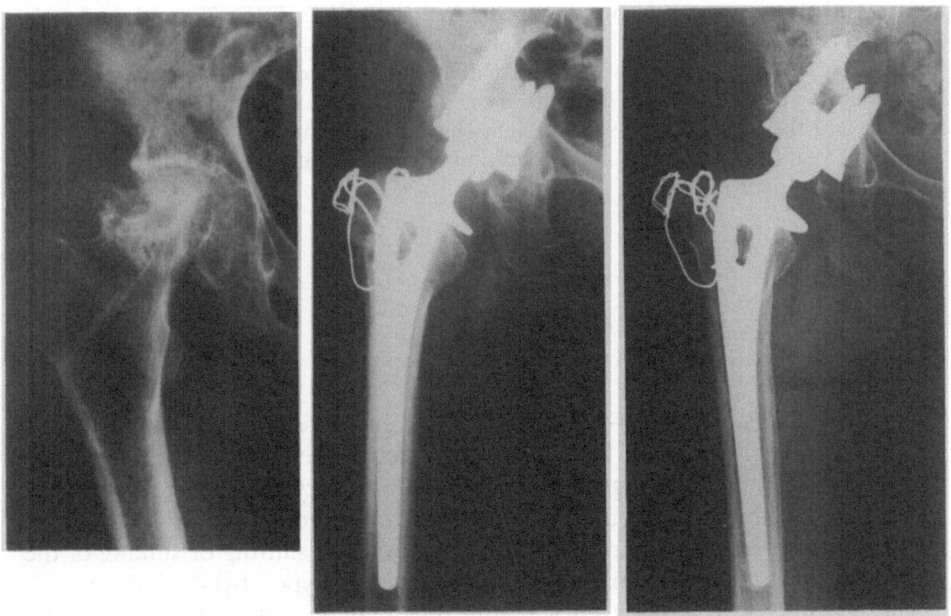

Fig. 4. Patient who was implanted by model Y 2 cementless THR (63-year-old female with aseptic necrosis of the right hip joint). Roentgenograph taken 5 weeks after surgery (*middle*) shows excellent press fit of the stem. The patient has not complained of thigh pain. The roentgenograph shows no shifting or sinking of the stem and no clear zone (type 1) around the stem. Clinically, neither pain nor limping was present 4 years and 2 months after surgery (*right*). Roentgenograph on the *left* was taken preopenatively

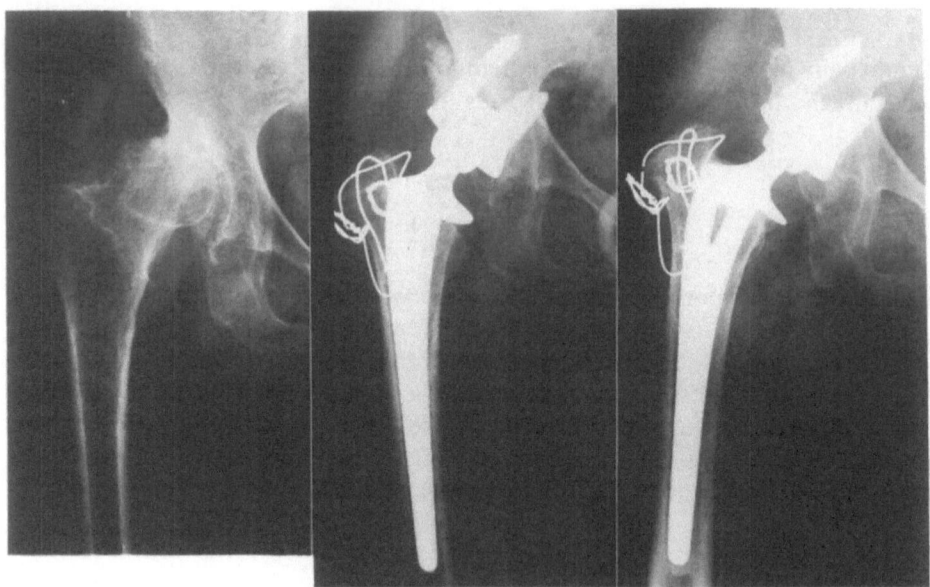

Fig. 5. Patient who was implanted by model Y 2 cementless THR (77-year-old female with osteoarthritis of the right hip joint). Roentgenograph taken 5 weeks after surgery (*middle*) shows excellent press fit. The patient has not complained of any pain or limping during the 2 years and 6 months of the postoperative course. There was no clear zone around the stem, nor was there stem shifting or sinking on the roentgenograph taken at 2 years and 6 months after operation (*right*). Roentgenograph on the *left* was taken preoperatively

Case 2

This patient was a 77-year-old female who underwent replacement with a model Y 2 cementless total hip system for osteoarthritis of the right hip. The roentgenographic AP findings immediately after surgery showed excellent press fit. For the 2 years and 6 months of the postoperative course, there has been no pain or limping; roentgenographic findings showed no shifting or sinking of the stem, and there was no clear zone around the stem (type 1) (Fig. 5).

These results were compared with the results in 15 out of 26 patients who was implanted by model Y and showed excellent or fair press fit. With regard to the absence or presence of a clear zone the stem, type 1 was observed in 6 of these 15 patients (40%), type 2 in 3 (20%), and type 3 in 6 (40%). Shifting or sinking of the stem were seen in 6 patients (40%). These findings showed that the model Y 2 cementless stem yielded better results than the model Y.

Discussion

There are two types of biomechanical stability displayed in regard to cementless stems; the initial biomechanical stability obtained immediately after surgery and the secondary stability provided by subsequent new bone formation. To achieve long-term secondary stability, it is essential to obtain favorable initial stability, which is provided by excellent press fit, as has been reported by Ide et al. [8], Amano et al. [9], Yamamoto [2], and the authors [10]. When we compared the currently used model Y 2 with the previously used model Y, in which the stem shape and size were the same as in the original JIAT stem, we found that the incidence of thigh pain, the shifting or sinking of the stem, and the frequency of the presence of a clear zone around the stem, which features reflect secondary stability achieved by new bone formation, were much lower in those patients who underwent the model Y 2 replacements than in those who underwent the model Y, even though these two models were compared in patients in whom press fit was acquired in the same manner. We believe that the difference in stem design contributed to the different results, although they may have been affected also by improvements in surgical procedure and surface design. That is, in the computer simulation using the RBSM, the model Y 2 showed uniform stress distribution and was mechanically stable, while the model Y was less stable (Fig. 1). The acquisition of favorable press fit, as well as the employment of a design which achieves excellent secondary stability via uniform stress distribution around the stem and subsequent new bone formation, may be important.

From this point of view, there are problems with both the graphic processing system and simulatin employing cadaveric specimens as means of assessing cementless stem design, since it is difficult to investigate secondary stability with these simulations. Biomechanical analysis with the RBSM, on the other hand, allows for comprehensive assessment of various factors, i.e., the degree of press fit, the resection angle of the femoral neck, and the length and width of the stem. That this is so is demonstrated in the clinical results obtained with the patients who underwent model Y 2 replacements that were developed on the basis of the results of computer simulation using the RBSM.

Conclusion

Computer simulation using the RBSM is a very useful method for the evaluation of an ideal design for a cementless stem. The model Y 2 cementless stem, whose development was based on the results of such a simulation, may indeed the ideal design at present.

References

1. Akamatsu N, Hamada Y, Nakajima I, Ide T, Yamamoto Y, Tatsugi S, Horiuchi T (1989) Basic research on the design of cementless total hip system. Jpn J Artif Organs 18:1626–1635
2. Yamamoto Y (1990) Clinical and basic research on the design of cementless total hip system. Jpn J Artif Organs 19:1476–1478
3. Engh CA, Bobyn JD, Glassman AH (1987) Porous-coated hip replacement. J Bone Joint Surg [Br] 69B:45–55
4. Burke DW, O'Connor DO, Zalenski EB, Jasty M, Harris WH, et al (1991) Micromotion of cemented and uncemented femoral components. J Bone Joint Surg [Br] 73-B:33–37
5. Sugiyama H, Murota K, Tomita Y, Ohtani T, Higo Y, Whiteside LA (1992) Rotational stability of cementless total hip system (abstract). Hip Biomechanics Symposium, Nov. 19, 1992, Fukui, p 41
6. Akamatsu N, Hamada Y, Fukushima H, Nakajima I, Tomita Y, Kushi M (1983) Cementless total hip system. Orthopaedic Surgery 3:42–48
7. Kawai T (1977) A new discrete model for analysis of solid mechanics problems. Seisan Kenkyu 29:208–210
8. Ide T, Akamatsu N, Hamada Y, Nakajima I, Yokoyama Y, Yamaguchi T (1987) Biomechanical study regarding fixation of cementless total hip system. Jpn J Artif Organs 16:1569–1572
9. Amano R, Ide T, Hamada Y, Tatsugi S, Akamatsu N (1989) Non-linear stress analysis of clinical case of cementless total hip system. Jpn J Artif Organs 18:325–355
10. Hamada Y, Akamatsu N, Ide T, Yamamoto Y, Tatsugi S (1992) The problems of cementless total hip system consisting of a socket with three spikes. Orthop Surg Traumatol 35:541–551

List of Contributors

Akamatsu, N. 265, 321
Amano, R. 265
Aoyama, T. 83
Arima, J. 225
Azuma, H. 187
Boh, A. 239
Doi, T. 187
Ebihara, K. 187
Fujii, G. 205
Funayama, K. 205
Furuya, I. 157
Gesso, H. 239
Hamada, T. 83
Hamada, Y. 321
Hara, Y. 11, 157
Hashimoto, T. 129
Hattori, T. 73, 105, 303
Hayashi, K. 225
Higo, Y. 313
Higuchi, F. 171
Himeno, S. 35
Hirohashi, K. 11, 157
Hirose, S. 73, 105
Hyodo, K. 277
Ichihashi, K. 239
Ide, T. 21, 265, 321
Iida, H. 139
Imura, S. 239
Inadome, T. 225
Inoue, A. 171
Ishinada, Y. 195

Iwata, H. 83
Izumida, R. 195
Kaneda, K. 123
Kasahara, F. 83
Kasai, R. 139
Kawate, K. 49
Kimura, S. 129
Konishi, N. 3
Kumeta, H. 205
Mashima, T. 225
Matsuda, Yasutaka 139
Matsuda, Yasumasa 303
Matsuno, S. 123
Matsuno, T. 123
Matsusue, Y. 139
Miura, H. 225
Miura, T. 83
Murota, K. 313
Niwa, S. 73, 105, 117, 303
Ohashi, H. 157
Ohneda, Y. 49
Ohtani, T. 313
Ohtsuka, H. 117
Oka, M. 255
Okumura, H. 149
Okumura, Y. 239
Okuno, M. 95
Oomori, H. 239
Otsuka, T. 95

Sakamaki, T. 289
Sakamoto, K. 11
Sakurai, M. 205
Sawai, K. 73, 105, 117, 303
Shiba, N. 171
Shimakage, K. 123
Shimazu, A. 11, 157
Sindo, H. 187
Sugioka, Y. 225
Sugiyama, H. 313
Suzuki, M. 61
Takasu, N. 95
Takedani, H. 239
Tamai, S. 49
Tanaka, S. 215
Tanaka, Y. 239
Tateishi, T. 277
Tatsugi, S. 21, 321
Teshima, R. 95
Tokura, A. 83
Tomihara, M. 215
Tomita, Y. 313
Watanabe, H. 321
Yamamoto, K. 95
Yamamoto, Y. 21
Yamamuro, T. 139
Yamashita, H. 171
Yamazaki, N. 61
Yanagimoto, S. 289
Yano, S. 129

Subject Index